AF577107

Experimental Methods in Orthopaedic Biomechanics

Experimental Methods in Orthopaedic Biomechanics

Edited by

Radovan Zdero

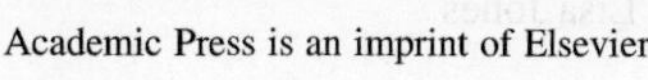

Academic Press is an imprint of Elsevier

Academic Press is an imprint of Elsevier
125 London Wall, London EC2Y 5AS, United Kingdom
525 B Street, Suite 1800, San Diego, CA 92101-4495, United States
50 Hampshire Street, 5th Floor, Cambridge, MA 02139, United States
The Boulevard, Langford Lane, Kidlington, Oxford OX5 1GB, United Kingdom

Library of Congress Cataloging-in-Publication Data
A catalog record for this book is available from the Library of Congress

British Library Cataloguing-in-Publication Data
A catalogue record for this book is available from the British Library

ISBN: 978-0-12-803802-4

For information on all Academic Press publications visit our website at https://www.elsevier.com

Publisher: Joe Hayton
Acquisition Editor: Fiona Geraghty
Editorial Project Manager: Maria Convey
Production Project Manager: Lisa Jones
Designer: Matthew Limbert

Typeset by TNQ Books and Journals

For my parents: Obrad and Ana Ždero
За моје родитеље: Обрад и Ана Ждеро

Contents

Acknowledgments

I am grateful to Fiona Geraghty, Maria Convey, and Lisa Jones of Elsevier Inc., who were efficient and energetic in bringing this project to fruition. My sincere thanks to all the contributing authors who took time out of their busy teaching and research schedules to add their highly valuable chapters to this book, as well as for being open to my editorial feedback. My appreciation also goes out to my past and present students and colleagues who inspire me with their enthusiasm for discovery. And, finally, I am beholden to the late great scientist and engineer Nikola Tesla, whose world-changing insights and innovations allow the rest of us to stand on the shoulders of a giant.

What Is Orthopaedic Biomechanics?

Radovan Zdero
Western University, London, ON, Canada

WHAT IS THE MUSCULOSKELETAL SYSTEM?

To understand what this book is really about and how to best use it, some basic background information must first be considered. The musculoskeletal system is a complex network of hard tissues, soft tissues, and biofluids that interact with each other to provide both form and function to the human body. Bones bestow shape and size to the body while also being involved in performing a variety of tasks. Muscles generate forces that move limbs and other body segments. Tendons connect muscles to bones, thereby transmitting force from the former to the latter. Ligaments connect bones to other bones, thus allowing appropriate relative positioning and/or motion between neighboring bones. Cartilage and meniscus are like shock absorbers, which modulate stresses transferred across the interfaces of articulating joints. Biofluids, such as blood, marrow, and synovial fluid, deliver nourishment to hard and soft tissues, as well as being involved in lubricating joints.

WHAT IS BIOMECHANICS?

Biomechanics is the field of study that uses the engineering tools of statics, dynamics, and strength of materials to analyze the kinetics (i.e., loads) and kinematics (i.e., motions) experienced by the musculoskeletal system. These loads and motions may be generated during a whole host of very different activities of daily living, disease conditions, or injury events. So, biomechanics can be divided into five or more primary subfields, although some overlap exists (Fig. 1). Occupational biomechanics addresses workplace tasks, ergonomic design of tools and workspaces, and injuries. Sports biomechanics deals with physical training, athletic activities, and injuries. Transportation biomechanics focuses on passenger comfort and safety during vehicle operation, ergonomic design of vehicle interiors, and accidents causing injury. Rehabilitation biomechanics concerns the recovery of musculoskeletal strength and function after disease, injury, or surgery. And, finally, orthopaedic biomechanics, which is the focus of this book, uses engineering analysis tools in order to (1) characterize the mechanical properties of bones, joints, and soft tissues; (2) develop new implants and biomaterials for artificial joint replacement, bone fracture fixation, and soft tissue repair; and (3) optimize orthopaedic surgical techniques.

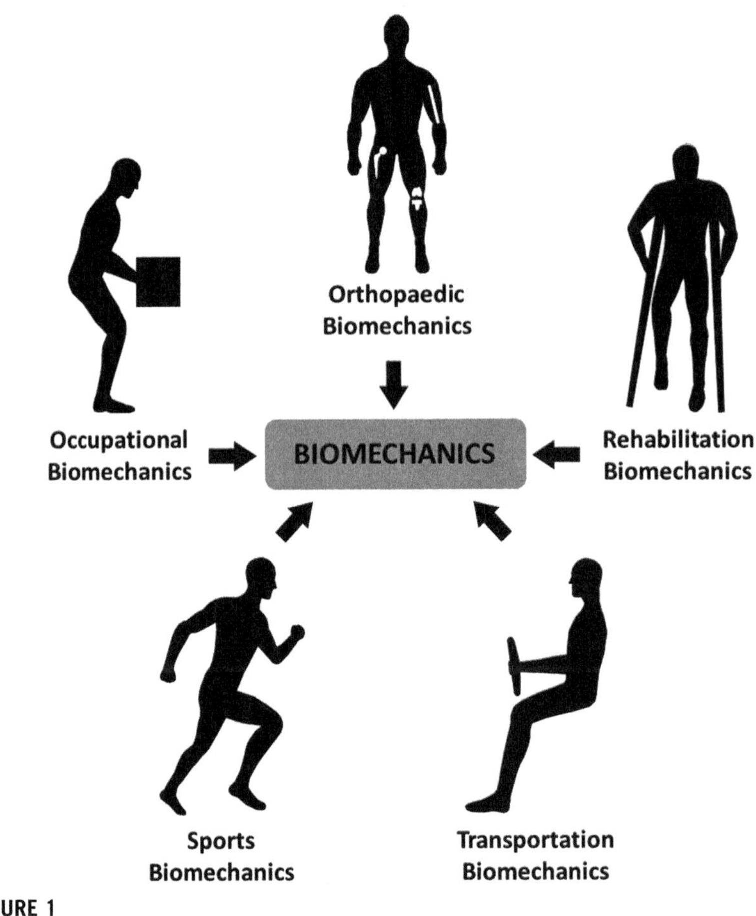

FIGURE 1

Biomechanics is a field of study dealing with loads and motions on the musculoskeletal system. The subfield of orthopaedic biomechanics is the focus of this book.

WHAT IS ORTHOPAEDIC BIOMECHANICS?

Orthopaedic biomechanics is about discovering and potentially optimizing the mechanical stresses experienced by normal, diseased, injured, or surgically treated bones, joints, and soft tissues. This subfield of study is particularly influenced by two groups of specialists, namely, orthopaedic surgeons and biomechanical engineers. Orthopaedic surgeons are on the "clinical frontline," as they treat patients by performing procedures like total or partial joint replacement, bone fracture repair, soft tissue repair, limb deformity correction, and bone tumor removal. Biomechanical engineers are on the "technological frontline," as they discover the basic mechanical properties of human tissues, design and test the structural stress limits of

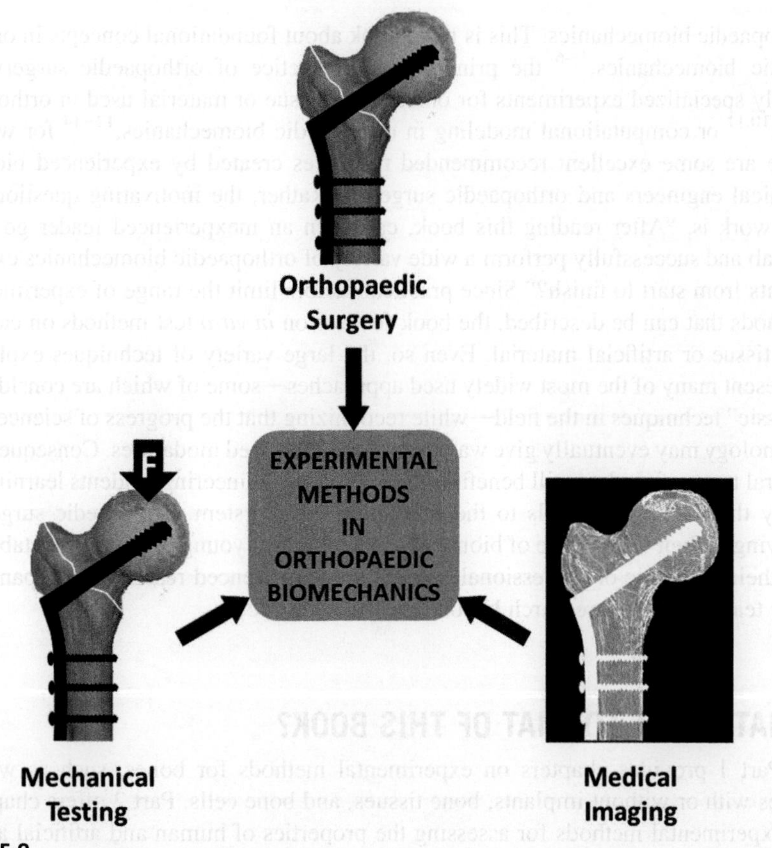

FIGURE 2

Experimental research activities in orthopaedic biomechanics.

orthopaedic implants, and develop new and improved biological and artificial biomaterials. Consequently, the strategy for conducting cutting-edge experimental research in orthopaedic biomechanics in hospitals, universities, and industry, includes a combination of orthopaedic surgery, mechanical testing, and medical imaging (Fig. 2).

WHAT IS THE PURPOSE OF THIS BOOK?

Much of the valuable material published in the field of orthopaedic biomechanics, including textbooks, peer-reviewed journal articles, and conference proceedings, is still of limited practical use in learning about laboratory experimentation. So, the aim of this book is to give the reader step-by-step, hands-on, do-it-yourself, practical instructions for performing experiments in orthopaedic biomechanics. This is not a book about foundational concepts in orthopaedic biomechanics,[1–6] the principles and practice of orthopaedic surgery,[7–9] highly specialized experiments for one type of tissue or material used in orthopaedics,[10,11] or computational modeling in orthopaedic biomechanics,[12–14] for which there are some excellent recommended resources created by experienced biomechanical engineers and orthopaedic surgeons. Rather, the motivating question for this work is, "After reading this book, can even an inexperienced reader go into the lab and successfully perform a wide variety of orthopaedic biomechanics experiments from start to finish?" Since practical factors limit the range of experimental methods that can be described, the book focuses on *in vitro* test methods on cadaveric tissue or artificial material. Even so, the large variety of techniques explored represent many of the most widely used approaches—some of which are considered "classic" techniques in the field—while recognizing that the progress of science and technology may eventually give way to new and improved modalities. Consequently, several types of readers will benefit from this book: engineering students learning to apply their analytical skills to the musculoskeletal system, orthopaedic surgeons growing in their knowledge of biomechanical methods, young researchers establishing their academic or professional careers, and experienced researchers expanding their teaching and/or research beyond their current focus.

WHAT IS THE FORMAT OF THIS BOOK?

Part 1 provides chapters on experimental methods for bones, such as whole bones with or without implants, bone tissues, and bone cells. Part 2 offers chapters on experimental methods for assessing the properties of human and artificial articulating joints. Part 3 gives chapters on experimental methods applied to soft tissues like cartilage/meniscus, ligament/tendon, and muscle. Obviously, some chapters could easily be classified as belonging to another part of the book, but the chosen groupings are sufficient for the aims of this volume. Each chapter presents content in a generic and rational way in order to emphasize underlying principles that can be easily adapted to a range of research questions. To this end, each chapter follows a standardized structure to make learning more effective and logical: Background, Research Questions, Methodology, Results, Discussion, Summary, Quiz Questions, and References. Note that the Methodology and Results sections deliberately provide only "mock" graphs and/or "empty" tables in order to illustrate the typical data trends and parameters measured during experiments; thus, the graphs and tables do not show actual numerical data. But, representative numerical data reported in the scientific literature are summarized in each Discussion section. And, finally, gray boxes, which highlight special content, are scattered throughout each chapter to make learning more memorable and enjoyable: Glossary, Safety First, Tips and Tricks, The Gold Standard, Engineer's Toolbox, and Alternatives and Adaptations.

WHAT IS THE BEST WAY TO USE THIS BOOK?

Some science textbooks sit on shelves collecting dust, or perhaps they are read but without making any practical difference. Other science textbooks, however, are used for the advancement of teaching, research, and technology. And so, there are many ways to get the most out of this book, rather than just reading it, which can be applied to university, industry, or hospital contexts:

- Design a new university-level course, using this book as the course textbook.
- Create undergraduate student lab tasks, homework assignments, and class projects.
- Formulate hypotheses and/or methodologies for graduate student thesis projects.
- Train lab managers, technicians, and other staff in experimental techniques.
- Expand the research programs of experienced researchers into new areas.
- Develop "gold standard" test methods that are recognized internationally.
- etc.

Another helpful idea for practical hands-on training and teaching using this book can be summarized by the acronym S.H.O.E (Fig. 3). S stands for "show," meaning that the mentor performs a task, while the mentee only watches carefully. H stands for "help," meaning that the mentor helps the mentee perform the task themselves. O stands for "observe," meaning that the mentee performs the task by themselves, while the mentor only watches and makes helpful verbal comments. E stands for "exit," meaning that the mentor now physically leaves the context, trusting that the mentee will be able to carry on with the task by themselves. While teacher-to-student transfer of theoretical information through lectures, books, videos, etc., has a pivotal role in modern education, the mentor-to-mentee

FIGURE 3

The S.H.O.E. system of mentoring.

apprenticeship approach has also been used effectively and extensively over the course of human civilization for many different endeavors. Now equipped with a basic background, may the reader expectantly turn to the content of the book!

REFERENCES

1. Fung YC. *Biomechanics: mechanical properties of living tissues*. New York, NY, USA: Springer-Verlag; 1981.
2. Johnson KD, Tencer AF. *Biomechanics in orthopedic trauma: bone fracture and fixation*. Philadelphia, PA, USA: JB Lippincott Company; 1994.
3. Poitout DG, editor. *Biomechanics and biomaterials in orthopedics*. London, UK: Springer; 2004.
4. Mow VC, Huiskes R, editors. *Basic orthopaedic biomechanics and mechano-biology*. 3rd ed. Philadelphia, PA, USA: Lippincott, Williams, and Wilkins; 2005.
5. Bartel DL, Davy DT, Keaveny TM, editors. *Orthopaedic biomechanics: mechanics and design in musculoskeletal systems*. Upper Saddle River, NJ, USA: Prentice Hall; 2006.
6. Winkelstein BA, editor. *Orthopaedic biomechanics*. Boca Raton, FL, USA: CRC Press; 2012.
7. Garino JP, Beredjiklian PK, editors. *Adult reconstruction arthroplasty: core knowledge in orthopaedics*. Philadelphia, PA, USA: Mosby Elsevier; 2007.
8. Sanders R, editor. *Trauma: core knowledge in orthopaedics*. Philadelphia, PA, USA: Mosby Elsevier; 2008.
9. Wiesel SW, editor. *Operative techniques in orthopaedic surgery*. Philadelphia, PA, USA: Lippincott, Williams, and Wilkins; 2010.
10. An YH, Draughn RA, editors. *Mechanical testing of bone and the bone-implant interface*. Boca Raton, FL, USA: CRC Press; 2000.
11. Saunders MM. *Mechanical testing for the biomechanical engineer: a practical guide*. San Rafael, CA, USA: Morgan and Claypool Publishers; 2015.
12. Zdero R, Bougherara H. Orthopaedic biomechanics: a practical approach to combining mechanical testing and finite element analysis. In: Moratal D, editor. *Finite element analysis*. Rijeka, Croatia: InTech; 2010 [chapter 7] Available free online at, http://cdn.intechweb.org/pdfs/11992.pdf.
13. Nedoma J, Stehlik J, Hlavacek I, Danek J, Dostalova T, Preckova P. *Mathematical and computational methods in biomechanics of human skeletal systems*. Hoboken, NJ, USA: Wiley and Sons; 2011.
14. Zhang M, Fan Y, editors. *Computational biomechanics of the musculoskeletal system*. Boca Raton, FL, USA: CRC Press; 2014.

1

Bones: Whole, Tissue, and Cell

CHAPTER 1

High-Speed Impact Testing and Injury Assessment of Whole Bones

Cheryl E. Quenneville
McMaster University, Hamilton, ON, Canada

1. BACKGROUND

Human bones are vulnerable to injury from a wide range of events, such as falls, automotive crashes, and sporting collisions, to name a few (Fig. 1.1). In order to evaluate protective measures, the injury limits of whole bones need to be established through impact testing, which is typically done on cadavers. This allows investigation of the factors that influence injury, and using statistical methods safety limits can be established, which are used by industry to assess risk during various events. Furthermore, the injury tolerance of whole bones must be identified and translated to measures obtained using surrogates, such as anthropomorphic test devices (ATDs, or "crash test dummies"), in order to be used outside of the research lab environment. Lastly, an understanding of the factors that influence fracture allows development of better protective measures, thus reducing the incidence and severity of these debilitating injuries. Therefore, this chapter explains high-speed impact testing of whole bone, as well as how to analyze, present, and interpret results.

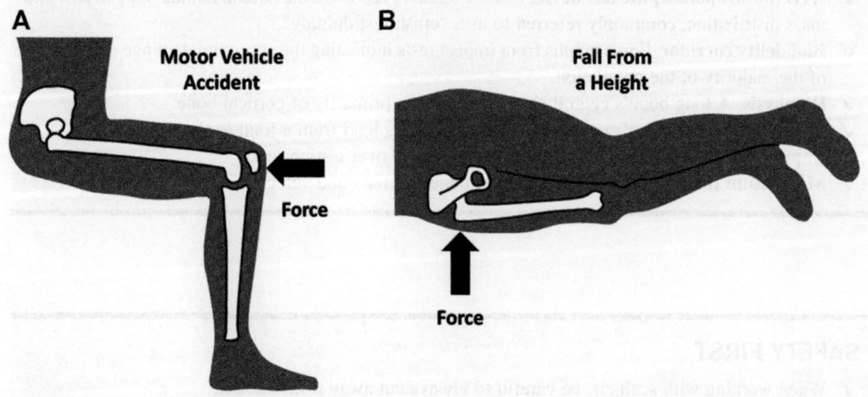

FIGURE 1.1

Typical whole bone injury mechanisms. (A) Motor vehicle accident, (B) fall from a height.

2. RESEARCH QUESTIONS

Typical research questions might include one or more of the following:

- What is the fracture tolerance of a specific whole bone?
- How do posture, impact angle, and impact duration affect the fracture tolerance of a bone?
- What is the most effective protective device for preventing injury of a bone?
- How can surrogates be developed to be used by industry for safety evaluations?
- etc.

3. METHODOLOGY

3.1 GENERAL STRATEGY

Cadaveric isolated whole bones are obtained from the relevant population (e.g., gender, age range, bone density, etc.) and then tested under high-speed impacts. As an example of how to perform such analysis, axial testing of the tibia is described here, but the methods can be applied to any bone, or even fully intact limbs. Impact testing is often done using a specialized impacting apparatus, such as pendulum-based[1] or pneumatic[2] systems. The impact parameters must be properly selected to replicate the relevant loading and may include impactor mass, velocity, acceleration, impact duration, and/or energy. Tibias are aligned according to anatomical landmarks to ensure consistency among tests, instrumented as required, and subjected to impacts of increasing intensity until failure occurs. Statistical methods are then used to establish injury tolerance curves and identify the factors that contributed to tibial fracture risk.

GLOSSARY

- ✓ **ATD (anthropomorphic test device)**. A surrogate to represent the natural human body in size and mass distribution, commonly referred to as a "crash test dummy".
- ✓ **Biofidelity corridor**. Force graphs from impact tests indicating the approximate range of response of the majority of the population.
- ✓ **Diaphysis**. A long bone's central shaft that is made primarily of cortical bone.
- ✓ **Epiphysis**. A long bone's rounded end that transmits load from a joint to the diaphysis.
- ✓ **Impulse**. The mathematical integral of impact force over testing time.
- ✓ **Momentum (linear)**. The mathematical product of mass and velocity.

SAFETY FIRST

- ✓ When working with scalpels, be careful to always cut away from oneself.
- ✓ Always follow standard operating procedures for the impacting device to ensure safety.
- ✓ Operate the device only when everyone is properly protected from the impact.
- ✓ Hearing protection is recommended, as impacts can cause high dB sounds.

3.2 MATERIALS AND TOOLS LIST

- alcohol wipes
- cement for potting
- high-speed camera (optional)
- human or animal bone
- impacting apparatus
- instrumentation (e.g., strain gage(s), accelerometer(s), etc.)
- levels (e.g., water or laser)
- PVC plastic pipe segment
- sandpaper
- scalpel
- strain gage glue

3.3 SPECIMEN PREPARATION

Step 1. Freeze the specimens. All isolated whole bones should be kept frozen prior to testing, preferably fresh–frozen within 24 h postmortem, and wrapped in plastic. Prior to testing, thaw isolated bones at room temperature for at least 8 h.

Step 2. Prepare the specimens. Strip isolated bones of all soft tissues, with particular attention to the end of the bone to be potted later, as well as locations of interest for instrumentation to be used later.

Step 3. Prepare for potting. In order to hold the specimen in the appropriate posture for impact testing, at least one end needs to be potted. The potting process ensures a consistent orientation of all specimens in the impacting apparatus and provides a secure place at which to hold the bone without slipping. Potting typically consists of embedding the end of a bone in a block of cement. This can be done by placing the specimen's end in a length of PVC plastic pipe and filling it with some sort of cement (e.g., dental cement).

Step 4. Anatomically align specimens. The orientation of a bone has a significant influence on the fracture tolerance and location under impact loading, and therefore it must be controlled to be consistent for all specimens in a given study. To achieve this alignment, support the bone in the required orientation using a frame structure. Select anatomical landmarks that allow consistent alignment. For example, to pot the proximal tibia, center the proximal epiphysis in the PVC plastic pipe segment with the anterior ridge defining one vertical axis and the center of the lateral malleolus defining a second vertical axis. Project laser levels along each specimen to fine-tune the orientation until satisfactory alignment has been achieved (Fig. 1.2A and B).

Step 5. Pour the cement. The cement level should be of sufficient depth to maintain a strong hold, even under impacts that could jar the specimen loose. For the proximal tibia, this is approximately 50 mm, but may vary depending on the bone and orientation of impact. To enhance the hold, make sure the specimen is well scraped of soft tissues, press the cement all around the contours of the specimen, and embed screws in the potted end, if fixation is a concern. Leave the specimen supported in the alignment frame for a minimum of 10 min while the cement sets, and at least 30 min further before testing (Fig. 1.2C).

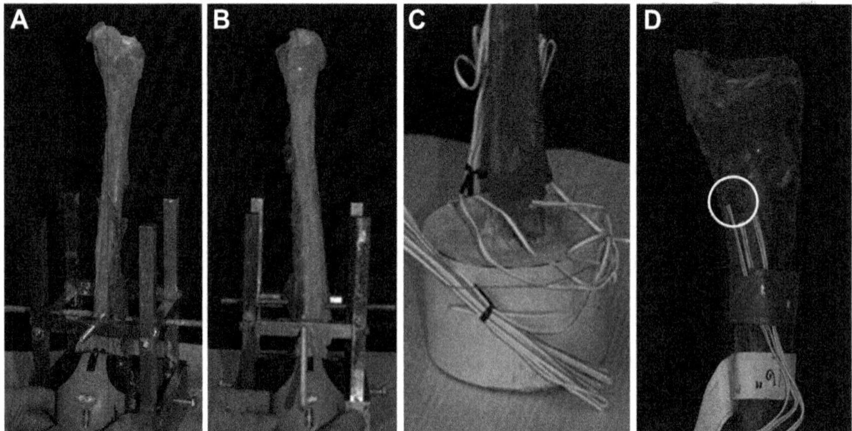

FIGURE 1.2

Whole bone preparation for impact tests. (A) Bone alignment using a laser and support frame (frontal view), (B) bone alignment using a laser and support frame (sagittal view), (C) the end of each bone is embedded in PVC plastic pipe using cement to provide a rigid method to grip specimens during impacts, (D) instrumentation such as strain gages can be affixed to the surface of bones using adhesives in order to characterize the load transmission along specimens during impact.

Step 6. Instrument the bones. Specimens can be instrumented at various locations to provide further information about load transmission and to provide data for developing finite element models. The most common types of instrumentation are strain gages and accelerometers. To provide accurate data, these sensors must be securely attached to the bone and, thus, should be placed above and below the expected location of fracture to provide the most useful data on stress propagation. Clean the surface of the bone by scraping it gently with a scalpel, and then rub it with sandpaper. Rub the surface with an alcohol pad to degrease the surface, and allow it to dry. Apply a thin layer of glue to the location of interest, and apply finger pressure through adhesive tape to form a smooth flat surface. After it dries, remove the adhesive tape, and let it cure for another 5 min. Lightly smooth any irregularities with sandpaper, apply the instrumentation (e.g., strain gage) with glue, and press evenly until it dries. Cover the sensor and lead wire attachments with a protective coating (e.g., clear nail polish) (Fig. 1.2D).

TIPS AND TRICKS

- ✓ Removing soft tissue is easier if bone is thawed for 1 h.
- ✓ Potting can be done before thawing of the whole bone specimen.
- ✓ To protect instrumentation, secure lead wires to the specimen with tape or elastic bands.
- ✓ Align strain gages or one element of strain gage rosettes with the long axis of the bone.
- ✓ Paired specimens can be used to directly compare different test conditions.

THE "GOLD STANDARD"

No known international standards exist specifically for performing high-speed impact tests on long bones. Thus, researchers should consult previously published peer-reviewed journal articles that have done experimental testing for this orthopaedic biomechanics application.

3.4 SPECIMEN TESTING

Step 1. Perform pre-test calibration. Once the relevant test parameters have been decided in order to accurately replicate the loading scenario being examined, the impacting apparatus must be calibrated prior to generating these impacts (Fig. 1.3). This ensures that known impact conditions are applied to the bone test specimen, and it allows for precise control over the increments used. This can be done using an ATD or even a potted section of PVC plastic pipe or wood so cadaveric bone specimens are not wasted during this process. However, remember that the output forces will vary depending on the stiffness of the object being struck. Select the appropriate input parameters (e.g., pressure, mass, velocity, force) for all calibration tests, which are conducted using the same steps 2–9 (below) that are employed for testing the whole bones. Next, generate the resulting calibration curve for the impacting device by testing at different conditions (e.g., increasing the pressure of a pneumatic system and then measuring the resulting projectile velocity immediately prior to impact) (Fig. 1.4).

Step 2. Mount the bone into the tester. Place an isolated whole bone into the impacting apparatus, attaching the pot securely (Fig. 1.3). If testing in a posture other than pure axial impacts, align one anatomical direction that was marked during the laser alignment process with the axes of the impacting apparatus (e.g., always place the anterior surface pointing upwards). Verify the alignment in both planes using levels typically held at the mid-diaphysis location.

Step 3. Ballast to a consistent weight. Add any required ballast weight to the bone specimen. Often it is desired to ensure that the overall mass of the specimen is equivalent to the segment mass *in vivo*, so this would be on the order of ~4.5 kg for the lower leg and foot, or ~12 kg to represent the full leg. Obtain these mass values from anthropometric tables for the population of interest. Weigh each potted specimen and any supporting components, and attach masses to the potted specimen until the desired mass is reached. Make sure these masses are mounted securely.

Step 4. Verify data acquisition. Connect the instrumentation (e.g., strain gages, accelerometers, etc.) to the data acquisition system, calibrate the sensors, and verify that all signals are collecting properly. High sampling frequencies are necessary for these tests, as they are quite rapid; thus, aim for a minimum of 10 kHz. Position any lights, and then adjust balance and focus of any high-speed cameras being used to record the event, which is typically recording at >1000 fps. The velocity of the projectile can be measured using reflective sensors or a photo gate at the location immediately prior to impact.

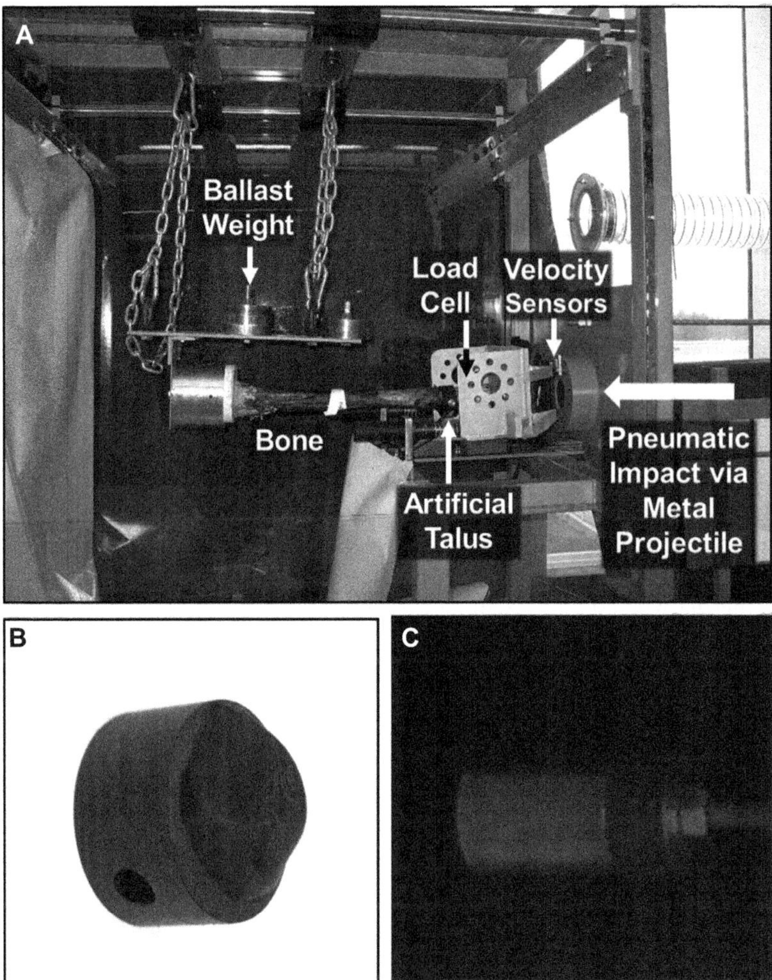

FIGURE 1.3

Impacting test setup. (A) Tibia loaded in axial alignment for testing, (B) artificial bone or joint (e.g., rapid prototyped plastic talus) used to transmit load to the articular surface of the bone, (C) metal projectile used to create the impact event.

Step 5. Position the bone for testing. Push the bone specimen up so it contacts the load transmission component (e.g., for distal tibia impacts, an artificial talus can be used to transfer the load evenly over the articular surface) (Fig. 1.3B). This load transmission component should be stronger than the specimens, and it can be made of a synthetic composite bone or a plastic (e.g., rapid prototyped plastic). Adjust this interface so the two components fit together as naturally as possible to ensure an even transfer of force. By mounting this load transmission component to an

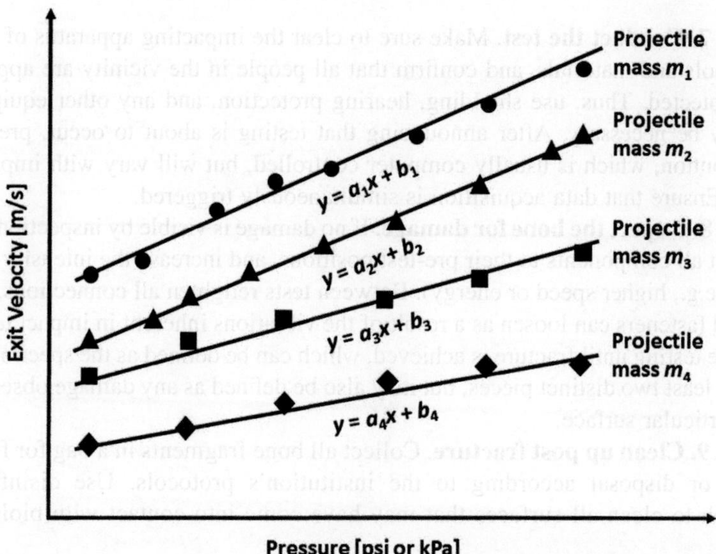

FIGURE 1.4

Typical calibration curves. Each impacting apparatus will be different and needs to be calibrated to ensure that the required impact velocity, energy, acceleration, and duration is achieved in order to be representative of the scenario under examination. These curves are representative of a pneumatic testing system, where the mass of the projectile and the pressure applied can be varied, resulting in a range of impact velocities. Curves for output force and any other required parameters can be similarly obtained. Each set of calibration trends can be fitted with a line of best fit represented by an equation, such as $y = ax + b$ for linear relationships, $y = ax^2 + bx + c$ for second-order polynomial relationships, etc.

in-line load cell, the forces and moments applied to the specimen during impact can be recorded. Make sure all components that move during a test are positioned in their pre-test configurations.

Step 6. Determine test parameters. Select parameters that accurately reflect the loading condition that is being simulated. These parameters may include impact speed, duration, energy, acceleration, etc. For example, for the lower leg in automotive collisions, typical speeds are on the order of 6 m/s with impact durations around 50 ms. Conversely, military blast conditions will reach 12 m/s with durations <10 ms. Furthermore, it is necessary to apply impacts in increasing intensity with the goal of targeting the fracture conditions without subjecting each specimen to a large number of repeated impacts, which can lead to cumulative damage. Ideally, one nonfracture test will be conducted, followed by a fracture test. The use of pilot specimens during calibration in step 1 may be necessary to determine the optimal test parameters, and the fracture conditions may vary from specimen to specimen, depending on the anatomical variations found within the population.

Step 7. Conduct the test. Make sure to clear the impacting apparatus of extraneous tools and materials, and confirm that all people in the vicinity are appropriately protected. Thus, use shielding, hearing protection, and any other equipment that may be necessary. After announcing that testing is about to occur, press the trigger button, which is usually computer controlled, but will vary with impacting device. Ensure that data acquisition is simultaneously triggered.

Step 8. Inspect the bone for damage. If no damage is visible by inspection or X-ray, reset all components to their pre-test positions, and increase the intensity of the impact (e.g., higher speed or energy). Between tests retighten all connections, since threaded fasteners can loosen as a result of the vibrations inherent in impact testing. Continue testing until fracture is achieved, which can be defined as the specimen being in at least two distinct pieces, but may also be defined as any damage observable on the articular surface.

Step 9. Clean up post fracture. Collect all bone fragments in a bag for freezer storage or disposal according to the institution's protocols. Use disinfectant or bleach to clean all surfaces that may have come into contact with biological tissue.

3.5 RAW DATA COLLECTION

Step 1. Record specimen details. Demographic factors such as donor age and gender must be included, since they have been shown to influence bone fracture tolerance[1] (Table 1.1). Obtain this information from the source of the cadaveric tissues at the time of purchase. Also, limb side, donor height, and donor weight should be entered, if known.

Step 2. Collect relevant load values. Both peak force and force at failure from the load cell should be examined, as these may differ (Table 1.1). For example, if for a given test, the specimen is very close to (but not exceeding) the fracture threshold, a very high force will be obtained. On the next test, where failure does occur, this releases a large amount of energy and reduces the peak force. Both factors should

Table 1.1 Raw data for whole bone impact tests.

Test	Age [Years]	Sex [M, F]	Limb Side [L, R]	Height [cm]	Weight [kg]	Peak Force [N]	Force at Failure [N]	Velocity [m/s]	Impact Duration [ms]
1									
2									
3									
etc.									
Avg		–	–						
SD		–	–						

Avg, *average;* SD, *standard deviation.*

be examined statistically for their influence. Accelerometers should be filtered as needed (e.g., according to Society of Automotive Engineers standards), and maximum principal strains can be calculated for rosettes. These values may be used for injury criteria development, but are more frequently used for validating finite element models.

Step 3. Create a test log. Any other test parameters that are varied (e.g., impact duration, impact angle, protective devices) should be similarly logged in the table (Table 1.1). Include a section in the table to write comments (Table 1.1). Denote any file naming structure used during data collection; for example, it is often helpful to include specimen numbers in the file name to avoid any posthoc analysis confusion.

3.6 RAW DATA ANALYSIS

Step 1. Compute statistical *P* values. Calculate average forces and standard deviations for all tests within a group for a given test condition. Statistical analyses are important in order to determine whether there is a statistical significance (e.g., $P < 0.01$ or < 0.05) between test conditions (e.g., axial vs. combined loading). For this purpose, an analysis of variance (ANOVA) can be used if an equal number of specimens are in each group and the distributions are assumed to be normal.

Step 2. Perform regression analysis. If examining a number of factors (e.g., Table 1.1) for their contribution to injury risk, a best subsets regression analysis can be performed to identify those that influence chance of fracture. This is typically done using a software package, which allows for investigating the best "model" to represent risk, starting with a single factor, and adding a factor at each stage until the best *adjusted* R^2 is achieved.

Step 3. Generate a fracture equation. Based on the best subsets regression analysis in step 2, an equation that represents this best "model" for fracture risk can be developed, so any input parameters (e.g., an older person with a higher applied force) can be used to predict the risk of injury, which is often a linear combination of the factors identified in step 2.

Step 4. Perform Weibull analysis. The final statistical technique used for analyzing injury risk data is the Weibull analysis. This is a survivability method used to make predictions for the lifespan of products in a population by extrapolating from a representative sample. The output from this analysis will take the form of a survivability curve with the x-axis representing the input value (e.g., impact force) and the y-axis representing the risk of fracture ranging from 0 (i.e., no risk) to 1 (i.e., guaranteed fracture).

Step 5. Compute statistical power. Power analysis can be done after the study to ensure there were enough specimens per test group to detect all statistical differences that were actually present (i.e., was type II statistical error avoided?).

Statistical power >80% is usually considered to indicate there were enough specimens per test group. Note that if good predictions of averages and standard deviations are available from prior studies, then the number of specimens and/or tests can be chosen before the study begins to ensure a power >80%.

ENGINEER'S TOOLBOX

Impact testing is essentially a collision that can be assessed using a momentum–impulse approach. Conservation of momentum states that the momentum prior to the collision is equal to the momentum after, which is calculated as $m_p v_{p1} + m_s v_{s1} = m_p v_{p2} + m_s v_{s2}$, where p is projectile, s is specimen, 1 is preimpact, 2 is postimpact, m is mass [kg], and v is velocity [m/s]. However, impulse is the integral of the force–time curve, which, according to Newton's second law, also represents the change in momentum from time t_1 to time t_2, such that $J = \int F dt = m\Delta v$, where J is impulse [N·s], F is force [N], t is time [s], m is mass [kg], and v is velocity [m/s]. Furthermore, the collision force will vary from specimen to specimen (i.e., dependent on the stiffness of the object being struck and any variations in total mass), and therefore, the impulse may change as well. This also makes it impossible to target specific force values for impact testing.

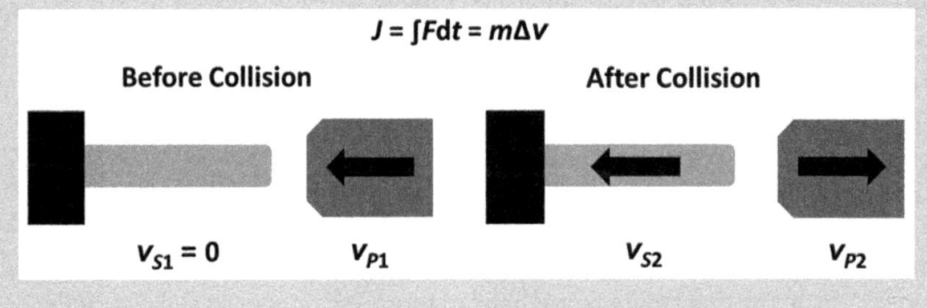

4. RESULTS

Once all impact testing data collection and analysis have been performed, it is then important to communicate and present the primary results in an understandable and concise manner to the reader of a journal article, conference paper, technical report, or book chapter.

Step 1. Plot relevant parameters. Force and moment graphs can be generated from an in-line load cell. These give an indication of peak force (or moment), impact duration, impulse (i.e., area under the force–time curve), and whether or not fracture occurred (Fig. 1.5A and B). Biofidelity corridors used to indicate the typical range of response for the population can be generated and shown by averaging all of the force curves for a given set of test parameters and presenting this average with ± 1 standard deviation error bars (Fig. 1.5C).

Step 2. Document injuries. Provide photos of each of the specimens and associated injuries so that they may be inspected and potentially categorized or

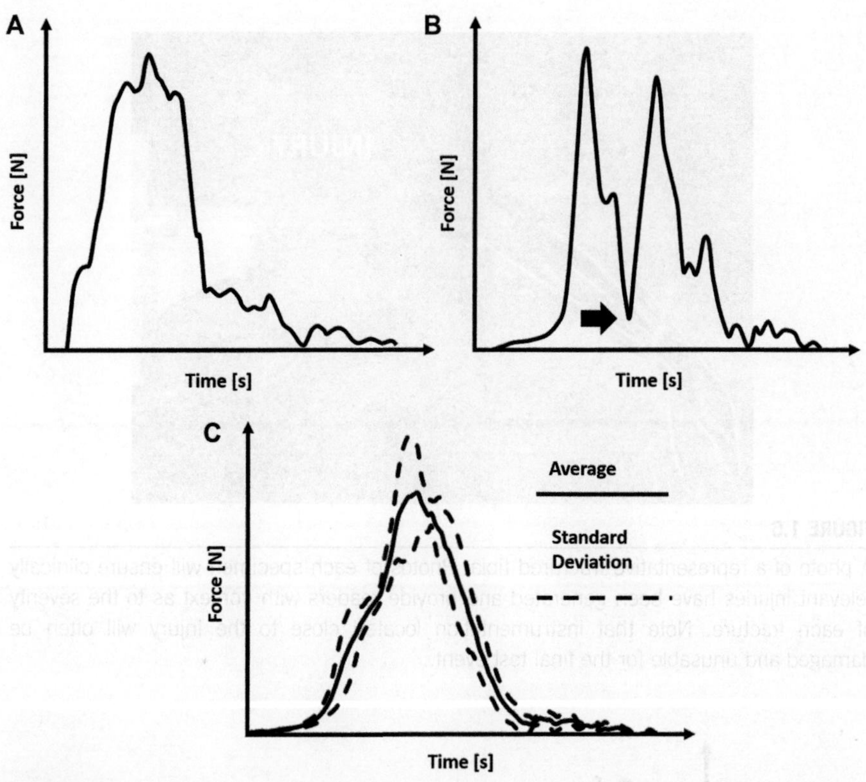

FIGURE 1.5

Output force curves. (A) Noninjurious event, (B) injurious (fracture) event, where the drop in load midway through the graph shown by the arrow is indicative of when the fracture occurs, (C) biofidelity corridors are calculated by averaging the force curves of multiple specimens to provide an average $\pm$ 1 standard deviation response.

correlated to the quantitative parameters that are measured, such as peak force, etc. (Fig. 1.6).

Step 3. Compare different impact conditions. Bar graphs can be used to compare peak force for various test groups (e.g., impact angle, impact speed, gender). These should be presented as average values with $\pm$ 1 standard deviation error bars (Fig. 1.7). Any statistical differences found during the analysis should be denoted with *P* values and symbols that are explained in the figure caption. Similar bar graphs can be shown for other measured parameters, if desired, such as impact velocity, acceleration, duration, etc.

Step 4. Create survivability curve. Plot the Weibull curve with the *x*-axis being the relevant factor that contributes to injury risk (e.g., force) and the *y*-axis being the risk of failure ranging from 0% to 100% (Fig. 1.8).

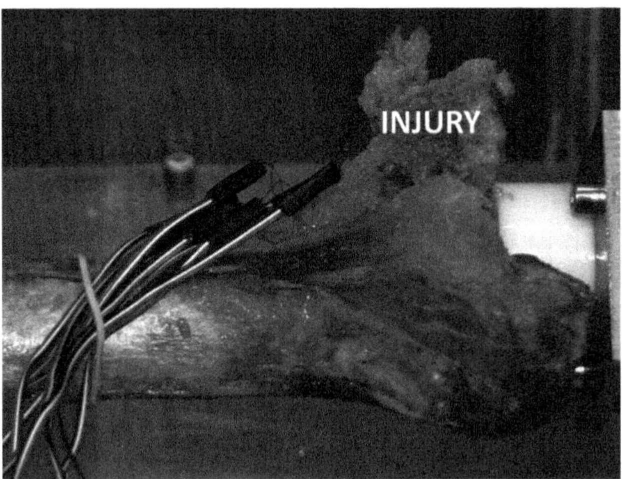

FIGURE 1.6

A photo of a representative fractured tibia. Photos of each specimen will ensure clinically relevant injuries have been generated and provide readers with context as to the severity of each fracture. Note that instrumentation located close to the injury will often be damaged and unusable for the final test event.

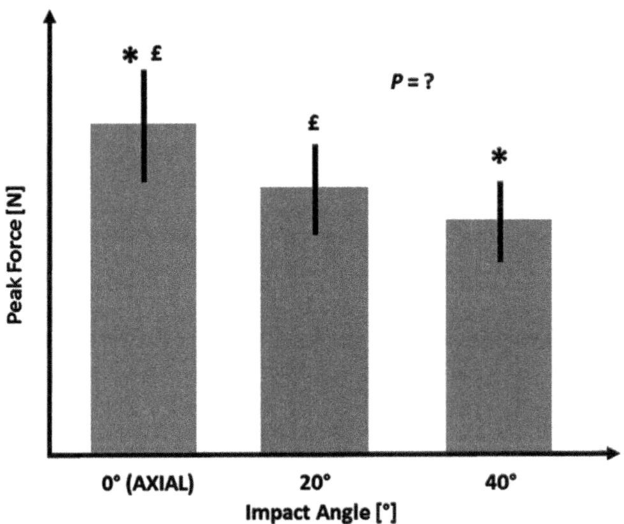

FIGURE 1.7

Typical peak force bar graphs. The average force is shown for each test condition (e.g., impact angle or impact parameters), while error bars indicate ± 1 standard deviation. Statistical *P* values and symbols like asterisks (*), pounds (£), etc., indicate statistical differences between test groups.

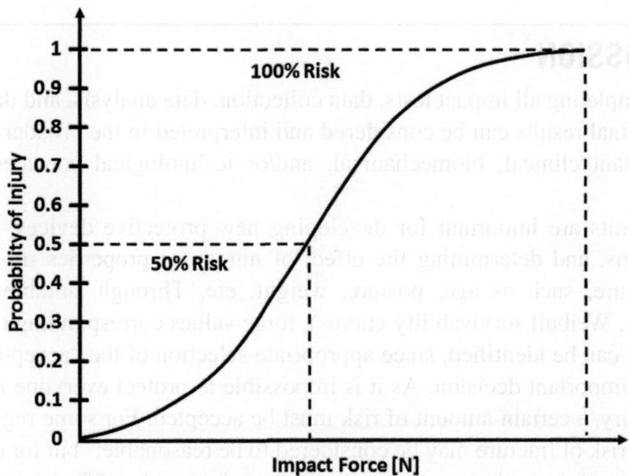

FIGURE 1.8

Weibull survivability curve. Impact force is plotted on the *x*-axis, and risk of injury is plotted on the *y*-axis ranging from 0 to 1. To predict the force associated with 50% risk of fracture, trace the line across at the 0.5 probability line until it intersects the graph, and then read the corresponding force.

ALTERNATIVES AND ADAPTATIONS

- ✓ **ATDs**. These devices are used in industry for evaluating injury risk and are commonly known as "crash test dummies." They are instrumented with accelerometers and load cells, readings from which are compared to injury risk thresholds during tests. ATDs do not match human biomechanical response well since they are often overly stiff with simplified joints and lines of action. Testing these components under the same conditions as cadaveric specimens allows a direct conversion of natural human body failure values to these standard tools. Be careful though, as subjecting them to the same loading conditions as those that cause fracture in the cadaveric specimens will often exceed the capacity of the load cells, potentially causing damage to these sensors.
- ✓ **Fully intact limbs**. Cadaveric intact limbs that include bones, joints, and soft tissues will more accurately replicate real-life *in vivo* conditions during impacts and injuries. Intact limbs can be assessed using laboratory impact testing with only slight modifications of the procedure used for isolated whole bones. For instance, once removed from freezer storage, intact limbs should be thawed for a minimum of 12 h, and the soft tissues are stripped off the end that will be secured during potting to ensure adequate adhesion. Appropriate positioning of all joints is extremely important to ensure repeatable testing of all specimens.
- ✓ **Synthetic bones**. Composite bones are widely used in biomechanical research for evaluating orthopaedic devices since they offer distinct advantages of reduced interspecimen variability and ease of handling and disposal. While they have been shown to not fracture at a comparable force or in a similar pattern to natural bones under axial impact loads, they have been shown to have similar properties to cortical bone, making them a potential surrogate for mid-diaphysis impacts.

5. DISCUSSION

After completing all impact tests, data collection, data analysis, and data presentation, then final results can be considered and interpreted in the broader context of some important clinical, biomechanical, and/or technological considerations, as follows.

Injury limits are important for developing new protective devices, evaluating safety systems, and determining the effect of numerous properties on a person's risk of fracture, such as age, posture, weight, etc. Through suitable statistical methods (i.e., Weibull survivability curves), force values corresponding to different levels of risk can be identified, since appropriate selection of the "acceptable" level of risk is an important decision. As it is impossible to protect everyone from every potential injury, a certain amount of risk must be accepted. For some regions of the body, a 10% risk of fracture may be considered to be reasonable,[3] but for other more critical regions of the body (e.g., head or spine), perhaps only a 5% risk is allowable.

Numerous studies have examined the fracture tolerance of the lower limb in an endeavor to identify the most appropriate injury limit. However, the 10% risk level has varied widely among studies: 5.4 kN,[1] 3.6 kN,[4] 2.1 kN,[5] 2.5 kN,[6] 5.8 kN,[7] and 7.9 kN.[8] This is likely the result of differing impacting conditions, and it highlights the importance of properly recreating in the lab the injurious event of interest. Furthermore, different studies have identified varying risk factors, including force, age, gender, body weight, height, and Achilles tension, that contribute to the potential for injury.

It is important to acknowledge that the injury limits identified using these impact techniques on cadaveric specimens cannot be directly applied to surrogates such as ATDs, as the stiffness and geometry of ATDs can vary from that of the natural human body.[9] Furthermore, the force response is not linear (i.e., the ratio of ATD loads to cadaveric loads is dependent on the impact speed), making establishing appropriate injury limits for use with an ATD especially challenging. By determining the impact parameters that would cause the injury limit and then subjecting an ATD to that impact, the correct injury threshold for this surrogate can be defined.

As cadaveric experimental tests are expensive and time-consuming, results from the instrumentation applied to the specimens can be used to validate computational finite element models. Strain gages have been shown to provide better validation agreement when placed over cortical bone regions than over cancellous regions since there are fewer local variations in strain gradient.[10,11] To obtain the geometry and material properties, the bone specimens should be scanned using computed tomography (CT) prior to testing.

Finally, impact testing can have important implications in the design of surgical fixation systems. While these devices tend to be designed and tested using specimens that are "fractured" using osteotomies, a more realistic and rigorous evaluation can be done by using the comminuted fractures created using impact testing techniques. Orthopaedic surgeons should be consulted to ensure that realistic fracture patterns are generated using impact tests to enhance the clinical applicability of any findings using these specimens.

6. SUMMARY

- Impact testing of bones can help develop appropriate injury limits for industry.
- Proper alignment of the bone is essential for consistent fracture response.
- Strain gages and accelerometers may help define injury limits or validate finite element models.
- Choose impact parameters carefully in order to replicate the injurious scenario appropriately.
- Minimize the number of impacts to reducing any cumulative loading effects.
- Weibull curves indicate the risk of fracture as a function of applied force or other parameter.

7. QUIZ QUESTIONS

1. What subject-specific factors will influence a person's risk of fracture?
2. Why can force during impact testing of whole bones not be predicted precisely in advance?
3. What information would strain gages or accelerometers provide for whole bone impact tests?
4. How could impact testing help the design and testing of better fracture fixation devices?
5. For a projectile of 9 kg and a lower leg specimen of 16 kg, if the initial velocity of the projectile is 6 m/s and it comes to rest after impact, what is the postimpact velocity of the specimen? Also, assuming the force−time curve assumes a triangular profile with duration of 15 ms, estimate what the peak force is for this collision (answer: 3.375 m/s, 7.2 kN).

REFERENCES

1. Yoganandan N, Pintar FA, Boynton M, Begeman P, Prasad P, Kuppa SM, et al. Dynamic axial tolerance of the human foot-ankle complex. *Society of Automotive Engineers* 1996; **962426**:207−18.
2. Quenneville CE, Fraser GS, Dunning CE. Development of an apparatus to produce fractures from short-duration high-impulse loading with an application in the lower leg. *Journal of Biomechanical Engineering* 2010;**132**(1):014502-1-4.
3. North Atlantic Treaty Organization (NATO). In: *Test methodology for protection of vehicle occupants against anti-vehicular landmine effects*, vol. 323. The Netherlands: Research and Technology Organisation; 2007. TR-HFM-090 Rijswijk.
4. Kuppa S, Wang J, Haffner M, Eppinger R. Lower extremity injuries and associated injury criteria. In: *17th international technical conference on the enhanced safety of vehicles in Amsterdam, The Netherlands*, vol. 4; June 4−7, 2001. Paper #457. p. 1−15.
5. Griffin LV, Harris RM, Hayda RA, Rountree MS. Loading rate and torsional moments predict pilon fractures for antipersonnel blast mine loading. In: *International IRCOBI conference on the biomechanics of impacts, Isle of Man, UK*; October 10−12, 2001.

6. Seipel RC, Pintar F, Yoganandan N, Boynton MD. Biomechanics of calcaneal fractures: a model for the motor vehicle. *Clinical Orthopaedics and Related Research* 2001;**388**: 218–24.
7. Funk JR, Crandall JR, Tourret LJ, MacMahon CB, Bass CR, Patrie JT, et al. The axial injury tolerance of the human foot/ankle complex and the effect of achilles tension. *Journal of Biomechanical Engineering* 2002;**124**(6):750–7.
8. Quenneville CE, McLachlin SD, Greeley GS, Dunning CE. Injury tolerance criteria for short-duration axial impulse loading of the isolated tibia. *Journal of Trauma* 2011;**70**(1): E13–8.
9. Quenneville CE, Dunning CE. Evaluation of the biofidelity of the HIII and MIL-Lx lower leg surrogates under axial impact loading. *Traffic Injury Prevention* 2012;**13**(1):81–5.
10. Burkhart T, Quenneville CE, Dunning CE, Andrews DM. Development and validation of a distal radius finite element model to simulate impact loading indicative of a forward fall. *Proceedings of the Institition of Mechanical Engineers (Part H): Journal of Engineering in Medicine* 2014;**228**(3):258–71.
11. Quenneville CE, Dunning CE. Development of a finite element model of the tibia for short-duration high-force axial impact loading. *Computer Methods in Biomechanics and Biomedical Engineering* 2011;**14**(2):205–12.

CHAPTER

Quasi-static Stiffness and Strength Testing of Whole Bones and Implants

2

Radovan Zdero[1], Mina S.R. Aziz[2], Bruce Nicayenzi[3]

Western University, London, ON, Canada[1]; University of British Columbia, Vancouver, BC, Canada[2]; Ryerson University, Toronto, ON, Canada[3]

1. BACKGROUND

The largest and heaviest human long bones are the femur, tibia, and humerus. They experience load transfer along their midshaft (i.e., diaphysis), but also at proximal and distal joints (i.e., shoulder, elbow, hip, knee, and ankle) during routine domestic, athletic, and workplace activities, as well as injury events. Diseases like osteoarthritis can lead to total joint replacement surgery; diseases like osteoporosis can lead to fragility fractures requiring surgical repair with plates, nails, screws, and cables; and diseases like bone cancer can lead to tumors or cavities, which change the bone's biomechanical properties. Consequently, orthopaedic surgeons and biomechanical engineers should understand the biomechanical properties of intact, instrumented, and diseased whole bones (Fig. 2.1).[1–4] Stiffness is a type of "ratio" between nondestructive load applied to the specimen with respect to the resulting displacement, for example, "toe touch" weight-bearing of a patient immediately after femur fracture repair surgery. Strength is the maximum load that a specimen can withstand before clinical or structural failure, for example, force needed for hip fracture. Therefore, this chapter explains how to perform quasi-static (i.e., low-speed) stiffness and strength tests on whole bones and whole bone–implant constructs, as well as how to analyze, present, and interpret results.

2. RESEARCH QUESTIONS

Typical research questions might include one or more of the following:

- What are the stiffness and strength for a particular whole bone?
- What is the effect of osteoarthritis or osteoporosis on stiffness and strength?
- Does artificial sawbone vs. biological bone have the same stiffness and strength?
- Which implant provides bone–implant mechanical properties closest to intact whole bone?

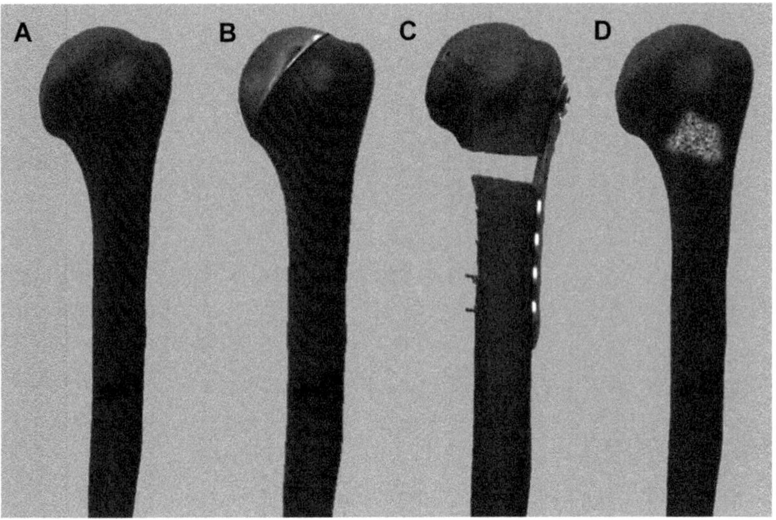

FIGURE 2.1

Photos of artificial sawbone humeri illustrating typical clinical conditions of whole bones. (A) Intact healthy humerus, (B) humerus with a shoulder joint replacement, (C) humerus with a fracture repair plate, (D) humerus with a simulated tumor.

- How do age, gender, limb side, nutrition, medication, etc., affect stiffness and strength?
- etc.

3. METHODOLOGY

3.1 GENERAL STRATEGY

Human whole bones are obtained and divided into equally-sized or different-sized groups, depending on the research topic. Specimens are intact, but may be instrumented or diseased depending on the research question. Quasi-static nondestructive stiffness tests are performed on all specimens, followed by quasi-static destructive strength tests. Raw data are then normalized by whole bone geometry. Quasi-static loading may not replicate many clinical conditions like impact injuries, but it is an acceptable standard approach, especially for comparative studies (i.e., test group 1 vs. 2 vs. 3, etc.), since all test groups undergo the same load regime. Statistical comparisons are then made between test groups for raw and normalized parameters. Finally, correlation coefficients are computed for raw and normalized data vs. bone demographics (i.e., age, sex, limb side, bone mineral density (BMD), and clinical T-score) to determine which factors are important.

GLOSSARY

✓ **Intact whole bones.** Whole bones that have not been cut, drilled, or damaged.
✓ **Quasi-static.** Load is applied slowly to minimize rate-dependent viscoelastic effects.
✓ **Sawbone.** Artificial bone substitute used for *in vitro* biomechanical research.
✓ **Stiffness.** Slope of the initial linear region of the load–displacement graph.
✓ **Strength.** Maximum load of the load–displacement graph, indicating failure.

SAFETY FIRST

✓ Remember to always wear goggles and gloves for protection.
✓ Place a protective plastic shield around the whole bone test area.
✓ Clean the work area and all tools with bleach or disinfectant after testing.

3.2 MATERIALS AND TOOLS LIST

- angle measurement gage
- cement for bone potting
- human whole bone
- mechanical tester
- steel cubes for bone potting
- tape measure
- Vernier calipers
- vices and clamps

3.3 SPECIMEN PREPARATION

Step 1. Store the bones. Fresh or fresh–frozen whole bones need to be wrapped in plastic strips or vacuum sealed in plastic bags for proper storage in a freezer prior to the study. Keep the freezer at −20°C or colder. Note that embalmed or dried/dehydrated bones do not need to be frozen and may simply be placed in a plastic box for storage at room temperature prior to tests.

Step 2. Thaw the bones. Remove specimens from the freezer, leave them in their plastic wrappings or vacuum-sealed bags, and place them on a suitable surface at ambient room temperature or in a warm water bath to thaw for at least 12 h. Remove bones from their plastic wrappings or bags, and soak or spray them with saline water solution to prevent dehydration. Remember to use a number tag or write a number on each bone to identify the specimens.

Step 3. Prepare the metal cubes. Cut a long hollow square or cylindrical metal tube into multiple smaller cube-like segments. For each cube, place several strips of

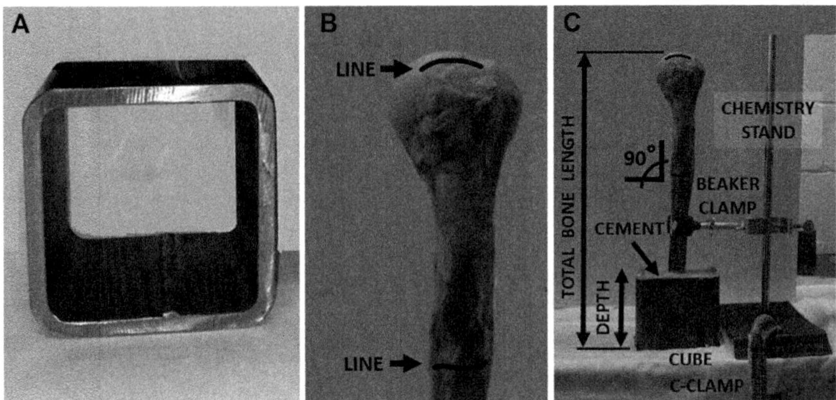

FIGURE 2.2

Specimen preparation for stiffness and strength tests. (A) Hollow metal cube, (B) pen lines drawn on the whole bone for alignment, (C) final whole bone potting setup.

duct tape on one open side in order to close it, which will prevent leakage during cement potting later on (Fig. 2.2A).

Step 4. Draw lines on bones. Use a pen to draw a line along the top surface (i.e., humeral head, femoral head, tibial plateau, etc.) of the whole bones in the coronal or sagittal direction and also at midlength (Fig. 2.2B). These will be used for specimen alignment in a later step.

Step 5. Mount the bones. Place the end of each whole bone into an empty metal cube (Fig. 2.2C). The end to be inserted (i.e., proximal or distal) will depend on the exact nature of the research question. In either case, because of geometric variability between biological whole bones, the relative depth of insertion into the metal cube must be the same with respect to the total length of each whole bone (e.g., % = relative dimension/reference dimension × 100 = insertion depth/total bone length × 100).

Step 6. Pot the bones. Insert and orient the whole bone vertically in coronal and sagittal planes using an angle measurement gage placed next to the previously drawn pen line at midshaft (Fig. 2.2C). Rotate the bone so the previously drawn pen line on the top surface is parallel to one side of the cube; this will aid consistent specimen-to-specimen mounting during mechanical tests. Next, fill the cube with casting material (e.g., anchoring cement, epoxy resign, nontoxic "field's metal," etc.), and allow for curing time according to the manufacturer's instructions.

TIPS AND TRICKS

- ✓ Whole bones can be cut to the same length, which will make potting go more quickly.
- ✓ Embalmed whole bones are easier to store and handle than fresh–frozen specimens.
- ✓ The same researcher(s) should perform all tests for consistency in results.

THE "GOLD STANDARD"

The American National Standards Institute (ANSI) document ANSI/ASAE S459 (Shear and three-point bending test of animal bone) provides a mechanical test method for biological whole bones, but only for midshaft shear and three-point bending. Otherwise, for other test modes that may be more clinically relevant for a particular research question, researchers are encouraged to refer to peer-reviewed journal articles.

3.4 SPECIMEN TESTING

Step 1. Determine testing order. Test the whole bones for nondestructive stiffness before destructive strength tests. An approach should be devised to deal with testing order effects (i.e., specimen order and test mode order) to avoid biased results, but this ideally requires changing both the test specimen and the test jig between each test. Alternatively, for a given test mode, individual whole bones can be randomly chosen (e.g., pulling numbers from a box) for all axial stiffness tests, then for all torsional stiffness tests, then for all cantilever stiffness tests, and then for all cantilever strength tests.

Step 2. Mount bones into test jig. Mount whole bones into the test jig for each test mode using the order decided above. For axial tests, mount the cement block into a vice attached to the base of the mechanical tester, position the proximal head into a smooth metal cup attached to the load cell, and decide if there should be any abduction, adduction, flexion, or extension angle to represent a clinical scenario (Fig. 2.3A). For torsional tests, mount the cement block into a vice attached to the load cell, and grip the head using four to six metal pins that are part of a ring

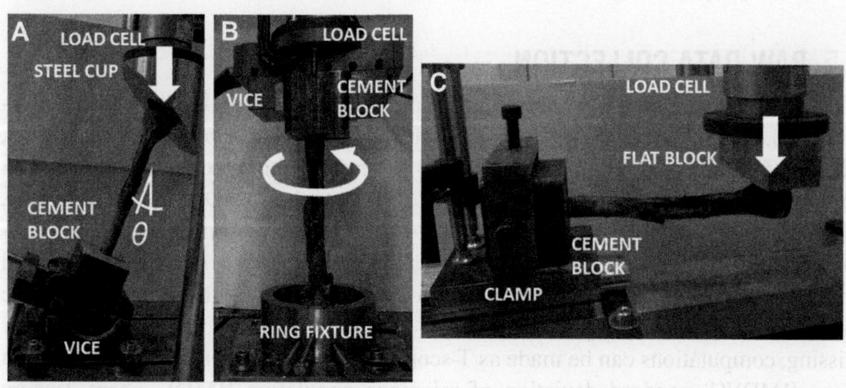

FIGURE 2.3

Experimental setup for whole bone stiffness and strength tests. (A) Axial loading, (B) torsional loading, (C) cantilever loading. White arrows indicate linear or torsional load, while θ represents abduction, adduction, flexion, or extension angle to replicate a clinical scenario.

fixture attached to the base of the mechanical tester (Fig. 2.3B). For cantilever tests, mount the cement block into a clamp or vice, and position the head under a smooth flat metal block attached to the load cell for load application (Fig. 2.3C). (Note: "Axial" mode produces axial compression along the long axis of the whole bone, plus some cantilever bending if an abduction, adduction, flexion, or extension angle is present.)

Step 3. Apply a preload. Before each stiffness or strength test, apply a small preload to eliminate any "mechanical slack" (i.e., hysteresis) in the system. For example, about 50–100 N (linear motion) or 0–0.5 N·m (torsional motion) are typical values for whole bones.[1,3,5,6]

Step 4. Perform stiffness tests. Apply 1–25 load cycles to precondition the whole bones so stiffness reaches "steady state," and then use the last one or two load cycles to compute stiffness.[7–9] For these load cycles, the specific research question(s) and the specimen's physical robustness may influence the choice of displacement or load control, displacement or load rate (e.g., 3–10 mm/min, 0.1–10 °/s, 50–100 N/s, and 0.25 N·m/s are common), and max displacement or load (e.g., 0.2–2 mm, 0.5–5°, 200–1500 N, and 12–30 N·m are common).[1,3,5–11] (Note: Such quasi-static tests may not replicate many clinical conditions like impact injuries, but they are an acceptable standard approach for comparative studies (i.e., test group 1 vs. 2 vs. 3, etc.), since all test groups undergo the same quasi-static load regime. Although preconditioning is sometimes not done, the specimens will have the same starting condition of not being preconditioned.)[1,5,6,10,11]

Step 5. Perform strength tests. Apply load until complete structural collapse of the specimen. Again, the choice of displacement or load control parameters will be informed by other factors.

Step 6. Inspect the specimens. Visually inspect the fractured whole bone, and take a sufficient number of photos for later analysis of the fracture mechanisms.

3.5 RAW DATA COLLECTION

Step 1. Record bone characteristics. Enter human or animal demographic information (i.e., age, sex, and left or right limb) and quantitative bone properties (i.e., BMD and clinical T-score), which will help determine which factors are statistically correlated with stiffness and strength (Table 2.1). Note that BMD is in 2-D units (i.e., g/cm^2 rather than g/cm^3), since DEXA (dual-energy X-ray absorptiometry) bone density scans only provide 2-D scans. DEXA scan reports often provide BMD values and corresponding clinical T-scores (i.e., normal T-score ≥ -1, $-1 >$ osteopenic T-score > -2.5, and osteoporotic T-score ≤ -2.5). When T-score is overlooked or missing, computations can be made as T-score = (bone BMD – reference population mean BMD)/(1 standard deviation of reference population BMD), where T-score reference populations are young healthy adults of the same sex that are aged 20–40 years.[12,13]

Step 2. Record bone geometry. Measure and record key aspects of whole bone geometry that will be used for data normalization later. Use a Vernier caliper to

Table 2.1 Demographic information for whole bones.

Whole Bone	Age [Years]	Sex [M, F]	Limb Side [L, R]	BMD [g/cm^2]	T-Score	Midshaft Outer Diameter [mm]	Total Length [mm]
1							
2							
3							
etc.							
Avg		–	–				
SD		–	–				

Avg, *average;* BMD, *bone mineral density;* SD, *standard deviation.*

measure the average midshaft outer diameter and a tape measure to determine the total length (Table 2.1). Due to the complex geometry of biological whole bones, measurements will only be approximations for each specimen and there will be great variability between specimens.

Step 3. Record raw stiffness. Stiffness is the slope of the average line of best fit of the load–displacement graph obtained from a linear or torsional stiffness test (Fig. 2.4A, Table 2.2). Any hysteresis (i.e., a gap between the load and unload curves) may be due to the nonlinear material behavior of bone caused by too much load being applied, mechanical slack in the test jig, slippage of bone under the loading platen, etc. This is acceptable if the line of best fit has a linearity of $R^2 \geq 0.9$, indicating no permanent damage has been done to the specimen. However, if $R^2 < 0.9$, then these other factors should be double-checked.

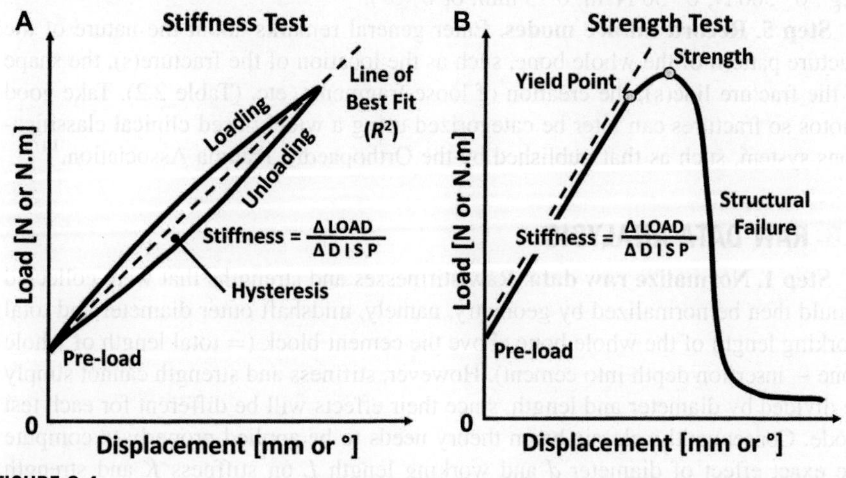

FIGURE 2.4

Raw stiffness and strength graphs. (A) Stiffness test showing a load–displacement profile, (B) strength test showing a load–displacement profile.

Table 2.2 Raw stiffness and strength for whole bones.

Whole Bone	Axial Stiffness [N/mm]	Torsional Stiffness [N·m/°]	Cantilever Stiffness [N/mm]	Cantilever Strength [N]	Fracture Pattern
1					
2					
3					
etc.					
Avg					–
SD					–

Avg, *average;* SD, *standard deviation.*

Step 4. Record raw strength. Strength is the maximum load of the load–displacement graph obtained from a linear or torsional strength test and is considered "mechanical failure" (Fig. 2.4B, Table 2.2). Alternatively, some clinical criterion for allowable load (e.g., 500 N or 50 N·m) or displacement (e.g., 5 mm or 5°) can be used to define "clinical failure." Also, if separate stiffness tests are not performed, stiffness can still be obtained from a strength graph (i.e., slope of the initial linear region of the load–displacement graph up to the linear yield point) (Fig. 2.4B). However, this stiffness depends on how the linear region is defined in a strength graph, since linear yield point is sometimes difficult to determine. One approach is to use the engineer's 0.2% offset displacement method. Another approach is to choose a load or displacement range based on a clinical criterion (e.g., 0–500 N, 0–50 N·m, 0–5 mm, or 0–5°).

Step 5. Record failure modes. Enter general remarks about the nature of the fracture pattern of the whole bone, such as the location of the fracture(s), the shape of the fracture line(s), the creation of loose fragments, etc. (Table 2.2). Take good photos so fractures can later be categorized using a widely used clinical classifications system, such as that published by the Orthopaedic Trauma Association.[14]

3.6 RAW DATA ANALYSIS

Step 1. Normalize raw data. Raw stiffnesses and strengths that were collected should then be normalized by geometry, namely, midshaft outer diameter and total working length of the whole bone above the cement block (= total length of whole bone − insertion depth into cement). However, stiffness and strength cannot simply be divided by diameter and length, since their effects will be different for each test mode. Consequently, classic beam theory needs to be applied properly to compute the exact effect of diameter d and working length L on stiffness K and strength S.[1,15] Thus, $K_{NORM} = K_{RAW}(L/d^2)$ for axial stiffness, $K_{NORM} = K_{RAW}(L/d^4)$ for torsional stiffness, $K_{NORM} = K_{RAW}(L^3/d^4)$ for cantilever stiffness, and $S_{NORM} = S_{RAW}(L/d^3)$ for cantilever strength.[1]

Step 2. Calculate correlation coefficients. Plot the measured result for each individual specimen (i.e., axial stiffness 1, 2, 3, etc.) (Table 2.2) vs. its corresponding specimen characteristic (i.e., age 1, 2, 3, etc.) (Table 2.1) to help visualize the interrelationship between measurements vs. characteristics. Then, calculate the correlation coefficient R for each measurement–characteristic pair to determine which characteristic has an influence on measurements, for example, $R > 0.8$ is a typical value considered to indicate a strong correlation. Make sure to do this for raw and normalized stiffness and strength measurements.

Step 3. Perform statistical comparisons. Choose the criterion for statistical difference (e.g., $P < 0.01$ or < 0.05). Use raw and normalized data to compare different patient groups (e.g., men ≤60 years old vs. men >60 years old, "normal" women vs. "osteoporotic" women, etc.), bone types (e.g., femur vs. tibia, left humerus vs. right humerus, etc.), etc. Use a statistical software program to compare two test groups (e.g., paired *t*-test) or two or more test groups influenced by multiple factors (e.g., analysis of variance, ANOVA).

Step 4. Compute statistical power. Perform power analysis after the study to ensure there were enough specimens per group to detect all statistical differences that were actually present (i.e., was type II statistical error avoided?). Statistical power > 80% is usually considered to indicate there were enough specimens per test group. Note that if good predictions of averages and standard deviations are available from prior studies, then the number of specimens and/or tests can be chosen before the study begins to ensure a power > 80%.

ENGINEER'S TOOLBOX

Raw stiffness K and strength S of a whole bone under standard test modes can be approximated using classic beam theory. Assume the whole bone is a solid circular cylinder with diameter d, length L, linear elastic modulus E, shear elastic modulus G, ultimate tensile stress σ_U, and abduction, adduction, flexion, or extension angle θ. The final forms of the formulas are axial stiffness $K = \pi E d^2 / (4L\cos\theta)$, torsional stiffness $K = \pi G d^4/(32L)$, cantilever stiffness $K = 3\pi E d^4/(64L^3)$, and cantilever strength $S = \pi \sigma_U d^3/(32L)$.

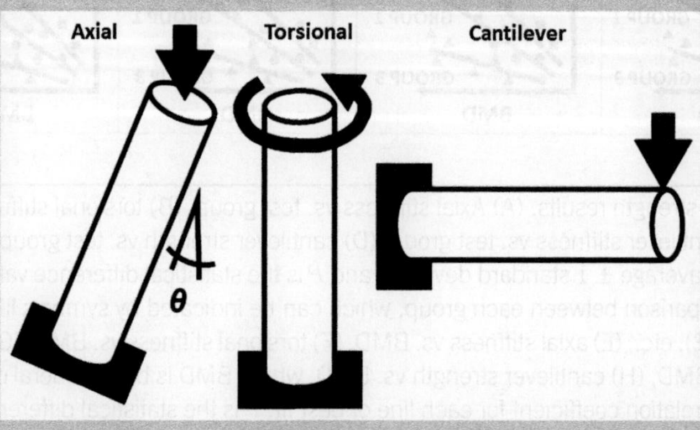

4. RESULTS

Once all stiffness and strength data collection and analysis have been performed, it is then important to communicate and present the primary results in an understandable and concise manner to the reader of a journal article, conference paper, technical report, or book chapter.

Step 1. Show main results. The main numerical findings of the study should be presented (Fig. 2.5). This includes raw and normalized stiffness and strength. All statistical pairwise comparisons should be made (i.e., test group 1 vs. 2 vs. 3, etc.) to generate multiple statistical P values (Fig. 2.5A–D). Also, any linear (or nonlinear) trends can be illustrated vs. BMD (Fig. 2.5E–H). For each line of best fit, several items can be generated: an equation $y = mx + b$ showing slope m and intercept b, a linear correlation coefficient R, and its own statistical P value to ensure that the slope (i.e., slope $= m$) is statistically different than a horizontal line (i.e., slope $= 0$).

Step 2. Show failure modes. A photo of a whole bone after strength tests helps visualize a typical fracture pattern and relates that to the type of stress (i.e., tension or compression) applied to the specimen (Fig. 2.6).

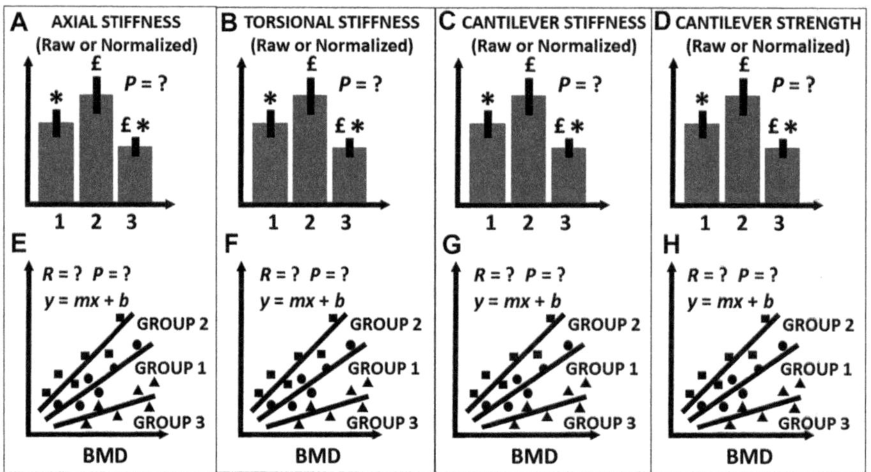

FIGURE 2.5

Stiffness and strength results. (A) Axial stiffness vs. test group, (B) torsional stiffness vs. test group, (C) cantilever stiffness vs. test group, (D) cantilever strength vs. test group, where bar graphs show average ± 1 standard deviation and P is the statistical difference value for each pairwise comparison between each group, which can be indicated by symbols like asterisks (*), pounds (£), etc., (E) axial stiffness vs. BMD, (F) torsional stiffness vs. BMD, (G) cantilever stiffness vs. BMD, (H) cantilever strength vs. BMD, where BMD is bone mineral density, R is the linear correlation coefficient for each line of best fit, P is the statistical difference value to ensure that the slope of each line of best fit (i.e., slope $= m$) is different than a horizontal line (i.e., slope $= 0$), and $y = mx + b$ is the equation of each line.

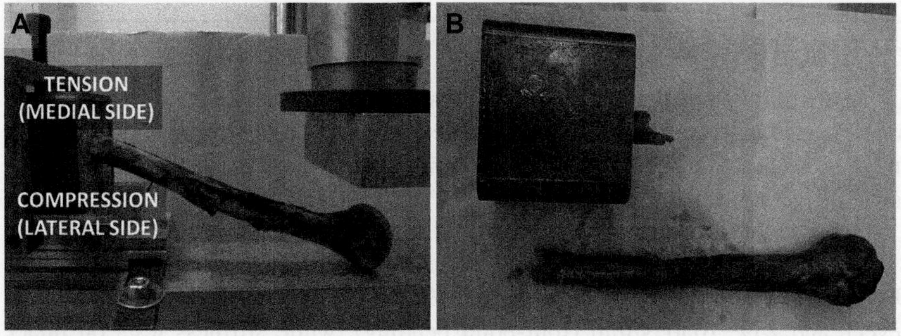

FIGURE 2.6

Fracture photos. (A) Whole bone position after completing the cantilever strength test, (B) whole bone fracture line near the cement block.

Table 2.3 Whole bone fracture types after cantilever strength tests.

Fracture Type	Whole Bone Group 1	Whole Bone Group 2	Whole Bone Group 3	etc.
Transverse				
Oblique				
Transverse + oblique				
Transverse + reverse oblique				
etc.				

Step 3. Show failure distribution. Tabulate the different fracture patterns to identify any trends. Statistically compare the test groups to see if they yielded the same distribution of fracture types (Table 2.3).

ALTERNATIVES AND ADAPTATIONS

- ✓ **Tibias and femurs**. For tibias, a swivel joint with two metallic spheres simulates axial load from femoral condyles. For femurs, an "offset" vertical load to the anterior or posterior side of the femoral head and a support block under the proximal shaft produces torsion around the shaft.
- ✓ **Three-point or four-point bending**. These are other very common test modes for biomechanical studies on intact, instrumented, or diseased whole bones. However, they often do not replicate clinical conditions as faithfully as axial, torsional, or cantilever loading modes.
- ✓ **Implants and bone–implant constructs**. Implants by themselves and whole bones with total joint replacements and/or fracture fixation plates or nails can be tested with only some minor modifications to the test procedures employed for intact whole bones.
- ✓ **Sawbones**. Artificial whole bone surrogates have similar mechanical properties to human whole bones and are anatomically correct, but have isotropic and homogeneous properties. 3-D density (i.e., g/cm^3), rather than 2-D BMD (i.e., g/cm^2), can be obtained from the manufacturer, since DEXA scans for BMD are only suitable for biological bone.

5. DISCUSSION

After completing all quasi-static stiffness and strength tests, data collection, data analysis, and data presentation, then final results can be considered and interpreted in the broader context of some important clinical, biomechanical, and/or technological considerations, as follows.

Quasi-static stiffness and strength of whole bones have a wide range of values.[1,3,16–18] For example, stiffness tests on intact human humeri yield 968–1499 N/mm (axial), 2.01–5.93 N·m/° (torsional), 7.5–27.5 N/mm (cantilever), and 259–1050 N/mm (four-point bending), while stiffness tests on intact sawbone humeri result in 179–1651 N/mm (axial), 2.72–4.47 N·m/° (torsional), 12.6–13.6 N/mm (cantilever), and 141–918 N/mm (four-point bending). Similarly, strength tests on intact human humeri yield 21–100 N·m (torsional) and 363–489 N (cantilever), while intact sawbone humeri show 2.72–3.22 N·m (torsional) and 294–300 N (cantilever). This kind of variability is common for the humerus, femur, and tibia, due to different bone geometry, bone density, and loading regime. Moreover, total joint replacements, fracture repair devices, or diseases can further influence whole bone stiffness and strength.[2–4,6,11]

BMD appears to be correlated to whole bone mechanical properties in some cases, but not others. One investigation showed that BMD for intact human humeri was moderately correlated with raw stiffness and strength in 9 of 12 instances (linear correlation coefficient $R = 0.49–0.80$), but only in 3 of 12 instances after data were normalized with geometry ($R = 0.43–0.46$).[1] This discrepancy may have been caused by the narrow age range of donors (i.e., 60–97 years), whereas a wider age range including younger adults (e.g., 20–60 years) may provide stronger correlations since it would be more representative of the general population.

Clinical fracture patterns of whole bones vary greatly and can occur due to many circumstances. This includes domestic, workplace, sports, or motor vehicle injuries, total joint replacements that create stress risers at bone–implant interfaces, diseases like bone tumors that weaken bone, and mechanical tests on whole bones performed in the laboratory. Fracture patterns differ depending on bone type, bone material properties, and loading conditions. The Orthopaedic Trauma Association has published a document that is used as a fracture classification system.[14] The document shows that whole bone injuries can occur for the proximal, shaft, or distal regions, but within each of these main zones, there are numerous fracture subtypes.

Biomechanical methodologies for *in vitro* laboratory research have not been fully standardized yet for whole bones, although some general considerations have been established.[19,20] Consequently, peer-reviewed journal articles must serve as a guide while considering the particular research question. For example, loading modes can include separate axial, torsional, cantilever, three-point bending or four-point bending tests, or simultaneous combinations of several test modes. Moreover, the use of quasi-static vs. dynamic variable vs. dynamic cyclic vs. high-speed impact loads may be based on practical limits of the research equipment available, the desire to minimize rate-dependent viscoelastic effects, and the clinical research question. Also, simplifications like quasi-static force application and the elimination of muscle, ligament, and tendon effects may be acceptable for cadaveric comparative lab tests (e.g., implant A vs. B vs. C for the same whole bone type), but do not fully replicate *in vivo* physiological conditions.

6. SUMMARY

- Intact, instrumented, or diseased whole bones can be tested biomechanically.
- Stiffness is the initial slope of the applied quasi-static load-displacement graph.
- Strength is the maximum load that causes complete structural collapse.
- Biomechanical testing may include various types of loading modes.
- Stiffness and strength can be predicted using engineering beam equations.
- Stiffness and strength for whole bones have a wide range of values.

7. QUIZ QUESTIONS

1. What are the definitions of stiffness and strength?
2. What simplifying assumptions are made when predicting stiffness?
3. Are stiffness and strength different for healthy vs. diseased whole bones?
4. Is immediate full weight-bearing allowed after tibial or femoral fracture repair?
5. Predict the torsional stiffness of an intact human whole bone. Assume shear elastic modulus = 4 GPa, midshaft outer diameter = 20 mm, and total length = 350 mm (answer: 3.13 N·m/°).

REFERENCES

1. Aziz MSR, Nicayenzi B, Crookshank MC, Bougherara H, Schemitsch EH, Zdero R. Biomechanical measurements of stiffness and strength for five types of whole human and artificial humeri. *Journal of Biomechanical Engineering* 2014;**136**(5):051006-1-10.
2. Merolla G, Nastrucci G, Porcellini G. Shoulder arthroplasty in osteoarthritis: current concepts in biomechanics and surgical technique. *Translational Medicine at UniSa* 2013;**6**: 16–28.
3. Lescheid J, Zdero R, Shah S, Kuzyk PRT, Schemitsch EH. The biomechanics of locked plating for repairing proximal humerus fractures with or without medial cortical support. *Journal of Trauma* 2010;**69**(5):1235–42.
4. Potter BK, Adams SC, Pitcher D, Malinin TI, Temple HT. Proximal humerus reconstructions for tumors. *Clinical Orthopaedics and Related Research* 2009;**467**(4):1035–41.
5. Papini M, Zdero R, Schemitsch EH, Zalzal P. The biomechanics of human femurs in axial and torsional loading: comparison of finite element analysis, human cadaveric femurs, and synthetic femurs. *Journal of Biomechanical Engineering* 2007;**129**(1):12–9.
6. Al-Jahwari A, Schemitsch EH, Wunder JS, Ferguson PC, Zdero R. The biomechanical effect of torsion on humeral shaft repair techniques for completed pathological fractures. *Journal of Biomechanical Engineering* 2012;**134**(2):024501-1-7.

7. Zdero R, Gallimore CH, McConnell AJ, Patel H, Nisenbaum R, Morshed G, et al. A preliminary biomechanical study of cyclic preconditioning effects on canine cadaveric whole femurs. *Journal of Biomechanical Engineering* 2012;**134**(9):094502-1-7.
8. Heiner AD, Brown TD. Structural properties of a new design of composite replicate femurs and tibias. *Journal of Biomechanics* 2001;**34**(6):773−81.
9. Cristofolini L, Viceconti M, Cappello A, Toni A. Mechanical validation of whole bone composite femur models. *Journal of Biomechanics* 1996;**29**(4):525−35.
10. Cristofolini L, Viceconti M. Mechanical validation of whole bone composite tibia models. *Journal of Biomechanics* 2000;**33**(3):279−88.
11. Fulkerson E, Koval K, Preston CF, Iesaka K, Kummer FJ, Egol KA. Fixation of periprosthetic femoral shaft fractures associated with cemented femoral stems: a biomechanical comparison of locked plating and conventional cable plates. *Journal of Orthopaedic Trauma* 2006;**20**(2):89−93.
12. Lewiecki EM, Borges JLC. Bone density testing in clinical practice. *Arquivos Brasileiros de Endocrinologia and Metabologia* 2006;**50**(4):586−95.
13. Pettersson U, Nordström P, Lorentzon R. A comparison of bone mineral density and muscle strength in young male adults with different exercise level. *Calcified Tissue International* 1999;**64**(6):490−8.
14. Marsh JL, Slongo TF, Agel J, Broderick JS, Creevey W, DeCoster TA, et al. Fracture and dislocation classification compendium − 2007: Orthopaedic Trauma Association classification, database and outcomes committee. *Journal of Orthopaedic Trauma* 2007;**21**(10 Suppl.):S1−133.
15. Norton RL. *Machine design: an integrated approach*. Upper Saddle River (NJ, USA): Prentice-Hall; 1996.
16. Grover P, Albert C, Wang M, Harris GF. Mechanical characterization of fourth generation composite humerus. *Proceedings of the Institution of Mechanical Engineers (Part H): Journal of Engineering in Medicine* 2011;**225**(12):1169−76.
17. Dunlap JT, Chong AC, Lucas GL, Cooke FW. Structural properties of a novel design of composite analogue humeri models. *Annals of Biomedical Engineering* 2008;**36**(11):1922−6.
18. Schopfer A, Hearn TC, Malisano L, Powell JN, Kellam JF. Comparison of torsional strength of humeral intramedullary nailing: a cadaveric study. *Journal of Orthopaedic Trauma* 1994;**8**(5):414−21.
19. Sedlin ED, Hirsch C. Factors affecting the determination of the physical properties of femoral cortical bone. *Acta Orthopaedica Scandinavica* 1966;**37**:29−48.
20. ANSI/ASAE S459. Shear and three-point bending test of animal bone. Washington (DC, USA): American National Standards Institute (ANSI); 1998. www.ansi.org.

CHAPTER

3 Surface Strain Gage Testing of Whole Bones and Implants

Radovan Zdero[1], Suraj Shah[2], Peter Goshulak[3]

Western University, London, ON, Canada[1]; Ryerson University, Toronto, ON, Canada[2]; University of Toronto, Toronto, ON, Canada[3]

1. BACKGROUND

Strain is the relative change per unit length of an object's physical dimensions caused by an applied mechanical stress. High strains in whole bones, implants, and whole bone–implant constructs may cause mechanical failure,[1] but low strains in whole bone may cause "stress shielding," which leads to bone atrophy, bone resorption, and implant loosening.[2] Consequently, a number of strain analysis techniques are commonly used in orthopaedic biomechanics applications, such as extensometers, fiber optic sensors, pressure sensitive film, thermography, and ultrasound.[2–4] However, strain gages are the "gold standard" tool for assessing whole bones and prostheses, as well as for validating other experimental and computational methods.[4–6] Strain gages are a sensitive, precise, inexpensive, widely used, and highly tested technology, although they can be physically fragile, inconsistent, and prone to drift.[3,6–8] Strain gages are composed of a metal alloy wire or foil laid out in a grid-like pattern, whose electrical resistance changes due to mechanical strain (Fig. 3.1).[7–11] Therefore, this chapter explains how to perform strain gage testing for whole bones, fracture fixation devices, total joint replacements, and biomaterials, as well as how to analyze, present, and interpret results.

2. RESEARCH QUESTIONS

Typical research questions might include one or more of the following:

- How high are surface strains on whole bones, implants, or whole bone–implant constructs?
- Do patient age, sex, limb side, or bone density influence a whole bone's surface strains?
- Do shape, size, density, and surface texture influence an implant's surface strains?

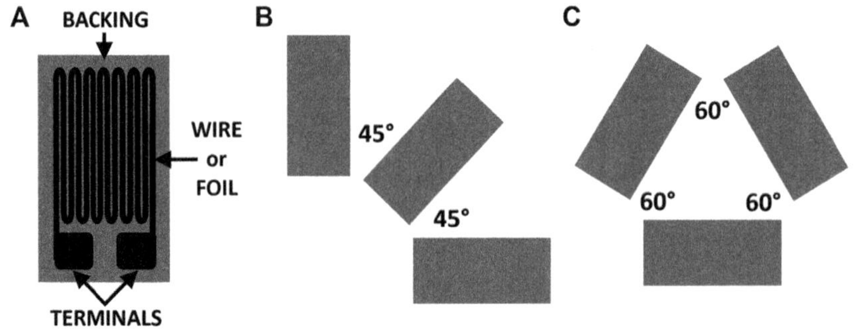

FIGURE 3.1

Strain gages for orthopaedic biomechanics research. (A) Linear strain gage, (B) rosette strain gage (45° rectangular pattern) composed of three linear gages, (C) rosette strain gage (60° delta pattern) composed of three linear gages. Rosette layouts do not need to be created by researchers using linear gages, since prefabricated rosettes are commercially available.

- How does the applied load on a bone or joint affect surface strain distribution?
- etc.

3. METHODOLOGY

3.1 GENERAL STRATEGY

Strain gage mounting, testing, and analysis are used to yield accurate surface strains at specific locations on artificial sawbones, implants, and/or artificial sawbone–implant constructs using a well-established protocol.[2,4,5,7–11] Strain gages are first mounted on specimens using a step-by-step process. Specimens are then mounted in a mechanical tester, which will generate loads. Simultaneously, strain readings are recorded using an analog-to-digital converter and dedicated computer software. Strains are analyzed to determine specimen areas with high strain (i.e., risk of failure) or low strain (i.e., risk of "stress shielding"). Finally, test groups composed of multiple specimens are statistically compared, and correlation coefficients are computed for strain vs. sawbone or implant characteristics to identify important factors.

GLOSSARY

✓ **Strain.** The relative change of an object's physical dimensions per unit length.
✓ **Strain gage.** A sensor that is able to detect changes in an object's surface strain.
✓ **Principal strain.** The maximum or minimum in-plane strain at a particular location.
✓ **Von Mises strain.** The magnitude, but not direction, of the resultant in-plane strain at a location.

SAFETY FIRST

✓ Always wear goggles and gloves for protection during preparation and testing.
✓ Clamp the specimen securely to a stable surface when mounting strain gages.
✓ Place a plastic protective shield around the test setup during strain gage experiments.

3.2 MATERIALS AND TOOLS LIST

- cleaning materials (i.e., liquid solvent, carbide paper)
- gluing materials (i.e., adhesive liquid, adhesive tape, sponge)
- lead wire materials (i.e., soldering iron, soldering wire, wire leads)
- mechanical tester
- sawbones and implants
- strain data acquisition system
- strain gages
- tweezers
- volt-ohm meter

3.3 SPECIMEN PREPARATION

Step 1. Complete preliminary tasks. Prior to mounting the strain gages, if possible, complete all preliminary tasks, which may be mechanically, thermally, or chemically detrimental to the strain gages. These tasks might include surgical procedures on sawbones and/or implants, potting of specimens into bone cement, anchoring cement, or liquid metal for later mounting into the test jig, and application of paints or coatings to the specimen surfaces.

Step 2. Choose the locations. Keep several things in mind when choosing locations for mounting strain gages. First, linear gages are most suitable for relatively flat surfaces when they are aligned lengthwise in the direction of primary tension or compression (if known) since they provide only 1-D data, while rosette gages can be used for curved surfaces and complex stress states since they can provide 2-D data. Second, for sawbone–implant constructs, it may be more optimal to mount gages on sawbone, rather than on implants, since sawbone often has more surface area to accommodate gages. Third, mount gages away from sharp ridges, fracture lines, plates, screws, etc., since gages may not always respond consistently to sudden changes in geometry or strain. Fourth, mount gages where high and low strains are expected in order to cover a wide range of potential strains. Finally, mount gages far away from each other to avoid any interference.

Step 3. Abrade the installation area. Remove loose debris, such as galvanized coatings, oxides, paint, rust, etc. This is done by sanding the surface by hand using silicon carbide paper with a grit size suitable for the material (i.e., sawbone, ceramic, metal, polymer) (Fig. 3.2A). (Note: In some situations, oils, organic contaminants,

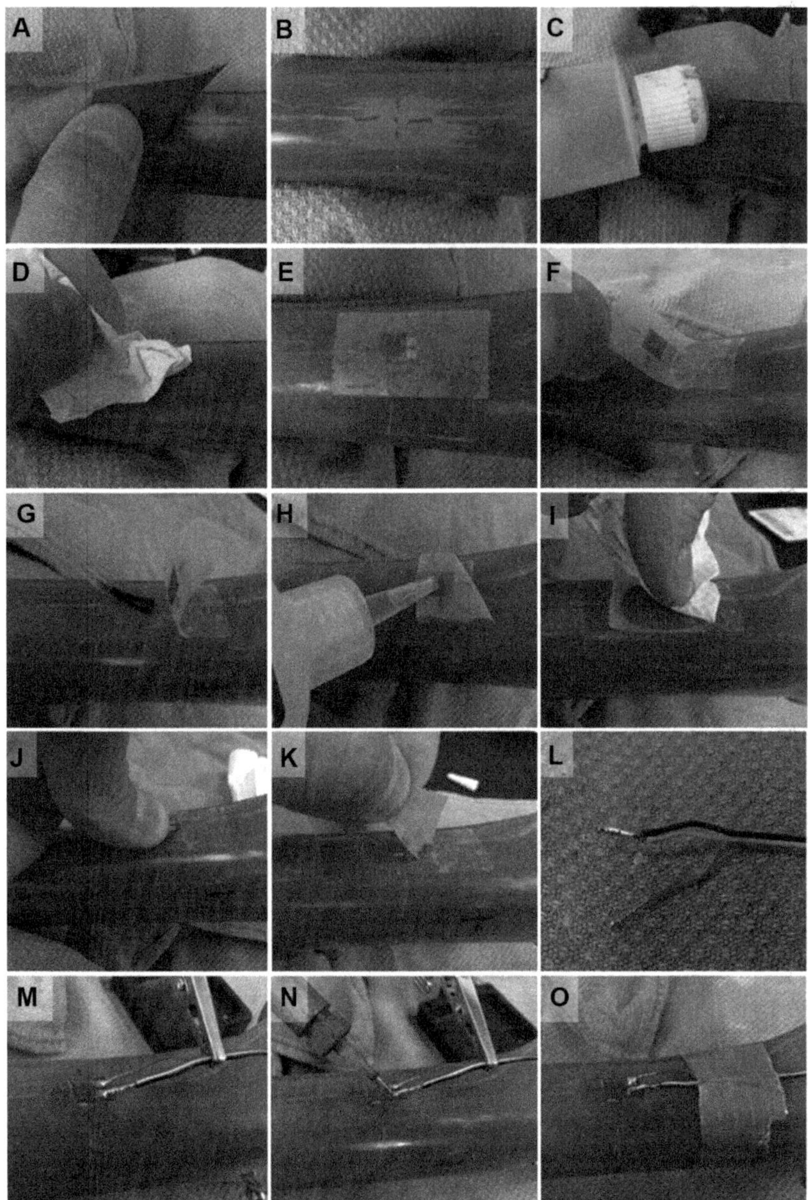

FIGURE 3.2

Strain gage mounting process. (A) Abrade, (B) draw, (C) condition, (D) dry, (E) tape, (F) lift, (G) catalyze, (H) glue, (I) wipe, (J) press, (K) remove, (L) strip, (M) stabilize, (N) solder, (O) tape. Linear and rosette gages are mounted using exactly the same process.

and soluble chemical residues may also need to be removed using a specialized "degreaser" just before abrading the surface.)

Step 4. Draw the installation area. Use a ruler to measure and a pen or pencil to draw layout lines onto or around the installation area (Fig. 3.2B). Small pen or pencil marks will not likely contaminate the surface appreciably.

Step 5. Condition the installation area. Repeatedly apply specialized liquid "conditioner" to the installation area while scrubbing with a cotton-tipped applicator (Fig. 3.2C). Conditioner should not be allowed to dry. Do this until the tip is no longer discolored, indicating the area is clean.

Step 6. Dry the installation area. Place a clean cloth or sponge inside the installation area, and wipe away from it using a single slow stroke (Fig. 3.2D). Then, with a fresh cloth or sponge, repeat this in the opposite direction. Never wipe the installation area back and forth, since this could bring in contaminants from outside the installation area.

Step 7. Tape the strain gage. Use tweezers to take the strain gage from its packaging and rest it (with bonding side facing down) onto a smooth clean surface, like glass, metal, or plastic. Obtain specialized, cellophane, clear adhesive tape that is, say, 10–20 mm longer than the strain gage. Place the tape (sticky side down) onto the strain gage, making sure it is centered on the tape. Then, lift the tape (with the strain gage attached) off the resting surface at an angle of about 45°. Next, place the tape (with the strain gage still attached) onto the installation area (Fig. 3.2E).

Step 8. Lift the strain gage. Lift one side of the tape (with the strain gage attached) completely off the installation area in order to fully expose the underside of the strain gage (Fig. 3.2F).

Step 9. Catalyze the installation area. Apply one to two drops of specialized liquid "catalyst" to the installation area. Use a clean brush to produce a thin uniform layer of catalyst (Fig. 3.2G). Allow the catalyst to dry for about 1 min.

Step 10. Glue the strain gage. Apply one to two drops of specialized liquid "adhesive" to the underside of the strain gage or, alternatively, at the tape and specimen juncture which should be about 10 mm away from the installation area (Fig. 3.2H).

Step 11. Wipe the strain gage. Make a slow and smooth wiping motion with a cloth or sponge across the top of the tape, thereby pressing the strain gage (with glue on its underside) onto the entire catalyzed installation area (Fig. 3.2I).

Step 12. Press the strain gage. Apply thumb pressure evenly across the top of the tape and strain gage for approximately 1–2 min while the adhesive dries (Fig. 3.2J).

Step 13. Remove the tape. Carefully lift one end of the tape. Then, slowly remove the tape from the top of the strain gage in order to fully expose the strain gage terminals (Fig. 3.2K).

Step 14. Strip the wire. Obtain insulated, stranded, tinned copper, three-conductor wire. Cut the wire to the appropriate length, but leave a little extra in case. Separate the three color coded conductors at both ends of the wire. Strip 10–15 mm of insulation from each conductor; a thermal stripper is recommended to prevent nicking the conductors. By hand, tightly twist together the exposed black and white conductors so they cannot be separated (Fig. 3.2L).

Step 15. Stabilize the wire. Apply an alligator clip (attached to a stand) onto the insulated part of the wire. Then, bring the conductors into contact with the strain gage terminals (Fig. 3.2M).

Step 16. Solder the wire. Solder the entwined black and white conductors onto one terminal of the strain gage and the red conductor to the other terminal (Fig. 3.2N).

Step 17. Tape the wire. Apply tape to the insulated part of the wire to prevent conductors from being accidentally pulled off the terminals during specimen handling and testing (Fig. 3.2O).

Step 18. Repeat the process. Repeat steps 3–17 until each strain gage has been prepared properly for the specimen. Typical, fully strain-gaged specimens are shown (Fig. 3.3).

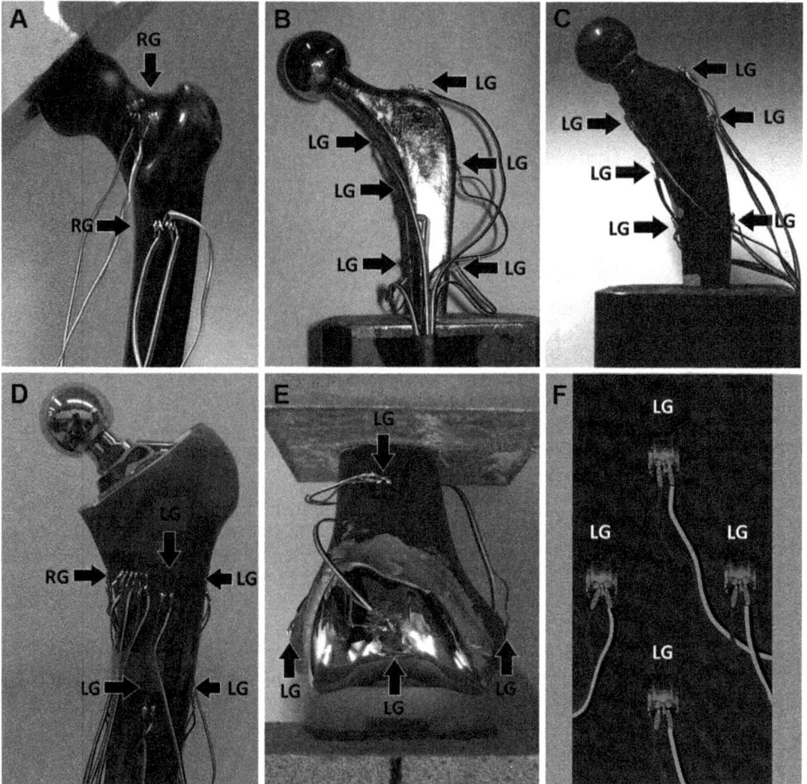

FIGURE 3.3

Orthopaedic test specimens with strain gages. (A) Sawbone femur, (B) metal hip implant, (C) polymer-based composite hip implant, (D) sawbone femur with a hip implant, (E) sawbone femur with a knee implant, (F) plate made from a polymer-based composite biomaterial. *LG*, linear gage; *RG*, rosette gage.

TIPS AND TRICKS

✓ The same orthopaedic surgeon should perform all surgeries for consistency.
✓ Mount gages after surgeries and potting are done to prevent damaging gages.
✓ Mount gages away from stress risers, like ridges, fracture lines, and implant edges.
✓ Rosette gages, rather than linear gages, are more suitable for complex geometries.
✓ Keep electronic/electrical gadgets away from strain gages during data collection.

THE "GOLD STANDARD"

The American Society for Testing and Materials (ASTM) has several documents that provide standardized guidelines for strain gages, namely, ASTM E1237 (Standard guide for installing bonded resistance strain gages) and ASTM E1561 (Standard practice for analysis of strain gage rosette data). However, for specific orthopaedic biomechanics applications, researchers should also consult peer-reviewed journal articles.

3.4 SPECIMEN TESTING

Step 1. Connect the wire. Connect the exposed conductors of each strain gage to complete an electronic "wheatstone" bridge circuit (i.e., quarter, half, or full bridge) inside a connector plug, which in turn is plugged into an analog-to-digital converter box, which, in turn, is connected to a computer that has dedicated software for strain gage data collection and analysis (Fig. 3.4).

Step 2. Double-check the strain gages. Apply a volt-ohm meter across the strain gage terminals to ensure the conductors are properly soldered to the terminals. First

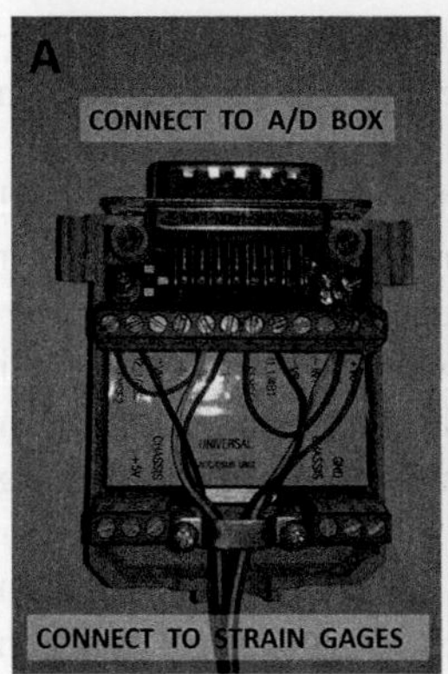

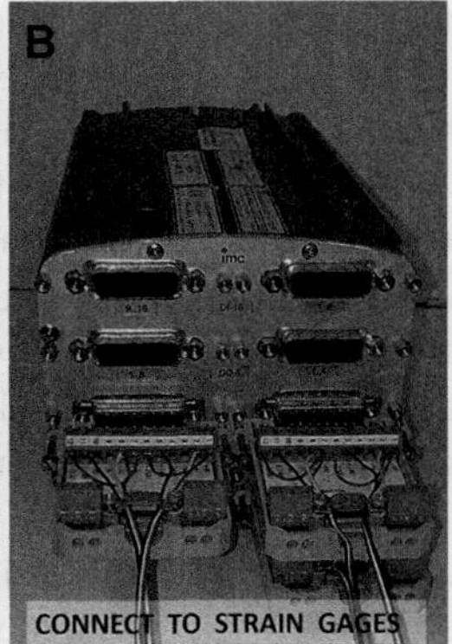

FIGURE 3.4

Strain gage electronics. (A) Connector plug showing internal "wheatstone" bridge circuit, (B) analog-to-digital (A/D) box showing the side which connects to the strain gages, (C) analog-to-digital (A/D) box showing the side which connects to the computer.

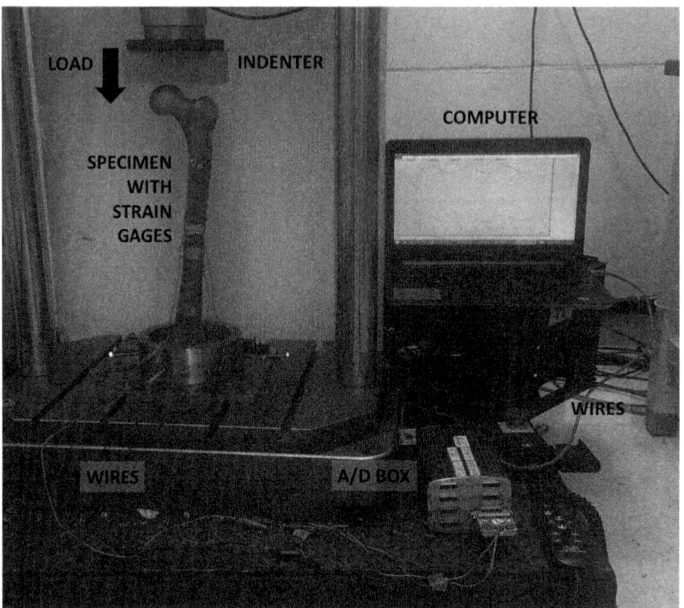

FIGURE 3.5

Strain gage test setup for orthopaedic biomechanics research.

check the electrical resistance of the unloaded specimen, and then see if resistance changes when a small load is applied by hand to the specimen. Reapply the soldering iron if there is a problem.

Step 3. Mount the specimen. Secure the sawbone, implant, or sawbone–implant construct into the test jig and the mechanical tester (Fig. 3.5). Orient the specimen appropriately. Lower the loading indenter to hover over the specimen.

Step 4. Position the safety shield. Place a transparent, protective, plastic shield around the specimen test area if there is a potential concern about specimen fracture and flying debris caused by accidental quasi-static or cyclic overloading.

Step 5. Configure the software. Adjust the computer software settings for strain gage data acquisition, such as gage factor, gage electrical resistance, data sampling duration, data sampling frequency, data processing algorithm, "wheatstone" bridge type, voltage level, etc.

Step 6. Start load application. Apply a small preload of 50–100 N to eliminate "mechanical slack" in the test setup (i.e., hysteresis), such as slippage between the specimen and loading indenter. Then, based on the research question being investigated, apply the appropriate nondestructive quasi-static or cyclic load for about 30–60 s to allow the strains to reach "steady state."

Step 7. Collect strain data. Trigger the computer to start acquiring strain data for an additional 30–60 s after the "steady state" mentioned above was reached.

Table 3.1 Raw linear and/or rosette strains.

Strain Gage	X-coordinate [mm]	Y-coordinate [mm]	Strain [μm/m]	Remarks
1				
2				
3				
etc.				
Avg	–	–		–
SD	–	–		

Avg, average; SD, standard deviation.

This should be enough to provide an accurate representative average strain value for each strain gage.

3.5 RAW DATA COLLECTION

Step 1. Record linear strain gage data. For each linear strain gage, record the dimensionless strain level (Table 3.1), but also record the X and Y coordinates of each strain gage location, relative to some reference point on each specimen. Since specimens can vary in shape and size (e.g., sawbone 1 vs. 2 vs. 3, implant 1 vs. 2 vs. 3, etc.), the coordinates should be documented so that the same or, at least, corresponding locations for strain gages are known between specimens. This ensures that reliable statistical comparisons between test groups can be done later.

Step 2. Record rosette strain gage data. For each rosette, record strain and location as described above for linear gages (Table 3.1). However, note that Von Mises strain (i.e., equivalent strain) of each rosette is computed using strain readings from individual linear gages comprising each rosette. This calculation is slightly different for rosettes with a rectangular vs. delta pattern.

Step 3. Record visual observations. Some remarks should be entered about cracking, bending, or movement that may have occurred at various strain gage locations or at any other locations in the sawbone, implant, or sawbone–implant interface (Table 3.1).

3.6 RAW DATA ANALYSIS

Step 1. Calculate correlation coefficients. Plot measured strains for each specimen or location of interest vs. quantified characteristics that were varied (e.g., average load, sawbone density, implant length, implant density, etc.) to visualize the interrelationship between strain vs. characteristics. Then, calculate several items: the linear equation $y = mx + b$ showing slope m and intercept b, the correlation coefficient R, and a statistical P value to ensure that the slope of each strain vs. characteristic line of best fit (i.e., slope = m) is statistically different than a horizontal line (i.e., slope = 0).

Step 2. Perform statistical comparisons. The criterion for statistical difference needs to be chosen (e.g., $P < 0.01$ or < 0.05). Then strain data can be used to statistically compare test groups representing different sawbone densities (e.g., low vs. medium vs. high, etc.), implant designs (e.g., design 1 vs. 2 vs. 3, etc.), implant materials (e.g., stainless steel vs. titanium vs. polymer-based composite), location on the same implant (location 1 vs. 2 vs. 3, etc.), and so forth. Various software programs exist for comparing two test groups (e.g., paired *t*-test) or two or more test groups influenced by multiple factors (e.g., analysis of variance, ANOVA).

Step 3. Compute statistical power. Power analysis can be done after the study to ensure there were enough specimens per group to detect all statistical differences that were actually present (i.e., was type II statistical error avoided?). Statistical power > 80% is usually considered to indicate there were enough specimens per test group. Note that if good predictions of averages and standard deviations are available from prior studies, then the number of specimens and/or tests can be chosen before the study begins to ensure a power > 80%.

ENGINEER'S TOOLBOX

Strain $\mathcal{E}$ is the relative change ΔL per unit length L of an object's dimensions at a particular location (i.e., $\mathcal{E} = \Delta L/L$) caused by an applied mechanical stress. A rosette strain gage can help evaluate how close to mechanical failure a specimen is at that location. To avoid failure, Von Mises stress σ_{VM} should be less than the specimen's ultimate tensile stress σ_U, which is often a known material property. Thus, $\sigma_{VM} = \mathcal{E}_{VM}E$, where σ_{VM} is Von Mises stress, $\mathcal{E}_{VM}$ is Von Mises strain $= (\mathcal{E}_{MAX}{}^2 + \mathcal{E}_{MIN}{}^2 - \mathcal{E}_{MAX}\mathcal{E}_{MIN})^{1/2}$, $\mathcal{E}_{MAX,MIN}$ is maximum and minimum principal strain, and E is modulus of elasticity for the material. For a 45° rectangular rosette, $\mathcal{E}_{MAX,MIN} = (\mathcal{E}_1 + \mathcal{E}_3)/2 \pm (1/\sqrt{2})[(\mathcal{E}_1 - \mathcal{E}_2)^2 + (\mathcal{E}_2 - \mathcal{E}_3)^2]^{1/2}$. For a 60° delta rosette, $\mathcal{E}_{MAX,MIN} = (\mathcal{E}_1 + \mathcal{E}_2 + \mathcal{E}_3)/3 \pm (\sqrt{2}/3)[(\mathcal{E}_1 - \mathcal{E}_2)^2 + (\mathcal{E}_2 - \mathcal{E}_3)^2 + (\mathcal{E}_3 - \mathcal{E}_1)^2]^{1/2}$. Note that $\mathcal{E}_1$, $\mathcal{E}_2$, $\mathcal{E}_3$ are raw linear strain readings from the three individual linear gages that comprise each rosette.

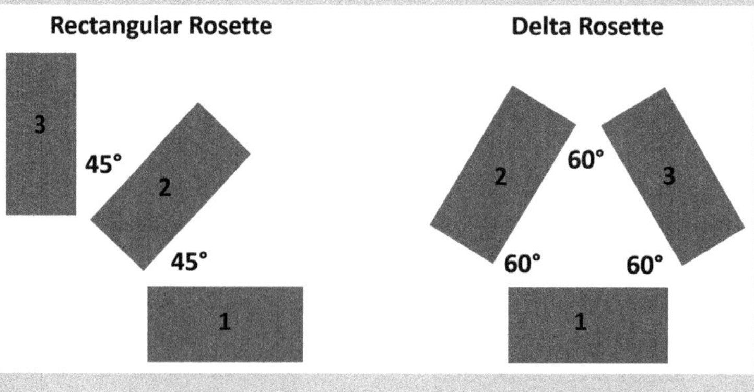

4. RESULTS

Once all strain gage data collection and analysis have been performed, it is then important to communicate and present the primary results in an understandable and concise manner to the reader of a journal article, conference paper, technical report, or book chapter.

Step 1. Show validation graphs. Strain gages are considered the "gold standard" method for assessing surface strains. Consequently, they are often used to validate other research techniques for a limited number of points on a specimen. The other techniques (e.g., finite element analysis, thermographic stress analysis, digital image correlation, etc.) can then be confidently used to perform more comprehensive strain analysis of the specimen.[2,4,5,12–17] If this is the case, then present the validation graph of the readings from strain gages vs. the other technique used (Fig. 3.6).

Step 2. Show comparison graphs. Sawbone strains, implant strains, and statistical *P* values for each relevant pairwise comparison between test groups should be depicted (Fig. 3.7). Test group comparisons may be for different sawbone densities (e.g., low vs. medium vs. high, etc.), implant designs (e.g., design 1 vs. 2 vs. 3, etc.),

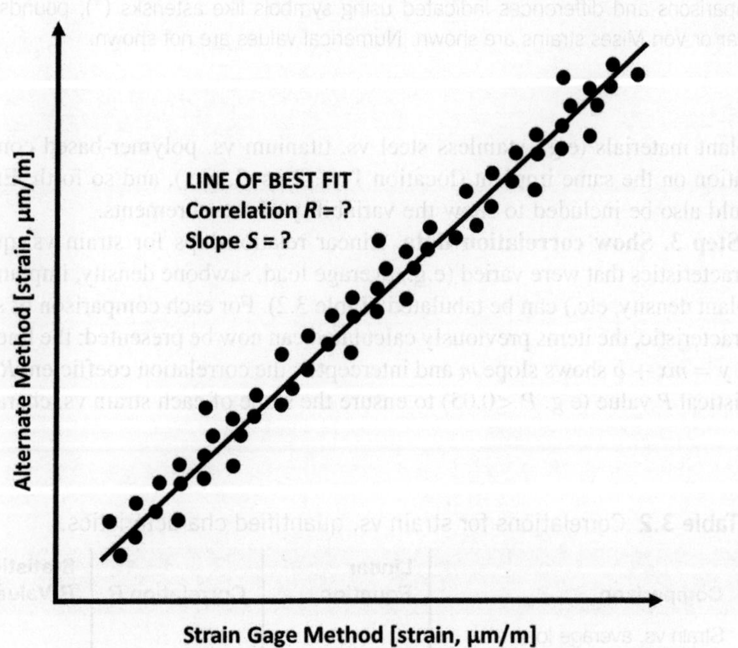

FIGURE 3.6

Validation graph of strain gages vs. an alternate strain testing method. *R* is the linear correlation coefficient, which is 1 for perfect precision. *S* is the slope, which is 1 for perfect accuracy.

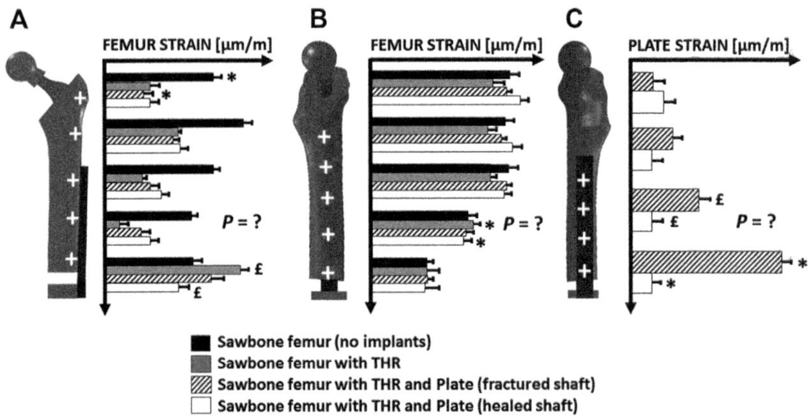

FIGURE 3.7

Strain graphs for sawbone–implant constructs. (A) Sawbone femur strain along the anterior surface, (B) sawbone femur strain along the medial surface, (C) plate strain along the lateral surface. *THR*, total hip replacement. White crosses are linear or rosette gage locations. Error bars indicate $\pm$ 1 standard deviation. *P* values should be calculated for all relevant statistical comparisons and differences indicated using symbols like asterisks (*), pounds (£), etc. Linear or Von Mises strains are shown. Numerical values are not shown.

implant materials (e.g., stainless steel vs. titanium vs. polymer-based composite), location on the same implant (location 1 vs. 2 vs. 3, etc.), and so forth. Error bars should also be included to show the variability of measurements.

Step 3. Show correlation data. Linear relationships for strain vs. quantified characteristics that were varied (e.g., average load, sawbone density, implant length, implant density, etc.) can be tabulated (Table 3.2). For each comparison of strain vs. characteristic, the items previously calculated can now be presented: the linear equation $y = mx + b$ shows slope m and intercept b, the correlation coefficient R, and the statistical P value (e.g. $P < 0.05$) to ensure the slope of each strain vs. characteristic

Table 3.2 Correlations for strain vs. quantified characteristics.

Comparison	Linear Equation	Correlation *R*	Statistical P-Value
Strain vs. average load			
Strain vs. sawbone density			
Strain vs. implant length			
etc.			

line of best fit (i.e., slope $= m$) is statistically different than a horizontal line (i.e., slope $= 0$). $R > 0.8$ is considered to indicate a strong correlation.

ALTERNATIVES AND ADAPTATIONS

- ✓ **Biological bone**. Strain gages can be used on whole biological bone, but only if surfaces are fully stripped of soft tissue, sanded for bone roughness, and cleaned of blood where the strain gage will be installed. However, strain gages may be more successful on bone coupons of standardized geometry, which have been machined to a good quality surface finish.
- ✓ **Interface strains**. Strain gages can be used to measure the strain at interfaces, such as between a metal implant and the underlying host bone. Strain gage mounting in these cases is much more challenging since strain gages are physically fragile and also because there may be limited access to the interface to correct any errors once the specimen has been prepared.

5. DISCUSSION

After completing all strain gage tests, data collection, data analysis, and data presentation, then final results can be considered and interpreted in the broader context of some important clinical, biomechanical, and/or technological considerations, as follows.

Strain gages can be used for diverse orthopaedic biomechanics applications. Human cadaveric forearms can have strain gages mounted on the ulna and radius to measure bone strains during axial force application on the hand.[18] Intact human and sawbone femurs can be tested to understand the kind of surface strains caused by realistic physiological loads during walking.[4,19] Human and sawbone femurs equipped with hip prostheses, knee prostheses, fracture plates, and/or intramedullary nails can be evaluated to determine strains on bones and implants, as well as near bone–implant interfaces.[2,5,15,20] Standard metal, modular metal, and polymer-based composite hip prostheses can be tested by themselves using strain gages to examine strains during walking-type loads.[12–14] Rectangular plates made from new polymer-based composite biomaterials can be equipped with strain gages and subjected to cyclic loads to obtain surface strains.[16] These various applications can produce strains ranging from 0–3500 μm/m, depending on the applied load, specimen material, specimen geometry, and location on the specimen surface.[2,5,14–16,19,20]

Strain gages are the "gold standard" technique for surface strain measurements. Consequently, they can be used to validate other strain analysis methods for orthopaedic biomechanics research. Linear or rosette strain gages vs. thermography tests yield excellent average linear correlation coefficients R for intact sawbone femurs ($R = 0.99$),[4] polymer-based composite hip implants ($R = 0.98$),[13] and polymer-based composite plates with biomaterials applications ($R = 0.93$).[16] Rosette strain gages vs. thermography tests for canine whole femurs demonstrate no statistical difference between the two methods ($P = 0.99$).[6] Rosette strain gages vs. digital image correlation show similar pelvic bone strain patterns for various acetabular cup implants.[17] Linear strain gages vs. computational finite element modeling for quasi-static and cyclic loads experienced by sawbone femurs, hip prostheses, knee prostheses, intramedullary nails, and/or fracture plates also yield good correlation coefficients ($R = 0.77–0.99$).[2,5,12,14,15]

Strain gage operating principles should be understood by researchers.[9–11] A strain gage is composed of a metal alloy wire or foil laid out in a grid-like pattern. Its electrical resistance changes when mechanical strain is applied to the underlying bone or implant on which the strain gage is mounted. This phenomenon can be expressed mathematically as the gage factor $K = (\Delta R/R)/(\Delta L/L)$, where R is original electrical resistance of the strain gage (units: ohms Ω) and L is original length of the strain gage. Obviously, K will depend on the particular material from which the strain gage wire or foil is commonly manufactured, such as constantan ($K = 2$), isoelastic ($K = 3.5$), and karma ($K = 2.1$). Also, note that strain $\mathcal{E} = \Delta L/L$ is the dimensionless strain value (i.e., μm/m) recorded during experiments and commonly reported in research articles and technical reports. The strain gage is typically connected to a "wheatstone" bridge circuit, which provides an excitation voltage, maximizes the sensitivity of the strain reading, and minimizes strain reading drift caused by temperature change. Then, the "wheatstone" bridge is connected to an analog-to-digital unit, which converts analog strains into digital format for computer data acquisition.

Strain gages have pros and cons.[3,9–11] They are a well-established and well-tested technology that is the "gold standard" method for surface strain measurements. They are extremely sensitive, thereby yielding strain data as small as several microns (i.e., μm/m). They are nondestructive, so they do not cause gross damage to test specimens. They provide real-time data for quasi-static, cyclic, or impact loads. They are relatively inexpensive compared to many other methods, thus making them ideal for researchers with limited budgets. However, strain gages have several cons. They require an extensive time-consuming step-by-step process for surface preparation and gage installation. They are physically fragile; thus, they can easily be damaged during mounting and cannot be reused for multiple specimens. They only provide one data point per strain gage, rather than a full-field stress map over the entire specimen surface like some other methods. They often give inaccurate readings near geometric discontinuities, such as edges. They can be temperature sensitive, so they must operate within a narrow temperature range for a given test.

6. SUMMARY

- Specimen loading, geometry, and material properties influence strain distribution.
- Strain gage methodology is an easy-to-use, nondestructive, "gold standard" approach.
- Strain gages can be used to obtain specimen surface strain distribution.

- Strain gages can identify high strain areas of bone or implant at risk of failure.
- Strain gages can locate low strain areas of bone at risk of "stress shielding."
- Strain gages can help optimize an implant's geometry and material properties.

7. QUIZ QUESTIONS

1. How do strain gages work to detect strain from a mechanics viewpoint?
2. What are the main steps to mount a strain gage on the surface of a bone or implant?
3. What are the main tasks required for quasi-static or cyclic strain gage measurements?
4. What are the advantages and disadvantages of using linear vs. rosette strain gages?
5. Does a repaired bone–implant construct fail under load at a particular location where a rosette gage is located? Assume the rosette gage has a 45° rectangular pattern with $\mathcal{E}_1$, $\mathcal{E}_2$, $\mathcal{E}_3 = 20, 300, 200$ μm/m, cortical bone's ultimate tensile stress $\sigma_U = 106$ MPa, and cortical bone's modulus of elasticity $E = 15$ GPa (answer: no failure, since Von Mises stress $\sigma_{VM} = 5.7$ MPa $< \sigma_U$).

REFERENCES

1. Zdero R, Walker R, Waddell JP, Schemitsch EH. Biomechanical evaluation of periprosthetic femoral fracture fixation. *Journal of Bone and Joint Surgery (American)* 2008;**90**(5):1068–77.
2. Bougherara H, Zdero R, Mahboob Z, Dubov A, Shah S, Schemitsch EH. The biomechanics of a validated finite element model of stress shielding in a novel hybrid total knee replacement. *Proceedings of the Institution of Mechanical Engineers (Part H): Journal of Engineering in Medicine* 2010;**224**(10):1209–19.
3. Roriz P, Carvalho L, Frazao O, Santos JL, Simoes JA. From conventional sensors to fibre optic sensors for strain and force measurements in biomechanics applications: a review. *Journal of Biomechanics* 2014;**47**(6):1251–61.
4. Shah S, Bougherara H, Schemitsch EH, Zdero R. Biomechanical stress maps of an artificial femur obtained using a new infrared thermography technique validated by strain gages. *Medical Engineering and Physics* 2012;**34**(10):1496–502.
5. Ebrahimi H, Rabinovich M, Vuleta V, Zalcman D, Shah S, Dubov A, et al. Biomechanical properties of an intact, injured, repaired, and healed femur: an experimental and computational study. *Journal of the Mechanical Behavior of Biomedical Materials* 2012;**16**:121–35.
6. Kohles SS, Vanderby Jr R. Thermographic strain analysis of the proximal canine femur. *Medical Engineering and Physics* 1997;**19**(3):262–6.
7. ASTM E1237. Standard guide for installing bonded resistance strain gages. West Conshohocken (PA, USA): American Society for Testing and Materials (ASTM). Available at: www.astm.org/Standards/E1237.htm.

8. ASTM E1561. Standard practice for analysis of strain gage rosette data. West Conshohocken (PA, USA): American Society for Testing and Materials (ASTM). Available at: www.astm.org/Standards/E1561.htm.
9. Wilson EJ. Strain-gage instrumentation. In: Piersol AG, editor. *Harris' shock and vibration handbook*. 5th ed. New York (NY, USA): McGraw-Hill; 2002.
10. Agilent Technologies. Practical strain gage measurement. Available free online at: www.omega.com/techref/pdf/StrainGage_Measurement.pdf.
11. Vishay Precision Group. www.vishaypg.com/micro-measurements.
12. Bougherara H, Zdero R, Dubov A, Shah S, Khurshid S, Schemitsch EH. A preliminary biomechanical study of a novel carbon-fibre hip implant versus standard metallic hip implants. *Medical Engineering and Physics* 2011;**33**(1):121–8.
13. Bougherara H, Rahim E, Shah S, Dubov A, Schemitsch EH, Zdero R. A preliminary biomechanical assessment of a polymer composite hip implant using an infrared thermography technique validated by strain gage measurements. *Journal of Biomechanical Engineering* 2011;**133**(7):074503-1-6.
14. Bougherara H, Zdero R, Shah S, Miric M, Papini M, Zalzal P, et al. A biomechanical assessment of modular and monoblock revision hip implants using FE analysis and strain gage measurements. *Journal of Orthopaedic Surgery and Research* 2010;**5**(34):1–12.
15. Bougherara H, Zdero R, Miric M, Shah S, Hardisty M, Zalzal P, et al. The biomechanics of the T2 femoral nailing system: a comparison of synthetic femurs with finite element analysis. *Proceedings of the Institution of Mechanical Engineers (Part H): Journal of Engineering in Medicine* 2009;**223**(3):303–14.
16. Saleem M. *A nondestructive study of a carbon fibre epoxy composite plate using lock-in thermography, cyclic loading, and finite element analysis* [M.A.Sc. thesis]. Toronto (Canada): Department of Mechanical and Industrial Engineering, Ryerson University; 2010.
17. Small SR, Berend ME, Howard LA, Tunc D, Buckley CA, Ritter MA. Acetabular cup stiffness and implant orientation change acetabular loading patterns. *Journal of Arthroplasty* 2013;**28**(2):359–67.
18. Shaaban H, Giakas G, Bolton M, Williams R, Scheker LR, Lees VC. The distal radioulnar joint as a load-bearing mechanism: a biomechanical study. *Journal of Hand Surgery (American)* 2004;**29**(1):85–95.
19. Cristofolini L, Viceconti M, Cappello A, Toni A. Mechanical validation of whole bone composite femur models. *Journal of Biomechanics* 1996;**29**(4):525–35.
20. Decking R, Puhl W, Simon U, Claes LE. Changes in strain distribution of loaded proximal femora caused by different types of cementless femoral stems. *Clinical Biomechanics* 2006;**21**(5):495–501.

CHAPTER

4

Thermographic Stress Analysis of Whole Bones and Implants

Zahra S. Bagheri[1], Habiba Bougherara[2], Radovan Zdero[3]

Toronto Rehabilitation Institute, Toronto, ON, Canada[1]; Ryerson University, Toronto, ON, Canada[2]; Western University, London, ON, Canada[3]

1. BACKGROUND

Experimental stress analysis of whole bones, implants, and whole bone–implant constructs is an important approach in orthopaedic biomechanics. High stresses in whole bones and implants may cause mechanical failure (Fig. 4.1),[1,2] but low stresses in whole bone may cause "stress shielding," which leads to bone atrophy,

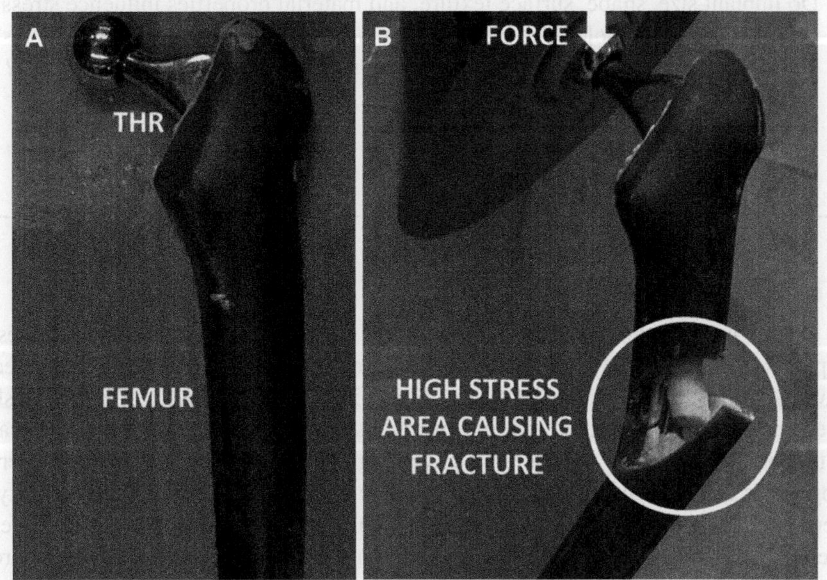

FIGURE 4.1

Testing an artificial sawbone femur implant. (A) Unloaded femur implant, (B) femur fracture caused by high stresses near the tip of a total hip replacement (THR).

bone resorption, and implant loosening.[3,4] Surface strain gages are the "gold standard" tool for strain (and stress) analysis in this field,[5,6] but strain gages are not fully reliable near high peak strains, provide only average strain at each location, cannot yield strain maps of the entire surface, are easily physically damaged, can permanently damage the surface during the mounting process, and have physical dimensions that limit where they can be mounted. Thermographic stress analysis (TSA) is an alternative, easy-to-use, point-and-shoot, noncontact, nondestructive, experimental stress analysis method based on thermoelasticity.[7,8] Put simply, thermoelastic theory states that changes in an object's mechanical stress state cause changes in its physical dimensions and, hence, its surface temperature. Therefore, this chapter explains how to perform TSA for orthopaedic biomechanics applications, such as whole bones, fracture fixation devices, total joint replacements, and biomaterials, as well as how to analyze, present, and interpret results.

2. RESEARCH QUESTIONS

Typical research questions might include one or more of the following:

- Do bone size, shape, surface texture, and material properties influence stress?
- Do implant size, shape, surface texture, and material properties influence stress?
- Which regions of whole bone are at risk of mechanical failure due to high stress?
- Which regions of whole bone are at risk of stress shielding due to low stress?
- What is the effect of loading level, duration, or frequency on stress?
- etc.

3. METHODOLOGY

3.1 GENERAL STRATEGY

TSA is used to yield accurate surface stress maps for artificial intact sawbones, implants, and/or sawbone–implant constructs using a well-established protocol.[3,5,7–11] Specimens are first dyed or painted with a nonglossy black finish to ensure optimal infrared emission. Specimens are then mounted in a mechanical tester for cyclic loading representing physiological forces. Simultaneously, a thermographic camera records surface temperature maps. To optimize image quality, thermographic image capture frequency is synchronized with cyclic loading frequency (i.e., "lock-in"). Dedicated computer software converts surface temperature maps to surface stress maps by employing user-inputted material properties for various specimen components. Stress maps are analyzed to determine sawbone and implant areas with high stress (i.e., risk of failure) or sawbone areas with low stress (i.e., risk of stress shielding). Finally, statistical comparisons are made between test groups, and correlation coefficients are computed for stress vs. sawbone or implant characteristics to identify important factors.

GLOSSARY

✓ **Adiabatic.** All heat is retained fully by an object with no heat lost to the surroundings.
✓ **Lock-in.** Synchronization of thermal image recording frequency with cyclic loading frequency.
✓ **Stress.** The ratio of applied mechanical force that is experienced per unit of surface area.
✓ **Stress map.** An image showing the stress distribution over an object's external surface.
✓ **TSA (thermographic stress analysis).** An experimental method for obtaining stress maps.

SAFETY FIRST

✓ Always wear goggles and gloves for protection during preparation and testing.
✓ Secure the sawbone and/or implant before applying load in the mechanical tester.
✓ Place a protective plastic shield around the mechanical test area to contain flying debris.
✓ Clean the work area and all tools with bleach or disinfectant if testing cadaveric bone.

3.2 MATERIALS AND TOOLS LIST

- artificial sawbone models
- black dye, paint, and/or tape
- electronic signal generator
- mechanical tester
- orthopaedic implants
- software for TSA
- thermography camera
- tripod to mount the camera
- various hand tools
- vice or clamp

3.3 SPECIMEN PREPARATION

Step 1. Clean the original surfaces. Surfaces to be thermographically imaged must be clean and smooth to avoid any image artifacts, which may incorrectly appear to be points of high or low stress. Loose debris should be removed using a nonabrasive cloth to prevent scratching the surface. However, if debris is stuck to the surfaces, sandpaper or a paint scraper may be used.

Step 2. Blacken the original surfaces. Surfaces to be thermographically imaged must also be as dark as possible to maximize their infrared emissivity constant (i.e., $\mathcal{E} = 1$ for a black body), which will result in good quality images. Consequently, if possible, all surfaces should be blackened. For artificial sawbones made from polymer, nonglossy black dye can be added by the manufacturer to the polymer mixture

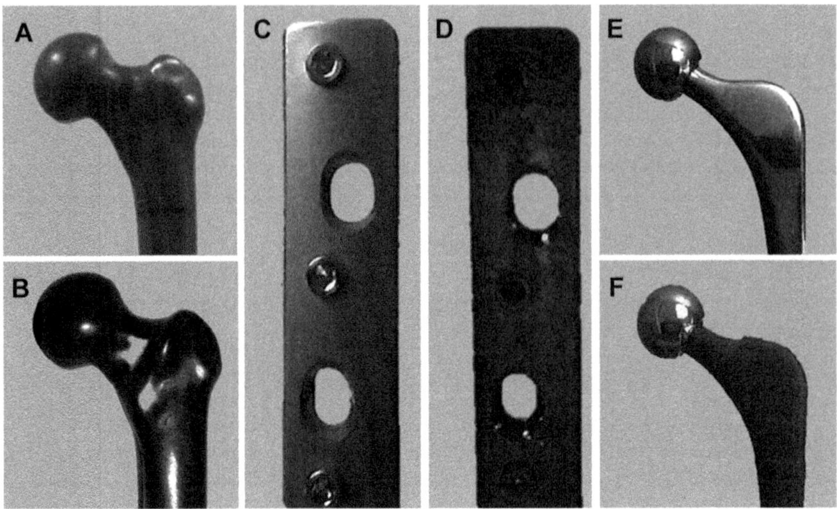

FIGURE 4.2

Typical blackened components for TSA. (A) Original sawbone femur, (B) sawbone femur dyed black, (C) original metal fracture plate, (D) metal fracture plate painted black, (E) original THR, (F) THR stem wrapped with black tape.

at the time of sawbone fabrication (Fig. 4.2A and B). For orthopaedic metal implants that are commonly gray in color, such as joint prostheses or fracture plates, a thin layer of nonglossy black paint needs to be sprayed or brushed onto the outer surface (Fig. 4.2C and D). Dye and paint have been used previously in this regard and do not substantially alter the material properties of the underlying components being investigated.[3,5,9,10]

Step 3. Sand the black surfaces. Use progressively finer grades of sandpaper to eliminate any paint globules, brush strokes, or air bubbles, which could introduce thermographic imaging artifacts. But also be careful not to leave scratches or debris from the sandpapering process.

Step 4. Apply black tape if needed. At times, some components of a sawbone–implant construct may not be thermographically analyzed (e.g., joint prosthesis), since other regions at risk of failure are of real concern to the researcher (e.g., surrounding sawbone). Moreover, these same components may need to be reused in their original unblackened condition for another study. In these cases, cover these components with black tape to eliminate light reflection, which could compromise image quality (Fig. 4.2E and F). Note that components covered in black tape cannot be accurately analyzed for stress, since deformation of the tape during load application to the sawbone–implant construct does not truly represent the stresses of the underlying surface.

Step 5. Perform surgeries if needed. Perform any necessary surgeries to represent clinical conditions. For example, one may want to investigate the biomechanical properties of a sawbone in its intact (Fig. 4.3A), implanted (Fig. 4.3B), injured and repaired (Fig. 4.3C), and/or healed (Fig. 4.3D) condition. Therefore, several sawbones should be prepared using typical surgical processes: (1) chamfering, drilling, rasping, reaming, and/or sawing the sawbone to receive a joint prosthesis that is then secured with bone cement and/or bone screw(s); (2) sawing the sawbone in an appropriate pattern to mimic a clinical-type fracture; (3) applying a metal fracture plate with the appropriate number of unicortical, bicortical, and/or lag screws; and/or (4) wrapping a metal cable or wire around the sawbone for fracture fixation.

Step 6. Pot the specimen if needed. In most cases, it is necessary to pot the specimen into anchoring cement, bone cement, or liquid metal in order to mount it securely during mechanical loading. Sawbones or implants can be potted as follows: (1) mount each specimen in a multiaxial clamp, (2) insert one end of the specimen into a hollow steel cube, (3) align the specimen in the appropriate orientation using a leveling gage, (4) pour the potting medium into the hollow steel cube, and (5) allow the potting medium to set properly until it is solid.

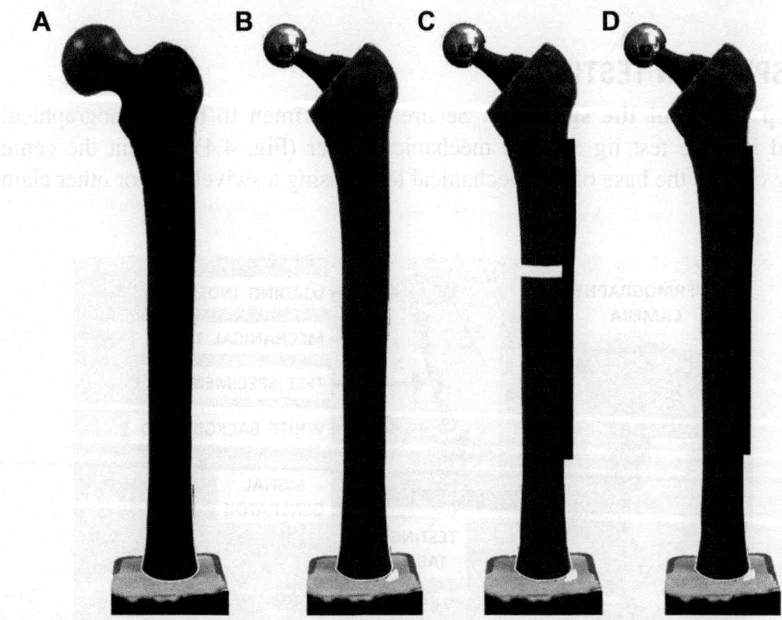

FIGURE 4.3

Typical blackened specimens for TSA. (A) Intact sawbone femur, (B) sawbone femur with total hip replacement (THR), (C) sawbone femur with THR and plate (fractured shaft), (D) sawbone femur with THR and plate (healed shaft). Sawbone femurs are potted in a hollow steel cube filled with anchoring cement.

TIPS AND TRICKS

✓ Apply nonglossy black paint or dye to the specimen surface for better image quality.
✓ Synchronize thermographic imaging and loading frequency for better image quality.
✓ Place a white sheet of cardboard behind the specimen to minimize image artifacts.
✓ Zoom into the area of interest to increase thermographic image resolution.
✓ Check that correct material properties are inputted into the computer software.

THE "GOLD STANDARD"

No known international standards exist specifically for using TSA on bones, implants, or bone−implant constructs. Thus, researchers should consult peer-reviewed journal articles that have used thermography for orthopaedic biomechanics. However, the American Society for Testing and Materials (ASTM) has a testing standard that may provide some guidelines, namely, ASTM E1934 (Standard guide for examining electrical and mechanical equipment with infrared thermography).

3.4 SPECIMEN TESTING

Step 1. Mount the specimen. Secure the specimen to be thermographically imaged into the test jig and the mechanical tester (Fig. 4.4). Mount the cement potting cube to the base of the mechanical tester using a swivel vice or other clamp.

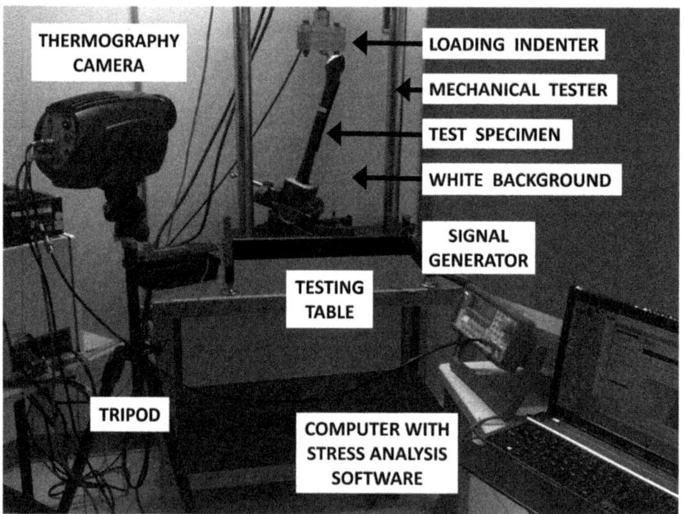

FIGURE 4.4

Photograph of a typical experimental setup used for thermographic stress analysis.

Orient the specimen to the angle of interest. Lower the loading indenter onto the area of the specimen to be directly loaded.

Step 2. Provide a white background. Move the test setup and specimen in front of a white wall without any other objects in the background, which could introduce image artifacts. This will provide excellent contrast with the blackened specimen and optimize images. Alternatively, place a white curtain, white foamboard, or white cardboard directly behind the specimen.

Step 3. Position the safety shield. Place a transparent, nonreflective, protective, plastic shield around the specimen if there is a potential concern about specimen fracture and flying debris. It is important to ensure that this shield does not interfere with thermographic image quality.

Step 4. Position the camera. It is not always necessary to image the entire specimen, but only a certain region. For instance, this region may be at particular risk for mechanical failure (i.e., excessively high stresses leading to sawbone fracture or sawbone–implant interface failure), or the sawbone region of interest may be at risk of stress shielding (i.e., excessively low stresses leading to bone atrophy, bone resorption, and implant loosening). Consequently, mount the thermographic camera on a tripod, and zoom into the region of interest so it is in the middle of the image widow. This technique will improve image detail and resolution.

Step 5. Enter material properties. Enter the thermoelastic coefficient and emissivity for the specimen into the TSA computer software (Fig. 4.5). Thermoelastic coefficient K may be provided by the manufacturer or computed from material property tables available in the literature for common orthopaedic materials: 1.16×10^{-5}/MPa (artificial cortical bone),[3,5,10] 1.10×10^{-6}/MPa (human cortical bone),[11] 2.19×10^{-6}/MPa (ceramic),[12] 3.44×10^{-6}/MPa (cobalt chrome),[12] 4.56×10^{-6}/MPa (316 L stainless steel),[3,10] 3.74×10^{-6}/MPa (titanium),[12] and 1.09×10^{-4}/MPa (ultra high molecular weight polyethylene).[12] Emissivity $\mathcal{E} = 1$ for an ideal black body, but a slightly lower value may represent the actual emissivity of the blackened specimen and, thus, may enhance image quality. Note that for artificial sawbones, it is common to use K only for the "cortical" material, while ignoring the "cancellous" material altogether. This simplification adequately represents the biomechanical behavior of sawbone, whose overall mechanical stiffness remains relatively constant, even over a six-fold change in "cancellous" material density.[2]

Step 6. Set the "lock-in" frequency. Loading and thermographic imaging frequencies must be synchronized (i.e., "lock-in") at a minimum frequency of 3–5 Hz, preferably using a sinusoidal waveform. This is required by many thermal cameras for specimens to reach adiabatic conditions (i.e., no heat loss to the surroundings). There are several possible ways to do this, depending on the capabilities of the particular mechanical tester and thermal camera, as follows.

The first option involves this setup: signal generator → thermal camera → computer, while the mechanical tester is independent.[3,10,13] The signal generator provides a sinusoidal waveform, which triggers the thermal camera to capture images at the same frequency, while the mechanical tester provides cyclic loading

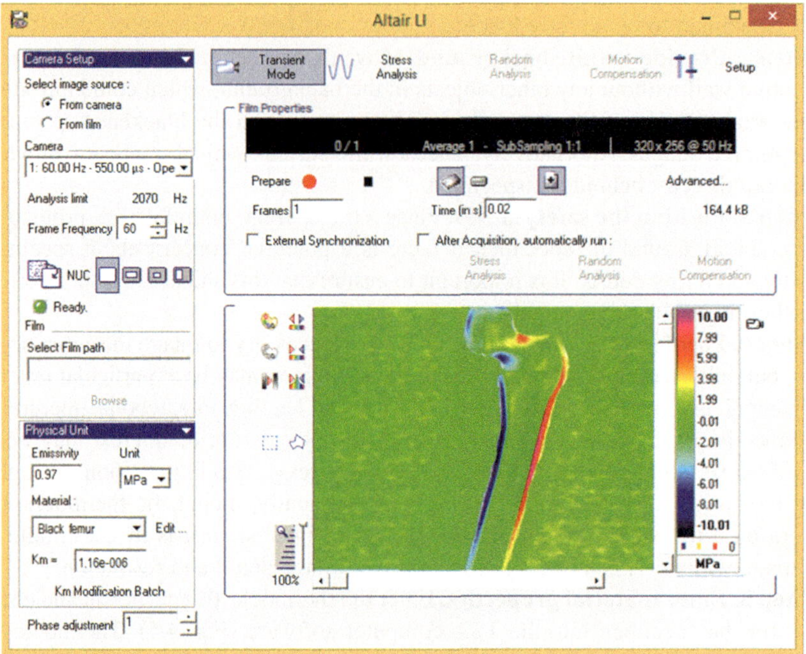

FIGURE 4.5

Typical TSA computer software window for inputting thermal properties. The user enters the name of the material, the thermoelastic coefficient, and the emissivity.

Software image used courtesy of www.flir.com.

independently. Thermal image quality is used as the criterion to ensure that imaging and loading are in phase. The computer stores the images.

The second option involves this setup: mechanical tester → thermal camera → computer. The mechanical tester provides a sinusoidal waveform, which triggers the thermal camera to capture images at the same frequency. The computer stores the images.

The third option involves this setup: mechanical tester → signal generator → thermal camera → computer.[5,6,14] The mechanical tester provides a sinusoidal loading waveform, which is read by the signal generator, which then sends a signal that triggers the thermal camera to capture images at the same frequency. The computer stores the images.

Step 7. Start loading and imaging. Apply a low vertical preload of 50–100 N to eliminate any "mechanical slack" in the test setup (i.e., hysteresis), such as slippage between the specimen and the loading indenter.[3,5,6,10] Then, set the minimum-to-maximum range for the applied cyclical load to a clinically realistic level if physiological loading is required, or to a subclinical level if permanent damage to specimens needs to be avoided. Specimens should then be cyclically loaded with a sinusoidal waveform for about 1000 cycles to reach the thermal steady state and adiabatic conditions required for imaging.[3,5,6,10] Finally, image the specimen for 2–3 min while the cyclic loading continues.[3,5,6,10] (Note: A sawbone–implant specimen is composed of different materials, e.g., artificial "cortical" material, polymer, stainless steel, titanium, etc. They cannot all be thermographically imaged at the same time, since each item has its own K and $\mathcal{E}$ values. Thus, repeat steps 5–7 for each new material of the sawbone–implant specimen to be imaged.)

3.5 RAW DATA COLLECTION

Step 1. Probe images for stress. In the TSA-dedicated computer software, open the stored thermal images in the correct file format (Fig. 4.6). Then, probe the image at each pixel of interest to obtain its stress and X and Y coordinates. Note that an artificially low or high stress for an individual pixel can occur due to image glare or surface roughness; this is not truly representative of the actual stress at that spot on the surface. To minimize this effect, an average stress can be calculated, say, for 10–20 adjacent pixels within a 3–5 mm diameter circle, the center point of which is considered to be the location of interest. Alternatively, the research question may require computing the average stress along a long line or a much larger region of the sawbone or implant, but a similar probing process is applied to obtain results.

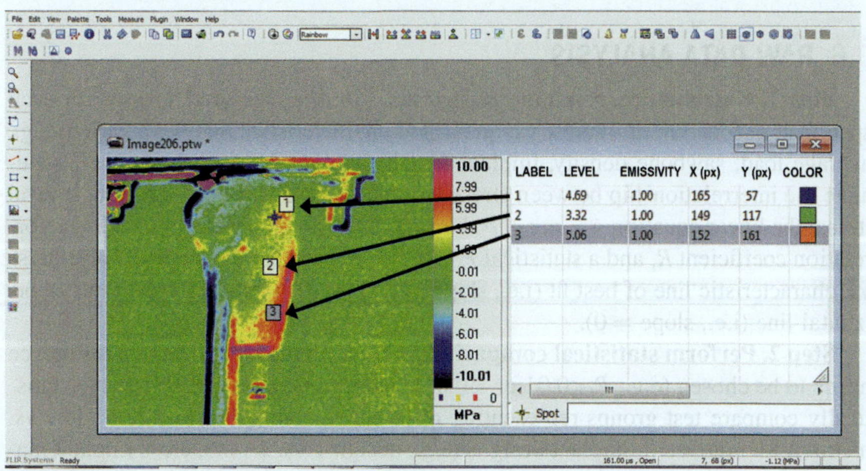

FIGURE 4.6

Typical TSA computer software window showing imaging results. For each point of interest chosen in the image, the software presents the label (i.e., point 1, 2, 3, etc.), level (i.e., stress), emissivity, X and Y pixel coordinates, and color.

Software image used courtesy of www.flir.com.

Table 4.1 Raw specimen surface stresses from a typical TSA image. The sum of the principal stresses is shown.

Location	X-coordinate	Y-coordinate	Stress [MPa]	Remarks
1				
2				
3				
etc.				
Avg	–	–		–
SD	–	–		–

Avg, *average;* SD, *standard deviation.*

Step 2. Record stress and coordinates. For each location of interest that was probed, enter not only the stress level, but also the X and Y coordinates into a data table (Table 4.1). Since specimens can vary in shape and size (e.g., sawbone 1 vs. 2 vs. 3, implant 1 vs. 2 vs. 3, etc.) coordinates should be recorded, so that the same or, at least, corresponding locations for stress probing are known. This ensures that reliable statistical comparisons between test groups can be done later.

Step 3. Record visual observations. Some final remarks should be recorded about any cracking, bending, or movement that may have occurred at the specific locations where stresses were probed, or at any other locations in the specimen (Table 4.1). These comments can confirm or deny the presence of high stress regions in the stress maps.

3.6 RAW DATA ANALYSIS

Step 1. Calculate correlation coefficients. Plot the measured stresses for each specimen or location of interest vs. quantified characteristics that were varied (e.g., average load, sawbone density, implant length, implant density, etc.) to help visualize the interrelationship between stress vs. characteristics. Then, calculate several items: the linear equation $y = mx + b$ showing slope m and intercept b, a linear correlation coefficient R, and a statistical P value to ensure that the slope of each stress vs. characteristic line of best fit (i.e., slope $= m$) is statistically different than a horizontal line (i.e., slope $= 0$).

Step 2. Perform statistical comparisons. The criterion for statistical difference needs to be chosen (e.g., $P < 0.01$ or < 0.05). Then stress data can be used to statistically compare test groups representing different sawbone densities (e.g., low vs. medium vs. high, etc.), implant designs (e.g., design 1 vs. 2 vs. 3, etc.), implant materials (e.g., stainless steel vs. titanium vs. polymer-based composite), location on the same implant (location 1 vs. 2 vs. 3, etc.), and so forth. There are software programs for comparing two test groups (e.g., paired t-test) or two or more test groups influenced by multiple factors (e.g., analysis of variance, ANOVA).

Step 3. Compute statistical power. Power analysis can be done after the study to ensure there were enough specimens per test group to detect all statistical differences that were actually present (i.e., was type II statistical error avoided?). Statistical power >80% is usually considered to indicate there were enough specimens per test group. Note that if good predictions of averages and standard deviations are available from prior studies, then the number of specimens and/or tests can be chosen before the study begins to ensure a power >80%.

ENGINEER'S TOOLBOX

TSA uses the thermoelastic equation under adiabatic conditions (i.e., no heat loss from the specimen to its surroundings) to relate stress and temperature for each image pixel. Specifically, $\Delta\sigma = \Delta T \rho C/(\alpha T_o)$, where $\Delta\sigma$ is change in specimen surface stress [MPa], ΔT is change in specimen surface temperature [°K], ρ is specimen material density [kg/m^3], C is specimen specific heat capacity [J/(kg°K)], α is specimen coefficient of thermal expansion [μm/(m°K)], and T_o is ambient temperature sensed by the camera [°K]. Alternatively, specimen properties can be combined into a single thermoelastic coefficient $K = \alpha/(\rho C)$, such that $\Delta\sigma = \Delta T/(KT_o)$.

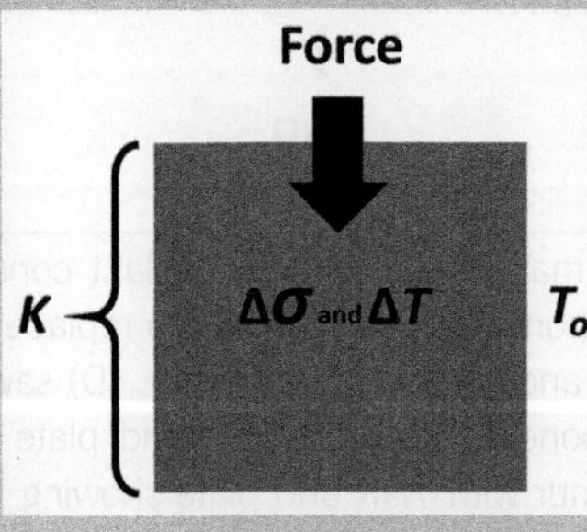

4. RESULTS

Once all TSA data collection and analysis have been performed, it is then important to communicate and present the primary results in an understandable and concise manner to the reader of a journal article, conference paper, technical report, or book chapter.

Step 1. Show raw stress maps. Present raw stress maps either with or without numerical values along the color scale (Fig. 4.7).[3,5,6,10,13,14] This helps visually identify locations of high and low stress, as well as any patterns. For most TSA systems, the color red represents areas under tension, the color blue represents areas under compression, and the color green represents no stress; this color scale may be different depending on the TSA system. Note also that TSA images commonly show the sum of the principal stresses, rather than Von Mises (i.e., equivalent) stresses.

Step 2. Show comparison graphs. Sawbone stresses, implant stresses, and statistical P values for relevant pairwise comparisons between test groups should be

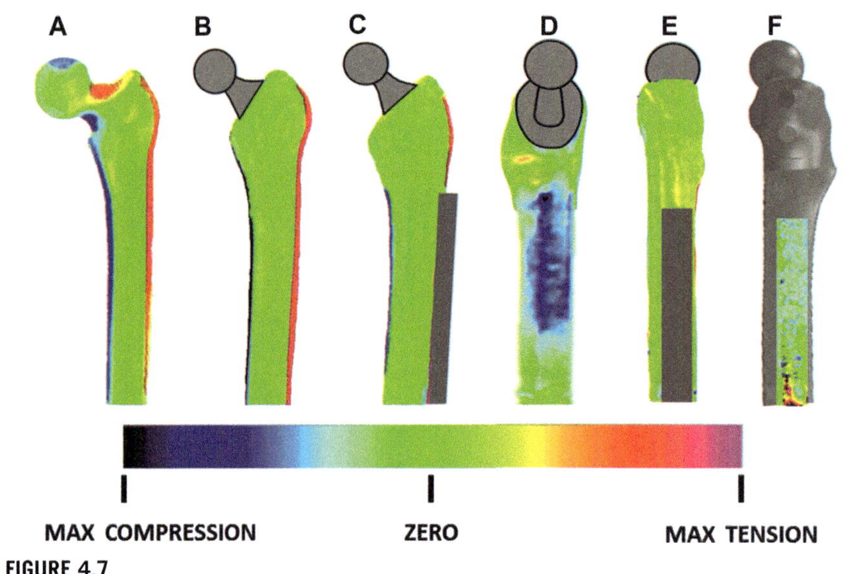

FIGURE 4.7

Typical thermographic stress maps for sawbone–implant constructs. (A) Intact sawbone femur (anterior view), (B) sawbone femur with total hip replacement (THR) (anterior view), (C) sawbone femur with THR and plate (anterior view), (D) sawbone femur with THR and plate (medial view), (E) sawbone femur with THR and plate showing sawbone stresses (lateral view), (F) sawbone femur with THR and plate showing plate stresses (lateral view). The THR is not analyzed. The sum of the principal stresses is shown. Numerical values are not shown.

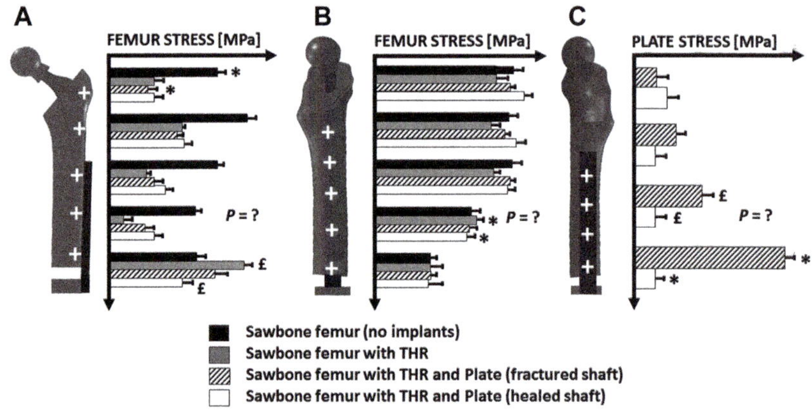

FIGURE 4.8

Typical TSA graphs for sawbone–implant constructs. (A) Sawbone femur stress along the anterior surface, (B) sawbone femur stress along the medial surface, (C) plate stress along the lateral surface. White crosses are locations where stresses were determined. Error bars indicate ± 1 standard deviation. *P* values are calculated for all relevant statistical comparisons, which can be indicated by symbols like asterisks (*), etc. The sum of the principal stresses is shown. Numerical values are not shown.

Table 4.2 Correlations for stress vs. quantified characteristics.

Comparison	Linear Equation	Correlation *R*	Statistical *P*-value
Stress vs. average load			
Stress vs. sawbone density			
Stress vs. implant length			
etc.			

depicted (Fig. 4.8).[3,5,6,10,14] Test groups compared via *P* values could represent different sawbone densities (e.g., low vs. medium vs. high, etc.), implant designs (e.g., design 1 vs. 2 vs. 3, etc.), implant materials (e.g., stainless steel vs. titanium vs. polymer-based composite), location on the same implant (location 1 vs. 2 vs. 3, etc.), and so forth. Error bars should also be included to show the variability of measurements.

Step 3. Show correlation data. Linear relationships for stress vs. quantified characteristics that were varied (e.g., average load, sawbone density, implant length, implant density, etc.) can be tabulated (Table 4.2).[3,5,6,10,14] For each comparison of stress vs. characteristic, the items previously calculated can now be presented: the linear equation $y = mx + b$ showing slope m and intercept b, the correlation coefficient R, and the statistical P value to ensure that the slope of each stress vs. characteristic line of best fit (i.e., slope $= m$) is statistically different than a horizontal line (i.e., slope $= 0$). $R > 0.8$ is considered to indicate a strong correlation.

ALTERNATIVES AND ADAPTATIONS

- ✓ **Biological bone.** TSA can be used on whole biological bone, but only if the surface is fully stripped of soft tissue, sanded for bone roughness, and cleaned of blood so it can be properly painted black. However, TSA may be more successful on bone coupons of standardized geometry, which have been machined from whole bone to a good quality surface finish.
- ✓ **Fatigue limit.** TSA can be a nondestructive alternative to conventional cyclic fatigue tests to determine the fatigue limit of biomaterials, implants, and bone–implant constructs. Specimens are stepwise cyclically loaded in the elastic region until there is an increase in the slope of the linear temperature vs. stress graph, which indicates microscopic crack nucleation and growth.

5. DISCUSSION

After completing all TSA tests, data collection, data analysis, and data presentation, then final results can be considered and interpreted in the broader context of some important clinical, biomechanical, and/or technological considerations, as follows.

TSA can be used for diverse orthopaedic biomechanics applications. Intact sawbone femurs, intact canine femurs, and standardized blocks of bovine cortical bone can be tested to provide stress maps caused by realistic physiological loads.[5,15–17]

Sawbone femurs equipped with a THR can be evaluated to determine if high stresses near the tip of the hip stem will cause fracture.[3,11] Sawbone femurs implanted with a standard metal fracture plate vs. a new polymer-based composite fracture plate can be analyzed to establish if low stresses at the bone–plate interface will lead to stress shielding and, hence, bone atrophy, bone resorption, and plate loosening.[3,10] Standard metal and polymer-based composite hip implants can be tested by themselves to examine stress maps under a variety of loads.[6,18] Rectangular plates made from new polymer-based composite materials can be subjected to uniaxial cyclic loads for thousands of cycles to obtain the *S–N* (i.e., stress–number of cycles) curve, thereby predicting the material's fatigue life.[13] These various applications can produce stresses ranging from 0–160 MPa, depending on the applied load, specimen material, specimen geometry, and location on the specimen surface.[3,5,6,11,13,15,17,18]

TSA can be validated against other stress analysis techniques applied to orthopaedic biomechanics. TSA vs. linear strain gage results yield excellent average linear correlation coefficients R for polymer-based composite hip implants ($R = 0.98$),[6] polymer-based composite rectangular plates with potential biomaterials applications ($R = 0.93$),[14] and bovine cortical bone blocks ($R = 0.95$).[17] TSA vs. rosette strain gage data show a similar excellent correlation for intact sawbone femurs ($R = 0.99$),[5] while canine whole femur tests demonstrate no statistical difference between the two methods ($P = 0.99$).[16] TSA for dynamic loads vs. computational finite element modeling for quasi-static loads using a sawbone femur implanted with a hip prosthesis and a midshaft fracture plate provide similar stress trends.[3]

TSA's operating principle involves obtaining four signal values, S_1, S_2, S_3, and S_4, for every image pixel at four different equidistant times.[3,5,6,10] The phase value is then calculated as $\Phi = \arctan[(S_1 - S_3)/(S_2 - S_4)]$ from which a phase image is generated. The phase image is used to obtain an average stress image from the four signal values, and it helps synchronize imaging and loading frequency. TSA assumes adiabatic conditions (i.e., no heat loss to the surroundings) in order to relate changes in temperature and stress for each image pixel. Specifically, $\Delta\sigma = \Delta T\rho C/(\alpha T_o)$, where $\Delta\sigma$ is change in specimen surface stress [MPa], ΔT is change in specimen surface temperature [°K], ρ is specimen material density [kg/m^3], C is specimen specific heat capacity [J/(kg°K)], α is specimen coefficient of thermal expansion [μm/(m°K)], and T_o is ambient temperature sensed by the camera [°K]. Alternatively, specimen properties can be combined into a single thermoelastic coefficient, $K = \alpha/(\rho C)$, such that $\Delta\sigma = \Delta T/(KT_o)$.

TSA has pros and cons.[3,5–8,10,11,13–18] First, TSA yields full-field surface stress maps. Second, TSA is noncontact and nondestructive and, thus, does not damage the specimen. Third, TSA requires the specimen to undergo cyclic loads, which are also generated during many real-world clinical conditions. Fourth, TSA is a simple point-and-shoot approach that needs only minimal experimental preparation. However, TSA also has several cons. Poor image quality, which can be difficult to interpret, may be caused by surface texturing of the original specimen, surface texturing caused by residual brush strokes or paint globules during application of black paint, and/or glare from nearby shiny surfaces or background objects. Additionally, a loading frequency of 3–5 Hz is needed to produce an adequate surface temperature gradient that can be detected by the camera, which does not represent common physiological conditions like walking (i.e., 1–2 Hz)[19] or impact injuries.[20] Finally, about 1000 loading cycles are required for the specimen to achieve thermal steady state; thus, some specimens may fracture before reaching steady state.

6. SUMMARY

- Specimen loading, geometry, and material properties influence stress.
- TSA is an easy-to-use, noncontact, point-and-shoot methodology.
- TSA can be used to obtain full-field specimen surface stress distributions.
- TSA can identify high stress areas of whole bone at risk of mechanical failure.
- TSA can locate low stress areas of whole bone at risk of stress shielding.
- TSA can help optimize an implant's geometry and material properties.

7. QUIZ QUESTIONS

1. What are the definitions of these terms: adiabatic, lock-in, and stress?
2. What are the pros and cons of TSA vs. strain gages for stress tests?
3. Why are regions of high stress of concern for whole bone or implant stability?
4. Why are regions of low stress of concern for whole bone or implant stability?
5. An unknown orthopaedic implant material is subjected to TSA under a 1 Hz cyclic load of 1–2 kN in a room kept at 22°C. The ratio of change in stress to change in surface temperature is 1548 MPa/°K. Compute the thermoelastic coefficient in order to determine the identity of the mystery material (answer: ceramic with $K = 2.19\text{x}10^{-6}$/MPa).

REFERENCES

1. Zdero R, Walker R, Waddell JP, Schemitsch EH. Biomechanical evaluation of periprosthetic femoral fracture fixation. *Journal of Bone and Joint Surgery (American)* 2008; **90**(5):1068–77.
2. Nicayenzi B, Shah S, Schemitsch EH, Bougherara H, Zdero R. The biomechanical effect of changes in cancellous bone density on synthetic femur behaviour. *Proceedings of the Institution of Mechanical Engineers (Part H): Journal of Engineering in Medicine* 2011; **225**(11):1050–60.
3. Bagheri ZS, Tavakkoli Avval P, Bougherara H, Aziz MSR, Schemitsch EH, Zdero R. Biomechanical analysis of a new carbon fiber/flax/epoxy bone fracture plate shows less stress shielding compared to a standard clinical metal plate. *Journal of Biomechanical Engineering* 2014;**136**(9):091002-1-10.

4. Bougherara H, Zdero R, Mahboob Z, Dubov A, Shah S, Schemitsch EH. The biomechanics of a validated finite element model of stress shielding in a novel hybrid total knee replacement. *Proceedings of the Institution of Mechanical Engineers (Part H): Journal of Engineering in Medicine* 2010;**224**(10):1209–19.
5. Shah S, Bougherara H, Schemitsch EH, Zdero R. Biomechanical stress maps of an artificial femur obtained using a new infrared thermography technique validated by strain gages. *Medical Engineering and Physics* 2012;**34**(10):1496–502.
6. Bougherara H, Rahim E, Shah S, Dubov A, Schemitsch EH, Zdero R. A preliminary biomechanical assessment of a polymer composite hip implant using an infrared thermography technique validated by strain gage measurements. *Journal of Biomechanical Engineering* 2011;**133**(7):074503-1-6.
7. Mix PE. *Introduction to nondestructive testing: A training guide*. 2nd ed. Hoboken (NJ, USA): Wiley and Sons; 2005 [Chapter 10].
8. Hellier C. *Handbook of nondestructive evaluation*. 2nd ed. New York (NY, USA): McGraw-Hill; 2012 [Chapter 9].
9. Robinson AF, Dulieu-Barton JM, Quinn S, Burguete RL. Paint coating characterization for thermoelastic stress analysis of metallic materials. *Measurement Science and Technology* 2010;**21**(8):5502–12.
10. Siddiqui FS, Shah S, Nicayenzi B, Schemitsch EH, Zdero R, Bougherara H. Biomechanical analysis using thermographic imaging of a traditional metal plate versus a carbon fiber/epoxy plate for Vancouver B1 femur fractures. *Proceedings of the Institution of Mechanical Engineers (Part H): Journal of Engineering in Medicine* 2014;**228**(1):107–13.
11. Hyodo K, Inomoto M, Ma W, Miyakawa S, Tateishi T. Thermoelastic stress imaging for experimental evaluation of hip prosthesis design. *JSME International Journal (Series C)* 2001;**44**(4):1065–71.
12. www.MakeItFrom.com.
13. Bagheri ZS, El Sawi I, Bougherara H, Zdero R. Biomechanical fatigue analysis of an advanced new carbon fiber/flax/epoxy plate for bone fracture repair using conventional fatigue tests and thermography. *Journal of the Mechanical Behavior of Biomedical Materials* 2014;**35**:27–38.
14. Saleem M. *A nondestructive study of a carbon fibre epoxy composite plate using lock-in thermography, cyclic loading, and finite element analysis* [M.A.Sc. thesis]. Toronto (Canada): Department of Mechanical and Industrial Engineering, Ryerson University; 2010.
15. Zanetti EM, Musso SS, Audenino AL. Thermoelastic stress analysis by means of a standard thermocamera. *Experimental Techniques* March–April 2007:41–50.
16. Kohles SS, Vanderby Jr R. Thermographic strain analysis of the proximal canine femur. *Medical Engineering and Physics* 1997;**19**(3):262–6.
17. Vanderby Jr R, Kohles SS. Thermographic stress analysis in cortical bone. *Journal of Biomechanical Engineering* 1991;**113**(4):418–22.
18. Harwood N, Cummings WM. Applications of thermoelastic stress analysis. *Strain* February 1986:7–12.
19. Bergmann G, Graichen F, Rohlmann A. Hip joint loading during walking and running measured in two patients. *Journal of Biomechanics* 1993;**26**(8):969–90.
20. Quenneville CE, Greeley GS, Dunning CE. Evaluation of synthetic composite tibias for fracture testing using impact loads. *Proceedings of the Institution of Mechanical Engineers (Part H): Journal of Engineering in Medicine* 2010;**224**(10):1195–9.

CHAPTER

Digital Image Correlation for Strain Analysis of Whole Bones and Implants

5

Kathryn Rankin, Martin Browne, Alex Dickinson
University of Southampton, Southampton, United Kingdom

1. BACKGROUND

Measurement of whole bone strain can give insights into fracture under traumatic loads and the influence of medical devices upon long-term bone maintenance, as bone adapts (or "remodels") in response to its mechanical loading (Fig. 5.1). An implant's long-term stability depends upon maintaining surrounding bone quality, and reduced bone density makes revision surgery more complicated and risky. Digital image correlation (DIC) uses image analysis for full-field shape, deformation, and strain measurement.[1] A region of interest (ROI) on a test specimen's surface is painted with a speckle pattern and imaged with digital cameras. Software tracks the relative displacement of the speckles during deformation between images by matching "subsets" (small groups of pixels with unique gray values), spaced center-to-center by a specified "step size." A strain tensor field is calculated by interpolating displacement data points at each subset center, which are smoothed to reduce noise. Stereo DIC employs two cameras with their relative angle calibrated using photographs of a target so that a common 3-D coordinate system can be determined. This allows deformation measurement across a nonplanar surface and can accommodate limited out-of-plane deformation. DIC requires appropriate experimental setup and analysis parameter values to produce valid results and a balance between measurement sensitivity and noise, bias, and systematic error. Therefore, this chapter explains how to perform DIC tests for whole bones, implants, and whole bone–implant constructs, as well as how to analyze, present, and interpret results.

2. RESEARCH QUESTIONS

Typical research questions could include one or more of the following:

- What strain does whole bone experience under quasi-static and dynamic loads?
- What strain does whole bone experience around implants under load?

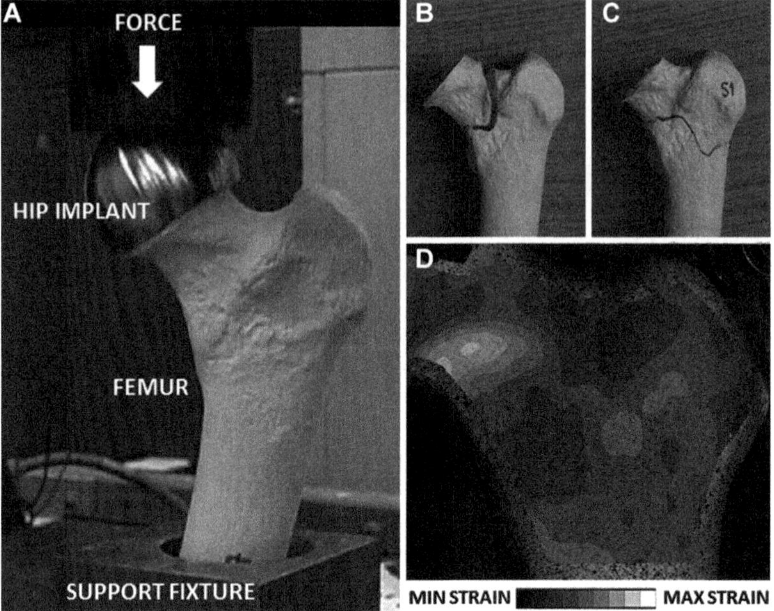

FIGURE 5.1

Testing an artificial sawbone femur–implant construct. (A) Femur–implant construct under load, (B) femur fracture through the neck caused by high strains around the implant, (C) alternative failure mode of femur fracture through the trochanter, (D) use of DIC for strain measurement to understand these failures.

- How does the strain change from intact to implanted conditions?
- What is the effect of implant factors such as geometry, fixation, and material?
- What is the effect of surgical factors such as implant positioning and sizing?
- What is the effect of patient factors such as bone shape and material quality?
- etc.

3. METHODOLOGY

3.1 GENERAL STRATEGY

DIC is used to obtain strain maps of analog femurs and human cadaveric femurs as an example of how to perform DIC tests on whole bones, implants, and whole bone–implant constructs. Whole bones are prepared for mechanical testing under loads representing a normal or traumatic scenario. An ROI on the bone's surface is identified, cleaned, and painted with a base coat and a contrasting speckle pattern. The specimen is attached to a mechanical tester using a support structure enabling the desired loading to be applied. Digital cameras are set up, focused on the illuminated ROI, and their relative position and angle calibrated. The speckle pattern, camera setup, and lighting are verified using trial images under nominally zero-strain conditions and by calculating the artificial strain to indicate noise and sensitivity. Finally, the strain field in the ROI under load is evaluated, and the measurement statistics are compared to other experimental cases.

GLOSSARY

✓ **Bias**. Systematic offset of a measurement from the true value (i.e., inaccuracy).
✓ **Noise**. Variation in measurement values (i.e., uncertainty).
✓ **Spatial resolution**. The smallest distance between independent measurement data points.
✓ **Step size**. The center-to-center spacing of pixel subsets and, thus, the spacing of data points.
✓ **Strain**. A measure of the local relative deformation of a structure in response to stress.
✓ **Subset**. A group of pixels with unique gray values used to track deformation.

SAFETY FIRST

✓ Ensure that there is sufficient ventilation while using spray paint.
✓ Be aware of manual handling, cable trip hazards, and hot lighting sources.
✓ Secure specimens in a vice before using powered surgical instruments.
✓ Keep fingers free of the test jig during specimen mounting and testing.
✓ Always wear the safety equipment required by the institution.
✓ After using cadaver bone, clean the work area and tools with disinfectant.

3.2 MATERIALS AND TOOLS LIST

- cadaveric or artificial bone
- camera calibration plate
- DIC analysis software and computer
- electromechanical tester
- lighting (i.e., LED and fiber optic lights)
- matte white spray paint and black enamel (or acrylic) paint
- orthopaedic implants and surgical instrumentation
- specimen container and support/potting medium
- tripod or rigid bar and corresponding clamps
- two identical digital cameras

3.3 SPECIMEN PREPARATION

Step 1. Store and thaw the bones. Fresh or fresh–frozen bones should be wrapped in plastic strips or vacuum-sealed in plastic bags for freezer storage at −20°C or colder. When ready for use, remove the bones from the freezer and thaw.

Step 2. Pot the bone material. Determine how the specimen shall be loaded,[2] and cut it to length if required. For stability and reproducibility, fix the specimen at a desired orientation relative to the mechanical tester axis using a pot and two-part epoxy or acrylic (PMMA, poly methyl methacrylate) potting medium. Using a pot and fixation medium enables closely reproducible positioning of different test specimens for comparable imaging. Support the bone at the desired angle using a retort stand, clamps, and bosses while the fixation medium cures.

Step 3. Prepare the specimen for painting. If using cadaver bone, remove soft tissues including the fibrous periosteum sheath, using scalpels to scrape. Dry the bone with absorbent paper, and degrease with 70% ethanol prior to painting (Fig. 5.2A and B). If testing an orthopaedic device, use the associated surgical instrumentation to prepare the bone and implant the device. Implantation prior to painting avoids damaging the speckle pattern. If comparing intact and implanted bones, use the intact specimen as its own control (i.e., test the same bone before and after implantation) so that the speckle pattern is the same, thus minimizing intraspecimen variability. In this case, paint the specimen first and take care not to damage the speckle pattern during implantation.

Step 4. Paint the specimen. Identify an appropriate ROI on the specimen surface. Spray the ROI area with thin uniform layers of matte white paint until the surface is completely covered (Fig. 5.2C). Several thin layers are preferable to one thick coat because a thin well-adhering layer of paint ensures that the speckle pattern deformation directly corresponds to the bone deformation.

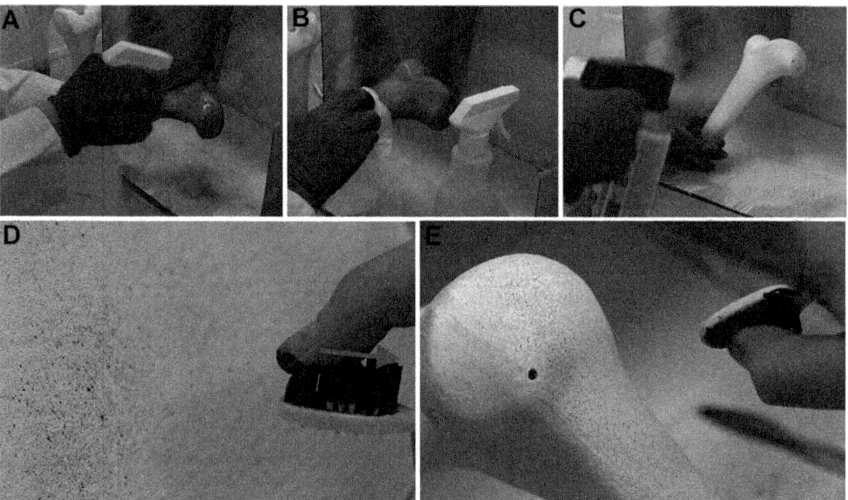

FIGURE 5.2

Surface preparation and speckle pattern application. (A) Degreasing the surface by spraying with ethanol, (B) wiping to remove excess ethanol, (C) applying a thin uniform base coat of paint, (D) applying a speckle pattern by dipping a toothbrush in contrasting colored paint and flicking the bristles with the thumb, (E) applying a speckle pattern to the specimen ROI.

Step 5. Apply a speckle pattern. Mix a small amount of black acrylic or enamel paint. Enamel is more difficult to work with and requires solvent thinner, but is more durable. This may be preferable if the protocol involves testing an intact bone and then performing implantation before retesting, where there is risk of damage to the speckle pattern. There are many methods for speckle pattern application, but a toothbrush flicking technique is flexible and inexpensive. Dip the brush bristles in the paint, hold the brush perpendicular to the surface, and draw a thumb across the bristles (Fig. 5.2D and E). Experiment on a trial surface, ideally with similar surface curvature to the bone sample, to find a brush–surface distance and amount of paint thinning/dilution to achieve the desired speckle size, pattern density, and shape. Alternatively, a fine marker pen can be used to create a speckle pattern, but this is time-consuming, so may not be feasible on cadaver specimens.

Step 6. Evaluate the speckle pattern. Take trial photographs of the speckle pattern with representative lighting and camera–specimen distance. For best image quality, speckles should be circular in shape, equal in size, and evenly distributed. Ideally, when the specimen is photographed, each speckle should be larger than two to three pixels across with high contrast to avoid aliasing. Any speckles smaller than one pixel will lead to interpolation bias of displacement toward integer–pixel values. With a fine speckle pattern (Fig. 5.3A), a small subset size may be used (15 × 15 pixels) to achieve high spatial resolution. However, sufficient speckle density is required to ensure that there are no sparse areas (Fig. 5.3B), which would require using a larger subset size (Fig. 5.3C) at the expense of spatial resolution and a loss of strain gradient information. High variability in speckle size should be avoided, as large speckles demand a larger subset size in order to avoid gaps in the data where there is insufficient unique information. This may be due to no unique grayscale information within a large speckle (Fig. 5.3D and E) or in sparser

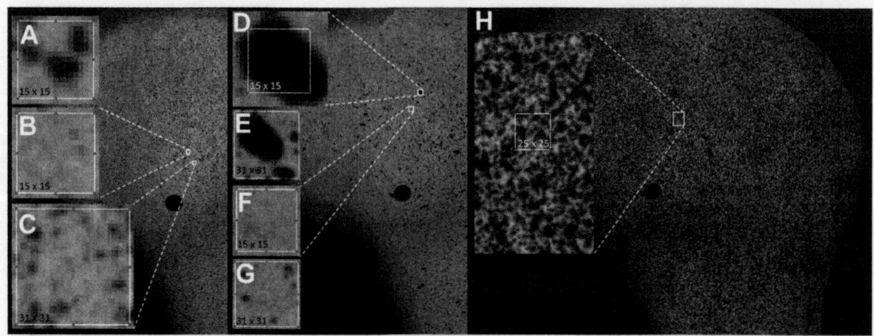

FIGURE 5.3

Typical speckle patterns and their influence on analysis subset size. (A–C) Sparse patterns may require a large subset to ensure successful correlation by avoiding featureless areas at the expense of resolution, (D, E) large speckles, (F, G) nonuniform speckle size also require increased subset size, (H) an optimal pattern is dense and uniform.

regions (Fig. 5.3F and G). The aim is a speckle pattern that is relatively uniform in size and dense (Fig. 5.3H).[3,4]

TIPS AND TRICKS

✓ A key advantage of full-field DIC strain data is that it may indicate the region of peak strain.
✓ This may inform the researcher of an appropriate location for a higher sensitivity strain gage.
✓ Always check that the calibration is still valid prior to each repeat test.
✓ To help with troubleshooting in DIC, check each stage of the testing process one by one.
✓ Avoid all causes of noise or compromised image sharpness in the experimental system.

THE "GOLD STANDARD"

No known international standards exist specifically for using DIC on bones, implants, or bone—implant constructs. Thus, researchers should consult the DIC manufacturer guidelines and also peer-reviewed journal articles that have used DIC for orthopaedic biomechanics applications.

3.4 SPECIMEN TESTING

Step 1. Set up the mechanical tester. An electromechanical tester is preferable to a servo-hydraulic tester due to its lower machine noise. Clamp the specimen support rigidly on the mechanical tester bed to ensure that it will not move under load (Fig. 5.4). Set control limits to maximum and minimum loads and displacements to protect the specimen from excessive loads.

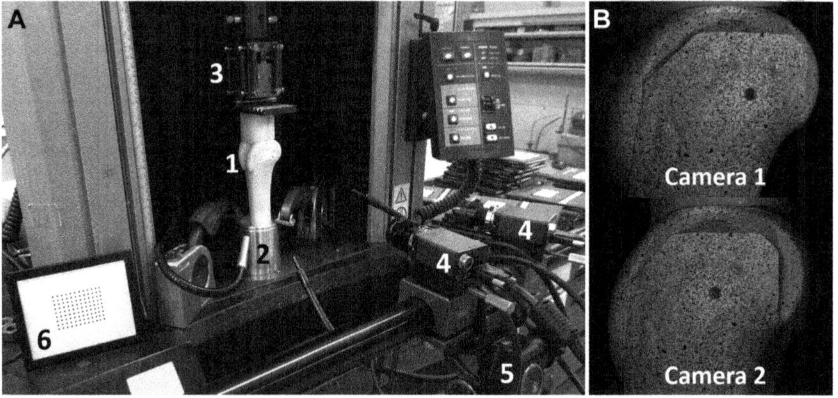

FIGURE 5.4

DIC test setup for the inverted distal femur. (A) Test setup showing (1) specimen, (2) mount, (3) planar bearing, (4) cameras, (5) beam camera mount, and (6) calibration target, (B) resulting images from both cameras.

Step 2. Set up the DIC cameras. Mount the two cameras on a rigid bar at a suitable height to capture images normal to the specimen surface ROI (Fig. 5.4). The ROI should fill around 80% of the field of view (FOV). Be aware that the edge of the analysis area will be lost during the analysis process, and results will not be reliable in highly curved areas. Lens selection is important for optimal imaging of the ROI. Short focal length lenses allow close-up images of the specimen and have low noise, but there may be practical limits to how close to the machine the cameras can be. Angle the cameras such that the ROI is in the center of each FOV. The intercamera angle should be from 15–45°, depending on the distance between the cameras and the specimen (Fig. 5.5). For a short focal length, a large stereo angle helps to minimize the standard deviation (SD) in *z* (out-of-plane) measurements. Once the camera position has been chosen, securely tighten the camera mounts to prevent movement during testing.

Step 3. Set camera focus and exposure. Open the camera apertures (small f-number, e.g., f/2.5), and focus each camera on the specimen surface, watching a live image on screen (Fig. 5.5). Zooming in until the image is pixelated shows when the camera is focused. Once both cameras are focused, reduce their apertures (large f-number, e.g., f/12) until the entire specimen ROI is in focus. Next, set a suitable exposure time for each camera. The required frame rate depends on the test strain rate and purpose. A low frame rate may be acceptable for quasi-static testing (e.g., 12 ms, manually triggered). However, if testing dynamically or trying to capture fracture, a higher frame rate (e.g., 100–9000 fps continuously triggered) is required with a high light level to allow adequate image exposure.[5–7]

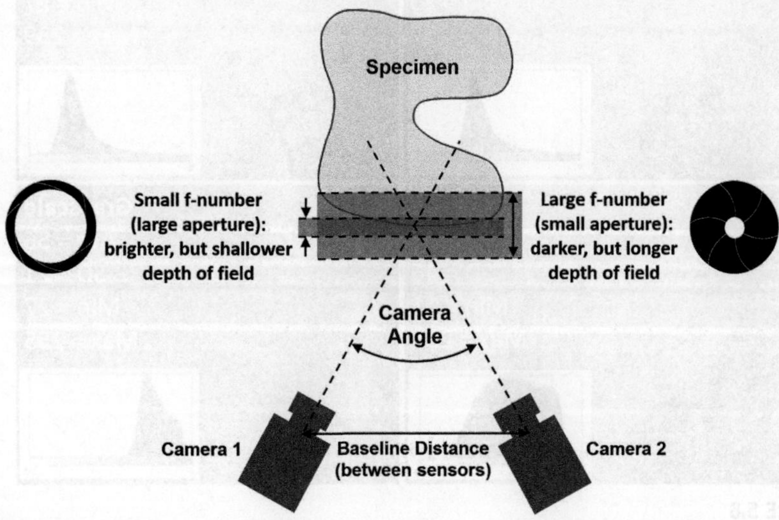

FIGURE 5.5

Focusing on the specimen. A simplified 2-D top view is shown. First, focus on the specimen surface with a large aperture (small f-number, left), and then close the aperture (large f-number, right) to increase the depth of field and bring the whole ROI into focus.

Step 4. Adjust the lighting. Check the raw image contrast by creating a histogram of grayscale pixel values. Within the speckle pattern region, the pixel gray values should lie between approximately 20–230 for a 0–255 8-bit image. Be sure to prevent saturation (i.e., peak values of 255), which may cause interpolation bias during the DIC analysis. The target is maximum raw image contrast with a broad histogram. If the histogram maximum value is too low (Fig. 5.6A), then increase the lighting (Fig. 5.6B). To obtain a broader histogram, add more speckles to the pattern (Fig. 5.6C). Avoid surface reflections, which can cause saturation, even with low lighting (Fig. 5.6D).

Step 5. Calibrate the DIC cameras. Remove the specimen from the test fixture, and replace it with a suitably sized calibration plate (DIC system–specific). Hold the calibration plate (Fig. 5.4) at varying angles, capturing an image pair in each position. Import 20–30 calibration image pairs into the DIC software and carry out a stereo triangulation calibration following the manufacturer's guidance. This will produce a calibration score for each image pair (average distance between the theoretical and actual positions where the point was found in each camera image) in pixels. The lower the score the better, and the software will highlight unacceptable scores. Place the specimen back on the mechanical tester without touching the

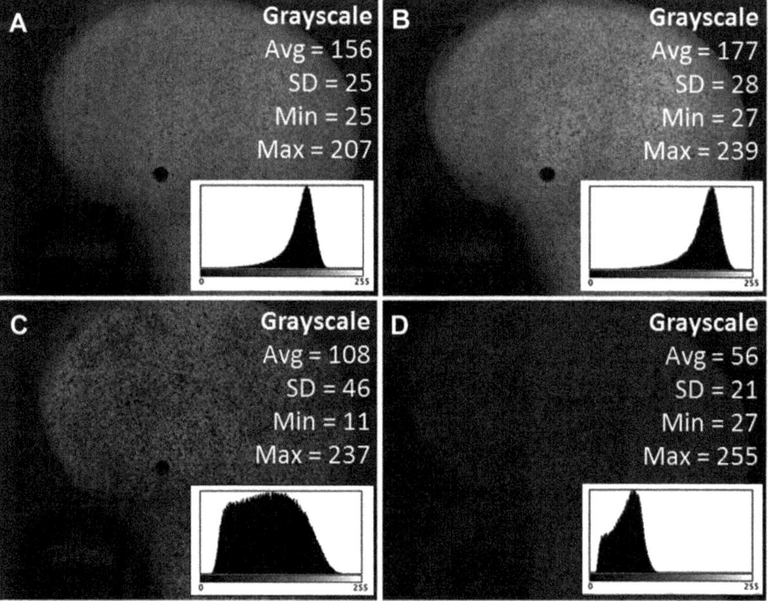

FIGURE 5.6

Typical images and grayscale histograms of the approximate analysis area. (A) Image with an overly low peak grayscale value, (B) image with increased lighting level, (C) denser speckle pattern used to increase grayscale spectrum, (D) surface reflections that cause image saturation (peak grayscale value 255), even with low light levels.

cameras, and acquire several speckle image pairs. Perform a quick correlation with these pairs to check that the projection error is sufficiently low. If the projection error is excessively high, as a first measure try repeating the calibration.

Step 6. Define the loading profile. Two approaches are available: quasi-static testing to a target structure displacement or load, or dynamic testing representing a biomechanical scenario such as gait or falling. If testing strain comparatively and nondestructively, a quasi-static load level can be determined by an appropriate signal-to-noise ratio rather than a biomechanical justification. Ramp the machine displacement until a target load is reached, and then hold while multiple images are obtained to account for noise. A displacement ramp pause also gives a stable image with low mechanical noise, but may capture viscoelastic effects. Load control with a pause for imaging would potentially result in creep or subsidence. Some systems allow image capture at a set frequency or triggering by an output from the mechanical tester when a target load is reached. If running a dynamic or impact test to investigate fracture, a high frame rate and continuous image acquisition will be required. Physiological loading rates should be applied, for example, by ramp loading (e.g., 15 mm/s)[6] or a drop tower with mass and speed as appropriate for a realistic impact energy.[5,7]

Step 7. Conduct the test. Start the mechanical tester, and collect raw data.

3.5 RAW DATA COLLECTION

Step 1. Collect verification images. Prior to testing under load, a set of unloaded, nominally zero-strain images should be collected to quantify the system noise and error in rigid body movement correction. Collect six unloaded images for the noise study, which identifies the measurement resolution due to sensor noise, the speckle pattern, and image contrast. Collect a further six images after translating the specimen to assess the ability of the software to recognize that the speckle pattern has translated without deforming. The rigid body translation correction also gives an indication of any noise introduced by distortion from the camera lenses.

Step 2. Collect strain images. Collect a set of reference unloaded images before initiating the test and image acquisition. Collect a final set of reference unloaded images after the test is complete. The image of the specimen in the unloaded state is normally the reference case for correlation vs. the loaded (i.e., strained) state. Provided the specimen is not damaged during the test, it may be more reliable to use post-test unloaded images in case loading causes the specimen to settle in the support structure. Alternatively, it is possible to carry out an incremental correlation to determine how the strain changes from one time point to the next. Carry out a suitable number of repeat tests to account for experimental error from variation in the test setup.

Step 3. Collect raw data speckle images. Along with their corresponding calibration images, the speckle images must be saved as suitably named files within an easy-to-follow folder system, such that each image set can be identified as corresponding to a specific test or repeat (e.g., femur1-unloaded-repeat1_0.tiff). The image capture software will add to the end of the filename to identify which camera captured the image (e.g., _0 or _1).

Step 4. Record conditions for each set of images. Record unique testing information for each image pair so that they are clearly identifiable (Table 5.1).

3.6 RAW DATA ANALYSIS

Noise and rigid body motion analysis

Step 1. Define the analysis area. Import the first test repeat speckle images. Select one of the unloaded specimen images as the reference image. Mark an analysis area on the speckle image. Remove areas where there are holes in the material or no speckle pattern (Fig. 5.7).

Table 5.1 Typical image recording information as a key to the raw data.

Time	Specimen ID	Test repeat #	Load [N]	Ramp rate [mm/s]	File name [*.tiff]

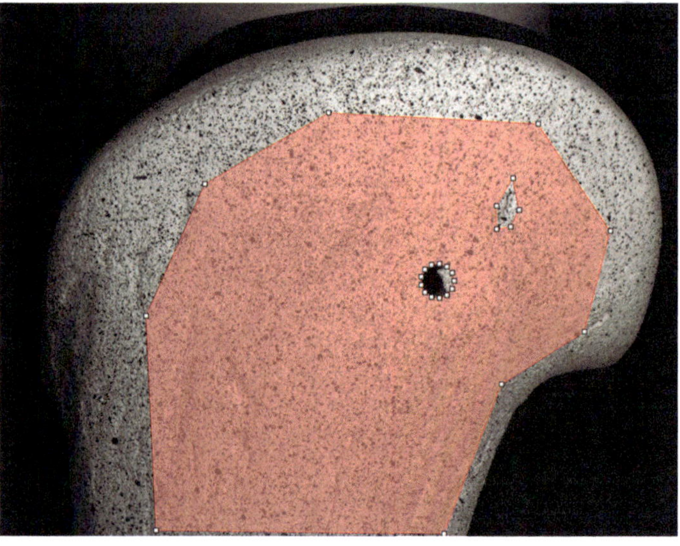

FIGURE 5.7

Defining the analysis area. Areas with holes or where the speckle pattern is not acceptable are removed.

Step 2. Select a correlation criterion equation. Choose a suitable correlation criterion equation. Choose the zero-normalized sum of squared differences since it is robust to both offset and scale in lighting during deformation, so it is recommended for most analyses. Choose the highest order interpolation filter possible to minimize bias and improve subpixel accuracy.

Step 3. Test a range of calculation parameters. Select a range of subset sizes, step size, and strain filter sizes. As a first estimate, choose a subset size, which contains approximately three speckle features (Fig. 5.8). Run the DIC calculation for the unloaded images as the "deformed" images (where a nominally zero strain state exists), using each combination of parameter values.

Step 4. Select the analysis parameters. Export the calculated strain statistics from the DIC software for each tested analysis parameter set. Plot a graph of strain standard deviation (SD) in the area of interest (Fig. 5.9), and select final analysis parameters, which balance noise and resolution.

Step 5. Evaluate the strain fields. Inspecting the calculated strain fields also indicates whether the noise with the selected parameters is acceptable (Fig. 5.10). Noise is higher around the edges of the analysis area, and these images indicate what proportion of the area can be trusted.

Step 6. Perform rigid body translation analysis. Repeat steps 1−5 and use the images taken after specimen translation as the "deformed" images.

Strain Analysis

Step 1. Import speckle images. Set an unloaded speckle image as the reference image.

Step 2. Define analysis area. It is important to use the same analysis area on a specimen where comparisons are being made, for example, between the intact and the implanted state.

Step 3. Perform the correlation calculation. Select the subset size, step size, and filter size determined from noise analysis (Fig. 5.9) and process the speckle

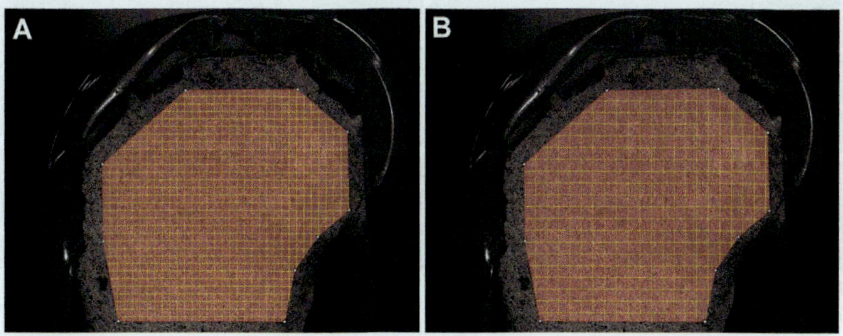

FIGURE 5.8

Comparison of the subset size (overlaid grid in the analysis area) and speckle size. (A) A first-estimate subset size (29 pixels) contains around three speckle features, (B) a larger subset (45 pixels) compromises spatial resolution and sensitivity for lower noise.

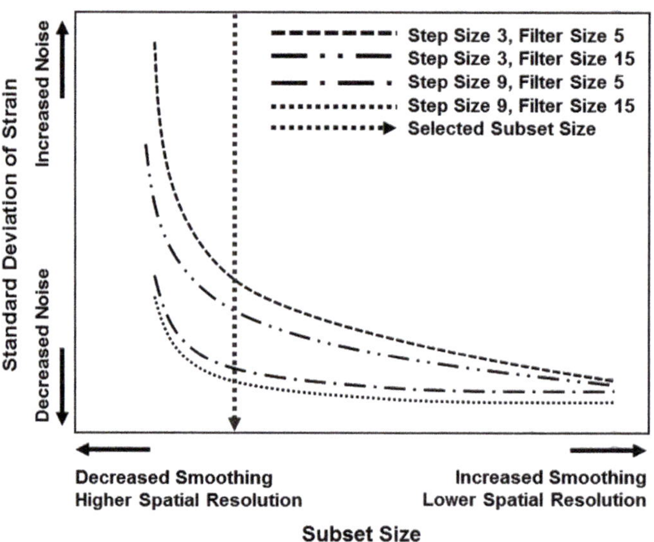

FIGURE 5.9

Influence of DIC subset, step, and filter size parameters on measurement sensitivity, noise, and spatial resolution. Calculate the strain standard deviation under nominally zero-strain conditions. Test a range of parameter sets, and select the smallest values before noise starts to increase substantially (e.g., the dashed arrow indicates a good compromise subset).

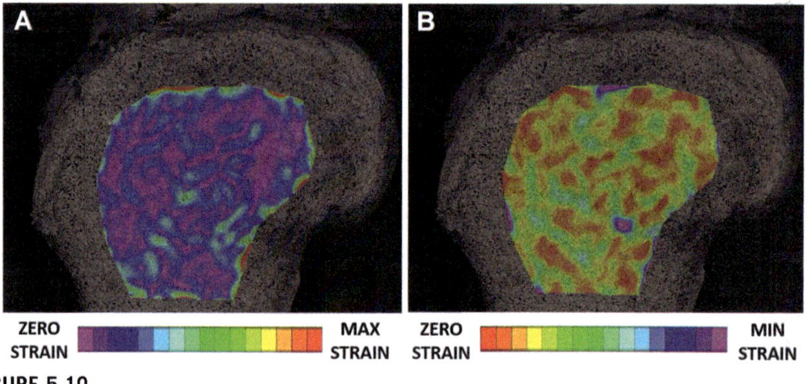

FIGURE 5.10

Unloaded nominally zero-strain maps. (A) Measurement noise for Von Mises strain, (B) measurement noise for minimum principal strain. Note the contour scaling.

images. Recheck the strain maps generated using a range of subset sizes to ensure that high strain gradients have not been lost due to excessive smoothing, and if necessary, adjust the subset size.

Step 4. Extract raw strain data. Overlay strain maps onto the test image, present them alongside unloaded nominally zero-strain (noise) data, and scale the maps within the same range so strain fields can easily be compared (Fig. 5.11). In addition to strain magnitudes, principal strain directions and surface displacement magnitude can be given.

Statistical Analysis

Step 1. Calculate correlation coefficients. Plot measured strain at key points for each specimen or location of interest vs. quantified characteristics that were varied (e.g., load, bone density, implant length, etc.) to visualize the interrelationship. Then, to establish, for example, whether a linear relationship exists, calculate several items: the linear equation $y = mx + b$ showing slope m and intercept b, and a linear correlation coefficient R.

Step 2. Perform statistical comparisons. Using statistical criteria (e.g., $P < 0.01$ or <0.05), strain data can be compared between test groups, such as bone densities (e.g., low vs. medium vs. high, etc.), implants (e.g., design 1 vs. 2 vs. 3, etc.), and so forth. A pairwise Wilcoxon signed-rank test will identify if there is a significant difference between averaged virtual strain gage values.

Step 3. Compute statistical power. Power analysis can be done after the study to ensure there were enough specimens per test group to detect all statistical differences that were actually present (i.e., was type II statistical error avoided?). Statistical power >80% is usually considered to indicate there were enough specimens per test group. Note that if good predictions of averages and standard deviations are available from prior studies, then the number of specimens and/or tests can be chosen before the study begins to ensure a power >80%.

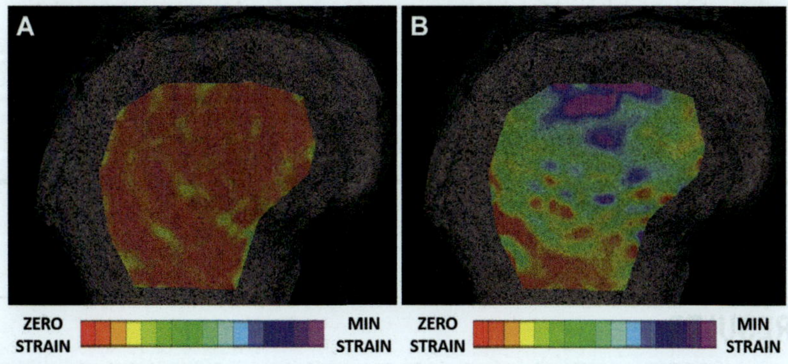

FIGURE 5.11

Maps of minimum principal strain. (A) Noise, (B) under load. Using the same scaling makes qualitative comparison easier and indicates the signal-to-noise ratio.

ENGINEER'S TOOLBOX

Speckle pattern displacement is tracked between image pairs by matching subsets, spaced center-to-center by a specified step size. A displacement data point is calculated at the center of each subset, and calculated strain is smoothed with a Gaussian decay filter over a group of surrounding data points. In the following image, the subset size is 5 pixels, the step size is 3 pixels (i.e., number of pixels between subset centers and, thus, between calculated data points), and the filter size is 5 (i.e., number of data points across the strain calculation zone for the highlighted data point). Hence, displacement resolution depends on the subset, step and filter sizes, as well as the number of pixels, the FOV, and the scale of the image relative to the speckle pattern. The spatial resolution of strain measurement is the width of the total area of pixels contributing to each data point's strain measurement. This represents the minimum distance between independently calculated data points, which can be approximated by the formula *Spatial Resolution of Strain* = {[*Step Size* × (*Filter Size* − 1)] + *Subset Size*}, equal to 17 pixels in the sample figure below. (Note: Different software packages may define the step size, subset size, and filter size differently, so the equation would be modified appropriately. Also, spatial resolution of strain is different from the strain resolution, which is the minimum change in strain magnitude that can be detected above noise. Both the strain resolution and spatial resolution influence the detection of strain gradients.)

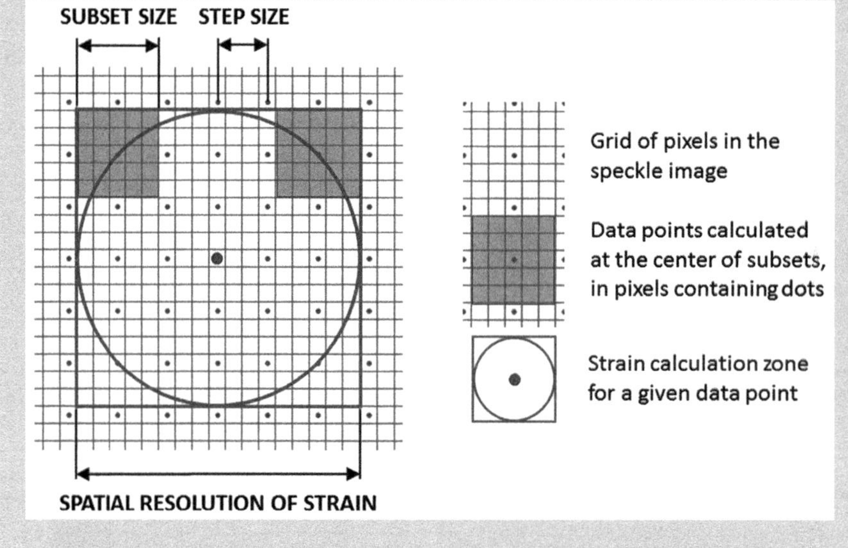

4. RESULTS

Once all DIC data collection and analysis have been performed, it is then important to communicate and present the primary results in an understandable and concise manner to the reader of a journal article, conference paper, technical report, or book chapter.

Step 1. Present general strain data. Plot the noise data (Fig. 5.12A), then extract and plot scaled strain contour plots of each test case (Fig. 5.12B and C) from the DIC software.

Step 2. Show strain data. Averaged values within subregions of interest may be reported as "virtual strain gage" values. Specifically, plot the corresponding virtual strain gage region values of each specimen case against one another, or measured against predicted strains, to form scatter plots for quantitative comparison of strain magnitudes and distributions (Fig. 5.12D). Also, a line chart allows strain along a

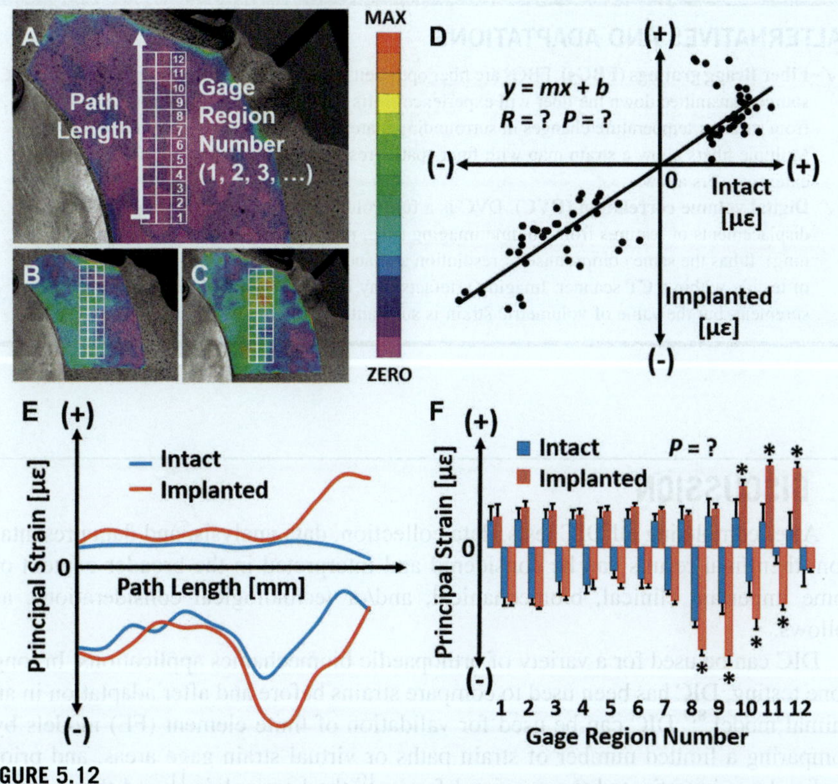

FIGURE 5.12

Typical results for comparing principal strains on intact and implanted bone. (A) Noise strain map, (B) intact bone strain map, (C) implanted bone strain map, (D) scatter plot with statistical analysis of general trend in strain change with implantation, showing the line-of-best-fit equation $y = mx + b$, the correlation coefficient R, and the P value result of a pairwise Wilcoxon signed-rank significance test, (E) path plot showing intact and implanted principal strains, (F) bar chart of principal strains in intact and implanted virtual strain gage regions showing averages $\pm$ 1 standard deviation, as well as statistical significance P values accompanied by asterisks (*), indicating which intact–implanted comparisons were different.

line path to be compared (Fig. 5.12E). Finally, a bar chart allows individual locations of strain changes to be inspected for each virtual gage region (Fig. 5.12F).

Step 3. Report all DIC verification data. This determines whether another researcher could replicate these tests and results. Include a graph of noise data vs. subset size for a range of potential step and filter sizes of the DIC calculation, and indicate the parameter values selected (Fig. 5.9).

Step 4. Report failure modes. If testing to failure, photos of typical fracture patterns help visualize failure modes in the implant, supporting bone, or at the implant–bone interface.

ALTERNATIVES AND ADAPTATIONS

✓ **Fiber Bragg gratings (FBGs)**. FBGs are fiber optic sensors. A known wavelength broadband light source transmitted down the fiber will experience shifts in wavelength due to axial deformations from strain or temperature changes in surrounding material, which can be measured in real time. Multiple fibers allow a strain map with finer spatial resolution than traditional strain gages and extensometers allow.

✓ **Digital volume correlation (DVC)**. DVC is a full-volume measurement technique that tracks displacements of features from volume imaging (e.g., micro-CT (computed tomography) scanning). It has the same compromise of resolution and accuracy, and practical and cost limitations of testing within a CT scanner. Imaging artefacts may interfere with peri-implant strain measurement, but the value of volumetric strain is substantial.

5. DISCUSSION

After completing all DIC tests, data collection, data analysis, and data presentation, then final results can be considered and interpreted in the broader context of some important clinical, biomechanical, and/or technological considerations, as follows.

DIC can be used for a variety of orthopaedic biomechanics applications. In long bone testing, DIC has been used to compare strains before and after adaptation in an animal model.[8,9] DIC can be used for validation of finite element (FE) models by comparing a limited number of strain paths or virtual strain gage areas, and prior studies have investigated the proximal femur,[10] the hemipelvis,[11] and the ankle.[12] DIC test data have also been compared to a subject-specific FE model by registering the DIC data to the FE mesh, enabling comparison of thousands of data points at a submillimeter spatial resolution.[13]

DIC can be applied to compare the effects of different implant materials upon surrounding bone strain, such as femoral total hip replacement stems[14] and acetabular cups.[15] These tests are most informative when comparing an implanted case to an intact control, such as for femoral total hip replacement [16] and total knee replacement implants.[17] Considering multi-individual analysis, DIC can also be used to analyze peri-implant bone strains around postmortem retrieved implants.[18] Finally,

DIC analysis can be performed to investigate bone fracture, such as in the proximal femur, under ramp loading[19] and under dynamic impact loading.[5–7,20]

DIC in long-bone biomechanical construct tests can achieve strain sensitivity in the range of 110–224 $\mu\varepsilon$ on relatively flat surfaces and up to 360 $\mu\varepsilon$ on curved geometries with an average $\pm$ 3 SD.[10,11,15] A technique may be precise (high resolution) and yet inaccurate (with bias from systematic errors) or vice versa, which indicates that the use of the mean noise as a measure of bias and SD noise as a measure of resolution (uncertainty) is preferable. Bias and resolution have been reported as 350 $\mu\varepsilon$ and 90 $\mu\varepsilon$, respectively, in a murine model.[9]

DIC is reported to correlate closely with "gold standard" strain gage measurements on composite bones under quasi-static conditions, with high correlation coefficient ($R = 0.99$), low standard error ($\sim$38 $\mu\varepsilon$), and near-unity regression slope ($m = 0.99$).[11] Under dynamic loading, DIC vs. strain gage root mean squared (RMS) errors of 127 $\mu\varepsilon$ (average) and 239 $\mu\varepsilon$ (SD) have been achieved in cadavers.[5]

As with any measurement technique, DIC has particular pros and cons. DIC has the advantage of providing full-field data, noninvasively, but only provides surface data and does not inform on volumetric strain. Application of adequate speckle patterns can be difficult, especially on cadaver material, and ultimately, strain measurements refer to the speckle pattern rather than the specimen itself. The drawback of the noncontact and optimization nature of DIC is that results are not necessarily true measurements of strain, but are smoothed estimates which depend upon user-selected calculation parameters. Thus, DIC could be regarded as providing qualitative information. However, the user may be able to optimize the parameters to achieve an acceptable signal-to-noise ratio and spatial resolution relative to strain gradients, and enable both reproducibility between tests and repeatability for comparison to other DIC and non-DIC studies. For this reason, speckle pattern details and correlation parameter verification must always be reported.

6. SUMMARY

- Bone strain indicates fracture risk, the stimulus for remodeling, and the influence of implants.
- Measurements may compare the pre- and postimplanted states and different implant options.
- DIC tracks displacement of painted speckles on a test specimen's surface to estimate strain.
- DIC calculation parameters must be verified, balancing resolution with sensitivity and noise.
- These parameters must be reported so that a study's results can be compared with others.
- DIC measurements depend fundamentally upon the quality of the speckle pattern and images.

7. QUIZ QUESTIONS

1. How do DIC image pixel subset, step, and filter size influence measurements?
2. What are the pros and cons of DIC vs. "gold standard" strain gaging?
3. What kinds of clinical data could help researchers interpret DIC results?
4. What are the characteristics of an optimal speckle pattern, and how can it be created?
5. Estimate spatial resolution for a 21-pixel subset size, a 7-pixel step size, and a 5-data point filter size. If one pixel corresponds to 0.05 mm, what is the spatial resolution? (answer: 2.45 mm).

REFERENCES

1. Sutton M, Orteu J, Schreier H. *Image correlation for shape, motion and deformation measurements: Basic concepts, theory and applications*. New York (NY, USA): Springer; 2009.
2. An Y, Draughn R. *Mechanical testing of bone and the bone-implant interface*. Boca Raton (FL, USA): CRC Press; 2000.
3. Lecompte D, Smits A, Bossuyt S, Sol H, Vantomme J, van Hemelrijk D, et al. Quality assessment of speckle patterns for digital image correlation. *Optics and Lasers in Engineering* 2006;**44**(11):1132–45.
4. Crammond G, Boyd S, Dulieu-Barton J. Speckle pattern quality assessment for digital image correlation. *Optics and Lasers in Engineering* 2013;**51**(12):1368–78.
5. Gilchrist S, Guy P, Cripton P. Development of an inertia-driven model of sideways fall for detailed study of femur fracture mechanics. *Journal of Biomechanical Engineering* 2013;**135**(12):12001.
6. Grassi L, Väänänen S, Amin Yavari S, Jurvelin J, Weinans H, Ristinmaa M, et al. Full-field strain measurement during mechanical testing of the human femur at physiologically relevant strain rates. *Journal of Biomechanical Engineering* 2014;**136**(11):111010.
7. Helgason B, Gilchrist S, Ariza O, Chak J, Zheng G, Widmer R, et al. Development of a balanced experimental-computational approach to understanding the mechanics of proximal femur fractures. *Medical Engineering and Physics* 2014;**36**(6):793–9.
8. Sztefek P, Vanleene M, Olsson R, Collinson R, Pitsillides A, Shefelbine S. Using digital image correlation to determine bone surface strains during loading and after adaptation of the mouse tibia. *Journal of Biomechanics* 2010;**43**(4):599–605.
9. Carriero A, Abela L, Pitsillides A, Shefelbine S. Ex vivo determination of bone tissue strains for an in vivo mouse tibial loading model. *Journal of Biomechanics* 2014;**47**(10):2490–7.
10. Dickinson A, Taylor A, Ozturk H, Browne M. Experimental validation of a finite element analysis model of the proximal femur using digital image correlation and a composite bone model. *Journal of Biomechanical Engineering* 2011;**133**(1):014504.
11. Ghosh R, Gupta S, Dickinson A, Browne M. Experimental validation of finite element models of intact and implanted composite hemipelvises using digital image correlation. *Journal of Biomechanical Engineering* 2012;**134**(8):081003.
12. Terrier A, Larrea X, Guerdat J, Crevoisier X. Development and experimental validation of a finite element model of total ankle replacement. *Journal of Biomechanics* 2014;**47**(3):742–5.

13. Grassi L, Väänänen S, Amin Yavari S, Weinans H, Jurvelin J, Zadpoor A, et al. Experimental validation of finite element model for proximal composite femur using optical measurements. *Journal of the Mechanical Behavior of Biomedical Materials* 2013;**21**: 86–94.
14. Tayton E, Evans S, O'Doherty D. Mapping the strain distribution on the proximal femur with titanium and flexible-stemmed implants using digital image correlation. *Journal of Bone and Joint Surgery (Br)* 2010;**9**(8):1176–81.
15. Dickinson A, Taylor A, Browne M. The influence of acetabular cup material on pelvis cortex surface strains, measured using digital image correlation. *Journal of Biomechanics* 2012;**45**(4):719–23.
16. Chanda S, Dickinson A, Gupta S, Browne M. Full-field in vitro measurements and in silico predictions of strain shielding in the implanted femur after total hip arthroplasty. *Proceedings of the Institution of Mechanical Engineers (Part H): Journal of Engineering in Medicine* 2015;**229**(8):549–59.
17. Rankin K, Dickinson A, Briscoe A, Browne M. Does a PEEK femoral TKA implant preserve intact femoral surface strains compared with CoCr? A preliminary laboratory study. *Clinical Orthopaedics and Related Research* 2016. Online, http://dx.doi.org/10.1007/s11999-016-4801-8.
18. Mann K, Miller M, Goodheart J, Izant T, Cleary R. Peri-implant bone strains and micromotion following in vivo service: a postmortem retrieval study of 22 tibial components from total knee replacements. *Journal of Orthopaedic Research* 2014;**32**(3):355–61.
19. Väänänen S, Amin Yavari S, Weinans H, Zadpoor A, Jurvelin J, Isaksson H. Repeatability of digital image correlation for measurement of surface strains in composite long bones. *Journal of Biomechanics* 2013;**46**(11):1928–32.
20. Op Den Buijs J, Dragomir-Daescu D. Validated finite element models of the proximal femur using two-dimensional projected geometry and bone density. *Computer Methods and Programs in Biomedicine* 2011;**104**(2):168–74.

CHAPTER

Force and Torque Measurements of Surgical Drilling Into Whole Bone

6

Radovan Zdero[1], Troy MacAvelia[2], Farrokh Janabi-Sharifi[2]

Western University, London, ON, Canada[1]; Ryerson University, Toronto, ON, Canada[2]

1. BACKGROUND

Repairing whole bone fractures requires surgical drilling to create pilot holes for easy insertion of cortical screws, cancellous screws, and locking bolts to align adjacent bone fragments, apply fracture plates, or insert fracture nails (Fig. 6.1).[1–14] Although drilling may be done using CO_2 pulsed lasers, haptic

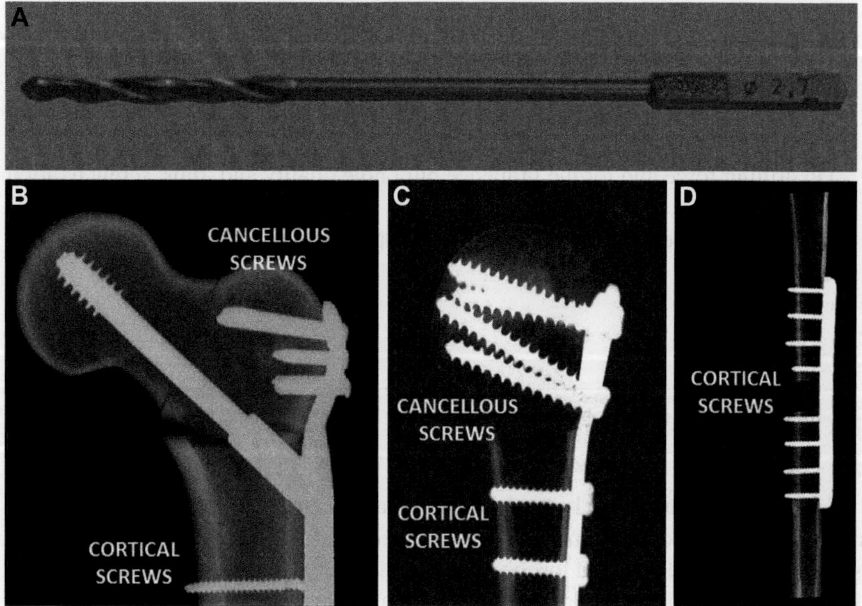

FIGURE 6.1

Surgical drilling for fracture repair. (A) Surgical drill bit, (B) proximal femur, (C) proximal humerus, (D) humeral diaphysis.

systems that offer tactile force and torque feedback, and teleoperation in which a surgeon conducts the procedure off-site, electrically-powered hand drills remain the common clinical practice. Force and torque generated at the drill bit–bone interface are influenced by feed rate (i.e., linear speed of the drill bit into bone), spindle speed (i.e., rotational speed of the drill bit), drill bit tip angle (i.e., angle of the pointed tip of the drill bit), drill bit size (i.e., outer diameter), and bone type (i.e., cortical vs. cancellous, normal vs. osteoporotic, human vs. animal, etc.). Nonoptimal force and torque can raise temperatures, causing bone necrosis, as well as poor pilot hole quality, causing poor screw fixation. Consequently, optimization of drilling force and torque is a clinically relevant topic. Therefore, this chapter explains how to measure force and torque during surgical drilling into whole bone, as well as how to analyze, present, and interpret the results.

2. RESEARCH QUESTIONS

Typical research questions might include one or more of the following:

- Do orthopaedic surgeons with varying experience generate different drilling force and torque?
- Do whole bone storage, thawing time, age, type, etc., affect drilling force and torque?
- Do drill bit feed rate, spindle speed, and geometry change drilling force and torque?
- Do drilling force and torque influence pilot hole quality and, thus, screw fixation strength?
- Do drilling force and torque alter bone temperature, potentially leading to bone necrosis?
- etc.

3. METHODOLOGY

3.1 GENERAL STRATEGY

Human or animal whole bones or segments are secured by a vice directly under a surgical drill bit whose feed rate and spindle speed are computer controlled. The vice is equipped with a force and torque sensor, which actively records data during drilling. Raw force and torque data are then normalized by the surface area of the pilot hole to eliminate geometric effects. Statistical comparisons are made between the test groups for raw and normalized drilling measurements. Finally, correlation coefficients are computed for raw and normalized data vs. biological bone demographics (i.e., age, sex, limb side, bone mineral density (BMD), and clinical T-score) to determine which factors are important.

GLOSSARY

✓ **Feed rate.** Linear speed of the drill bit along its long axis during drilling.
✓ **Force.** The linear load experienced by the bone along the long axis of the drill bit.
✓ **Spindle speed.** Rotational speed of the drill bit around its long axis during drilling.
✓ **Torque.** The angular load experienced by the bone around the long axis of the drill bit.

SAFETY FIRST

✓ Remember to always wear goggles and gloves for protection.
✓ Secure the vice and bone before using the drill press to conduct drilling tests.
✓ Use a fume hood or do tests in an open area, since bone may emit a burning odor when drilled.
✓ Clean the work area and all tools with bleach or disinfectant after testing bone specimens.

3.2 MATERIALS AND TOOLS LIST

- computer-controlled drill press
- force and torque sensors
- human or animal bone
- leveling gage
- surgical drill bits
- tape measure
- thickness gage
- Vernier calipers
- vice or clamp

3.3 SPECIMEN PREPARATION

Step 1. Store the bone. Fresh or fresh−frozen whole bones or segments initially need to be wrapped in plastic strips or vacuum-sealed in plastic bags for proper storage in a freezer prior to the study. The freezer should be held at −20°C or colder. Note that if using embalmed or dried/dehydrated bones, these do not need to be frozen and may simply be placed in a plastic box or bag for storage at room temperature prior to tests.

Step 2. Thaw the bone. Bone specimens should be removed from the freezer, left in their plastic wrappings or vacuum-sealed bags, and placed on a surface at ambient room temperature or in a warm water bath to thaw for at least 12 h. Bone specimens are then removed from their plastic wrappings or bags and soaked or sprayed with saline water solution to prevent dehydration.

Step 3. Determine the drilling site. Mount bone specimens into a vice or clamp them onto a table, make measurements for the drilling site, and mark the location with a pen. Because of geometric variability among biological bone, the relative location of the drilling site with respect to a common reference dimension needs to be computed and used for all bone specimens (e.g., % = relative dimension/reference dimension × 100 = distance from one end of the bone to the drilling site/total length of bone × 100).

TIPS AND TRICKS

- ✓ Whole bones can be cut into segments with a band saw for easier handling.
- ✓ Place a leveling gage on the bone surface to ensure drill direction is perpendicular.
- ✓ Insert a new drill bit when needed to avoid biasing data with dull drill bits.
- ✓ The same researcher should perform all tests for consistency.

THE "GOLD STANDARD"

The International Organization for Standardization (ISO) provides guidelines for drill bit design factors, such as materials selection, mechanical properties, and labeling, in its document, ISO 9714-1 (Orthopaedic drilling instruments – Part 1: drill bits, taps and countersink cutters), but no standards exist for surgical bone drilling itself. Thus, researchers should use peer-reviewed journal articles and/or medical manufacturer's surgical technique manuals as guidelines.

3.4 SPECIMEN TESTING

Step 1. Assemble the test setup. An experimental test setup for surgical bone drilling research must first be assembled (Fig. 6.2). Several components are vital. A computer-controlled drill press will adjust drill bit feed rate and spindle speed over the typical ranges used in the research literature or clinically in orthopaedic surgery. A vice will rigidly hold the bone specimens in proper position. A biaxial force and torque sensor should be attached under the vice to monitor surgical drilling load, but it must be of sufficient resolution and accuracy. An LVDT (linear variable displacement transducer) can be attached to the drill bit chuck to measure drilling depth. Vibration isolation pads should be placed under the drill press and vice to minimize mechanical vibration. A heavy work bench will be used to mount the entire assembly, but will also further minimize mechanical vibration. Water, liquid coolant, or compressed air may be used to cool the drill bit and bone, as well as to unclog debris from the drill bit and clear away debris from the bone specimen. Orthopaedic surgical drill bits that are unused need to be procured in sufficient quantity.

Step 2. Create a "pecking" hole. Create a preliminary pecking hole, which will later be used to guide the real surgical drill bit during the actual drilling test. To do this, place the bone into the vice under the center of the drill bit chuck. Use a leveling

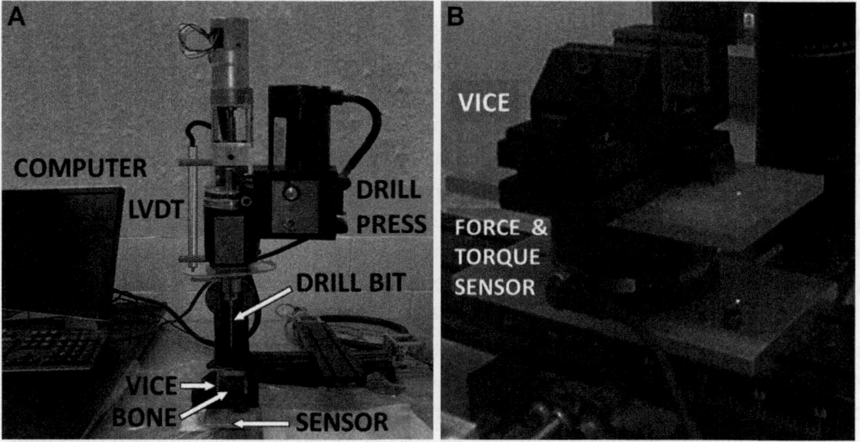

FIGURE 6.2

Experimental setup for drilling into bone. (A) Overall test setup, (B) close-up of the force and torque sensor.

gage to ensure the bone site is perpendicular to the drill bit direction. Insert a "dummy" center drill bit (e.g., size #3 with a 2.8-mm diameter) into the chuck, and center it about 20 mm above the bone drilling site. Set the desired spindle speed, but not the feed rate. Then, manually, slowly, and incrementally lower the spinning dummy center drill bit until the tip superficially punctures the bone to the correct pecking depth (e.g., 0.25–0.50 mm). Pecking depth can usually be detected by a grinding sound or slight vibration or visual observation, but the LVDT may be more reliable. Note that lowering the dummy center drill bit too quickly during pecking could cause the tip to slip overtop of a slippery bone surface, so the wrong location is punctured, as well as potentially bending or breaking the drill bit. Then, retract the dummy center drill bit and remove it from the chuck.

Step 3. Choose the drill bit. The correct surgical drill bit for the real drilling test then needs to be chosen. This is based on the surgical procedure that will make use of the pilot hole (e.g., cortical or cancellous screw insertion during fracture plating, mounting an acetabular cup during total hip arthroplasty, etc.), the type of bone being drilled (e.g., cortical vs. cancellous, normal vs. osteoporotic, etc.), and/or the particular research question being asked about the drill bit (e.g., effects of cutting flute geometry, drill bit material, drill bit diameter, etc.). Typical drill bits used in the research literature and/or clinically have outer diameters of 1.98–4.76 mm and lengths of 60–200 mm.[1,2,4,5,9,11,12,14]

Step 4. Do the drilling test. Perform the real drilling test that will create the final pilot hole. To do this, insert the real unused surgical drill bit of appropriate size into the chuck and position it approximately 20 mm above the bone specimen. Then, start the automated computer-controlled drilling process by setting the desired feed rate (e.g., 10–132 mm/min) and spindle speed (e.g., 40–3300 rpm) within typical ranges used in the research literature and/or clinically.[1,2,5,9,11,12,16] Ensure that the drill and the force and torque sensors are started simultaneously for data collection. Allow drilling to continue until the drill bit tip reaches the desired depth. Then, return the drill bit to its home position, turn off the drill press, and remove the bone from the vice. Clear the drill bit flutes of clogged material in preparation for the next test. Cooling the bone can be done using water, saline solution, liquid coolant, or compressed air if the research question requires it; however, irrigation may not always be done during actual orthopaedic surgery.

Step 5. Measure pilot hole length. Remove the bone from the drill press and hold it firmly on a work surface, whereas large specimens may be secured using a vice or clamp on a work surface. To measure cortical wall thickness at the drilling site, insert a clinical or mechanical depth gage that has a hooked end, pull the hooked end until it engages the intramedullary underside of the cortical wall, and then measure this length using a ruler or Vernier caliper (Fig. 6.3A–C). To measure cancellous pilot hole depth at the drilling site, insert a thin, rigid guide wire until it feels

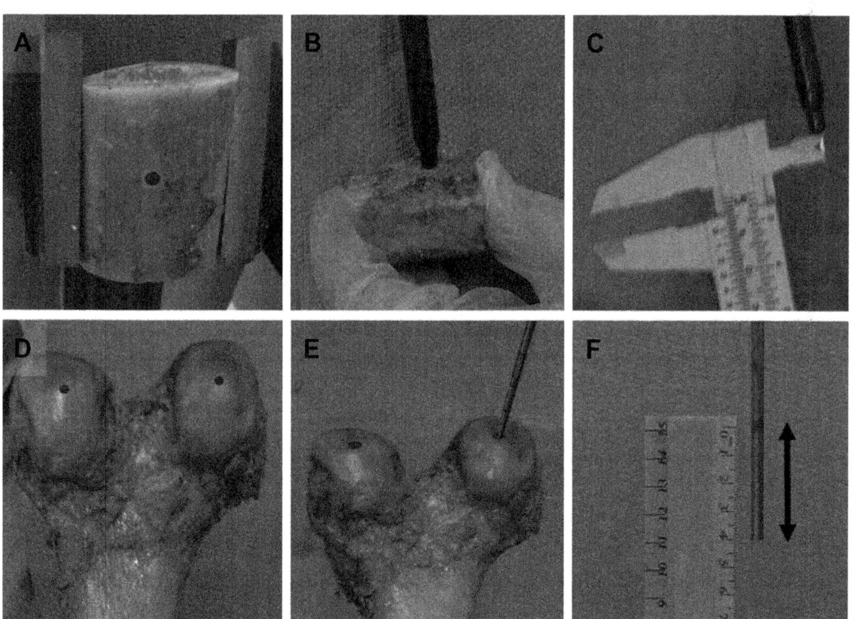

FIGURE 6.3

Measurement of pilot hole geometry. (A) Cortical bone specimen with a pilot hole, (B) depth gage insertion into a cortical pilot hole, (C) depth measurement of a cortical pilot hole using a depth gage and Vernier calipers, (D) cancellous bone specimen with a pilot hole, (E) guide wire insertion into a cancellous pilot hole, (F) depth measurement of a cancellous pilot hole using a guide wire and ruler.

like it has reached the bottom of the hole, use a pen to mark the guide wire's surface that is at the entrance of the pilot hole, remove the guide wire, and then measure the length of the guide wire's tip-to-pen mark using a ruler or Vernier calipers (Fig. 6.3D–F). Guide wires can also be used to clear away unwanted bone debris from the pilot hole.

Step 6. Photograph or draw the specimen. Photograph or draw the pilot hole that was created, the bone debris that was generated during drilling, and the drill bit for later data analysis.

3.5 RAW DATA COLLECTION

Step 1. Record bone characteristics. Enter human or animal demographic information (i.e., age, sex, and left or right limb) and quantitative bone properties (i.e., BMD and clinical T-score), which will help determine the factors that are statistically correlated with drilling force and torque (Table 6.1). Note that BMD is in 2-D units (i.e., g/cm^2 rather than g/cm^3), since DEXA (dual-energy X-ray absorptiometry) bone density scans are only 2-D. DEXA scan reports often provide BMD values and corresponding clinical T-scores (i.e., normal T-score ≥ -1, $-1 >$ osteopenic T-score > -2.5, and osteoporotic T-score ≤ -2.5). When such analysis is overlooked or missing, computations can be made as T-score = (bone BMD − reference population mean BMD)/(1 standard deviation of reference population BMD), where T-score reference populations are young healthy adults of the same sex that are aged 20–40 years.[15,16]

Step 2. Record pilot hole geometry. Input the pilot hole diameter (i.e., drill bit diameter) into the summary table (Table 6.2). Enter the pilot hole length, which is the wall thickness for unicortical or bicortical specimens, whereas it is the pilot hole depth for cancellous specimens (Table 6.2).

Step 3. Record drilling parameters. Feed rate, spindle speed, and drill bit tip angle should be noted (Table 6.2). Raw force and torque should have been recorded during drilling and then smoothed for noise artifacts using a running average function or other appropriate filtering algorithm. Then, peak force and torque can be identified (Table 6.2).

Table 6.1 Bone characteristics.

Bone	Age [years]	Sex [M, F]	Limb Side [L, R]	Bone Mineral Density [g/cm^2]	T-score
1					
2					
3					
etc.					
Avg		–	–		
SD		–	–		

Avg, *average;* SD, *standard deviation.*

Table 6.2 Drilling test parameters.

Bone	Feed Rate f [mm/min]	Spindle Speed ω [rpm]	Drill Bit Tip Angle θ [°]	Drill Bit Diameter D [mm]	Pilot Hole Length L [mm]	Peak Force F [N]	Peak Torque T [N·mm]	Remarks
1								
2								
3								
etc.								
Avg	–	–	–	–				–
SD	–	–	–	–				–

Avg, average; SD, standard deviation.

Step 4. Record visual observations. Inspect prior photographs or drawings and record remarks on the pilot hole (e.g., smooth? circular? surface cracking? etc.), the bone debris that has been removed (e.g., shape, size, texture, color), and the drill bit (e.g., wear, bending, breaking, etc.) (Table 6.2).

3.6 RAW DATA ANALYSIS

Step 1. Normalize raw data. Raw peak forces and torques should be normalized by geometry by taking into account the surface area of the pilot hole that is in contact with the drill bit. This can be computed as $F_{NORM} = F_{RAW}/A = F_{RAW}/(\pi DL)$ and $T_{NORM} = T_{RAW}/A = T_{RAW}/(\pi DL)$, where F is force, T is torque, A is surface area, D is pilot hole diameter, and L is bone wall thickness (cortical specimen) or hole depth (cancellous specimen).

Step 2. Calculate correlation coefficients. Plot the measured raw and normalized result for each individual specimen (i.e., peak force 1, 2, 3, etc.) (Table 6.2) vs. its corresponding specimen characteristic (i.e., age 1, 2, 3, etc.) (Table 6.1) to help visualize the interrelationship between measurements vs. characteristics. Then, calculate the correlation coefficient R for each measurement–characteristic pair to determine which characteristic has an influence on measurements (e.g., $R > 0.8$ is a typical value considered to indicate a strong correlation).

Step 3. Perform statistical comparisons. The criterion for statistical difference needs to be chosen (e.g., $P < 0.01$ or < 0.05). Then, data can be used to compare different patient groups (e.g., men ≤ 60 years old vs. men > 60 years old, "normal" women vs. "osteoporotic" women, etc.), bone types (e.g., femur vs. tibia, left humerus vs. right humerus, etc.), drilling sites (e.g., anterior vs. lateral surface), drill bit designs (e.g., small vs. large outer diameter), etc. Various software programs exist for comparing two test groups (e.g., paired t-test) or two or more test groups influenced by multiple factors (e.g., analysis of variance, ANOVA).

Step 4. Compute statistical power. Power analysis can be done after the study to ensure there were enough specimens per group to detect all statistical differences that were actually present (i.e., was type II statistical error avoided?). Statistical power >80% is usually considered to indicate there were enough specimens per test group. Note that if good predictions of averages and standard deviations are available from prior studies, then the number of specimens and/or tests can be chosen before the study begins to ensure a power >80%.

ENGINEER'S TOOLBOX

Force and torque for bone drilling can be estimated using engineering formulas. Assume no effects from bone isotropy, homogeneity, and viscoelasticity, or from bone chip clogging of drill bit cutting flutes, bone temperature, and drill bit wear. So, for each revolution of the drill bit around its long axis, the force $F = 2.5\sigma_U(f/\omega)D\sin(\theta/2)$ and torque $T = 0.625\sigma_U(f/\omega)D^2$, where F is force [N], T is torque [N·mm], σ_U is ultimate tensile stress of bone [N/mm^2], f is drill bit feed rate [mm/min], ω is drill bit spindle speed [rpm or revolutions/min], D is drill bit outer diameter [mm], and θ is drill bit tip angle [°].

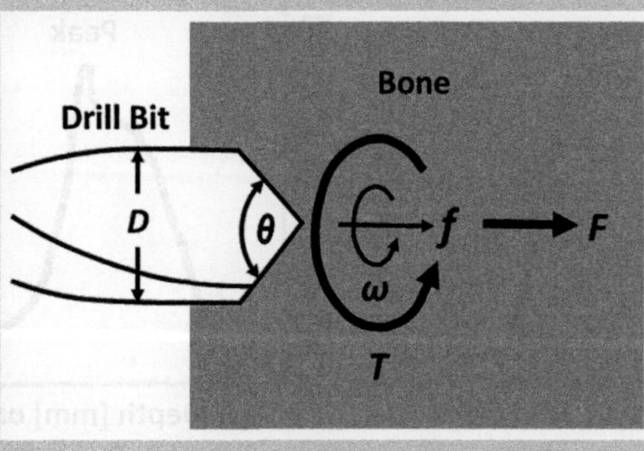

4. RESULTS

Once all surgical drilling data collection and analysis have been performed, it is then important to communicate and present the primary results in an understandable and concise manner to the reader of a journal article, conference paper, technical report, or book chapter.

Step 1. Show raw data profile. Begin with typical raw force and torque profiles vs. drilling depth or time for both unsmoothed and smoothed data (Fig. 6.4). Initially, force and torque are zero before the drill bit tip contacts the bone. Then, there is a small "step" in force and torque as the drill bit cutting flutes enter the "pecking" hole to make superficial contact with bone. Following this, there is a gradual increase in force and torque to some peak value as the drill bit tip becomes fully immersed into bone. Next, a gradual drop in force and torque occur as the drill bit exits through the other side of the bone wall (i.e., cortical drilling) or reaches the final drilling depth (i.e., cancellous drilling), but a small amount of torque is maintained as the drill bit

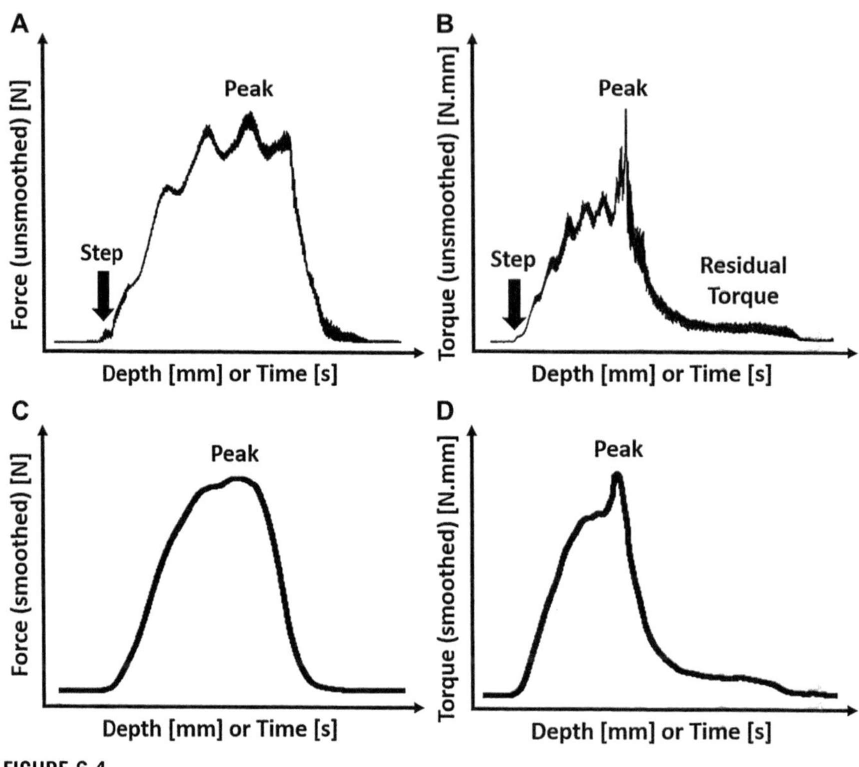

FIGURE 6.4

Raw data profiles of surgical drilling through one contiguous segment of cortical or cancellous bone. (A) Force (unsmoothed for noise), (B) torque (unsmoothed for noise), (C) force (smoothed for noise), (D) torque (smoothed for noise).

continues to spin and make superficial contact with surrounding bone. For the unsmoothed data, note that there are usually small-scale high-frequency fluctuations superimposed on top of large-scale low-frequency fluctuations in force and torque, which may be due a combination of factors: (1) bone anisotropy, porosity, and viscoelasticity; (2) mechanical flexing of the bone as the drill bit thrusts forward; (3) clogging of drill bit cutting flutes with bone debris; (4) drill bit vibration caused by drill bit bending or a loose bone specimen; and (5) load alterations with each circumferential pass of a cutting flute past a given point on the bone. Although noise smoothing algorithms eliminate these small-scale high-frequency fluctuations, some large-scale low-frequency fluctuations remain.

Step 2. Show main results. The main numerical findings of the study can be presented, namely, raw and normalized peak force and torque (Fig. 6.5). For each of these parameters, all statistical pairwise comparisons should be made (i.e., bone type 1 vs. 2 vs. 3, drill bit type 1 vs. 2 vs. 3, etc.) to generate statistical *P* values

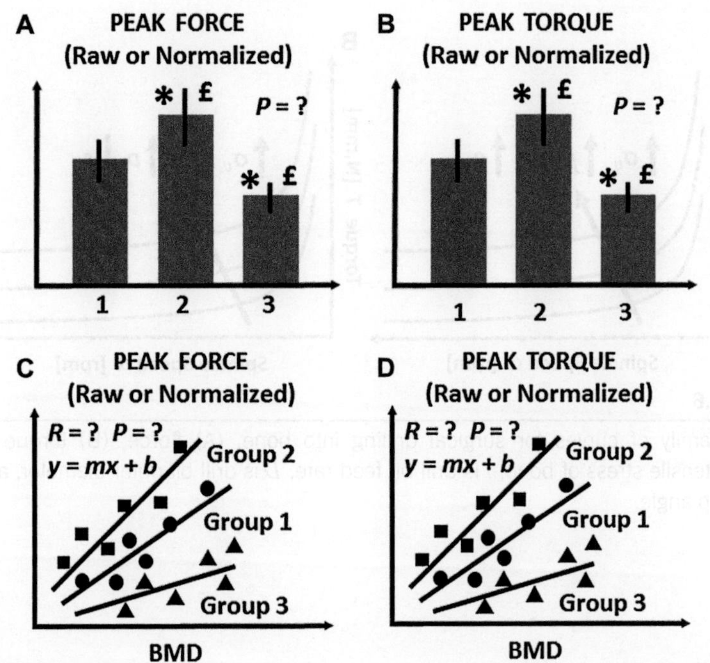

FIGURE 6.5

Surgical drilling test results. (A) Force vs. test group, (B) torque vs. test group, where bar graphs show average $\pm$ 1 standard deviation and P is the statistical difference result for each pairwise comparison between groups which can be indicated using symbols like asterisks (*), pounds (£), etc. (C) Force vs. BMD, (D) torque vs. BMD, where BMD is bone mineral density, R is the linear correlation coefficient for each line of best fit, P is the statistical difference value ensuring the slope of the line of best fit (i.e., slope $= m$) is statistically different than a horizontal line (i.e., slope $= 0$), and $y = mx + b$ is the equation of each line.

(Fig. 6.5A and B). Any linear (or nonlinear) trends can be illustrated for force and torque vs. BMD (Fig. 6.5C and D). For each line of best fit, several items can be generated: an equation $y = mx + b$ showing slope m and intercept b, a linear correlation coefficient R, and its own P value to ensure the slope of each line of best fit (i.e., slope $= m$) is statistically different than a horizontal line (i.e., slope $= 0$). Also, if a substantially wide range of drilling parameters has been examined, then results can be presented as a family of curves that show how various parameters affect peak force and torque (Fig. 6.6).

Step 3. Show specimen photos. Present photos to help visualize the quality of drilling, such as the pilot hole (e.g., circular? symmetric? surface cracking? etc.), bone debris (e.g., shape, size, texture, etc.), and drill bit cutting flutes (e.g., bending, breaking, wear, etc.) (Fig. 6.7).

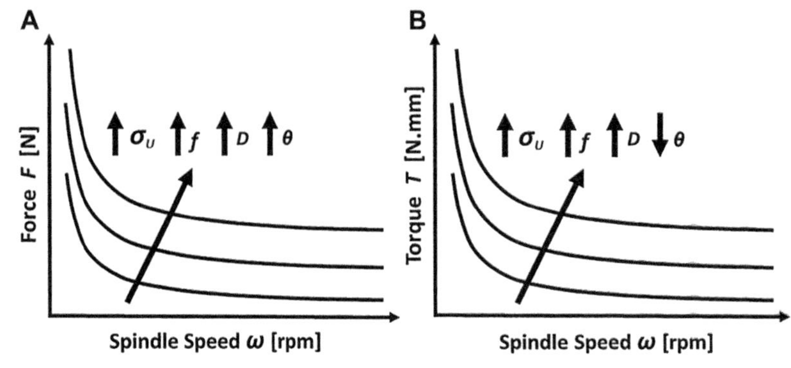

FIGURE 6.6

Typical family of curves for surgical drilling into bone. (A) Force, (B) torque. σ_U is ultimate tensile stress of bone, f is drill bit feed rate, D is drill bit outer diameter, and θ is drill bit tip angle.

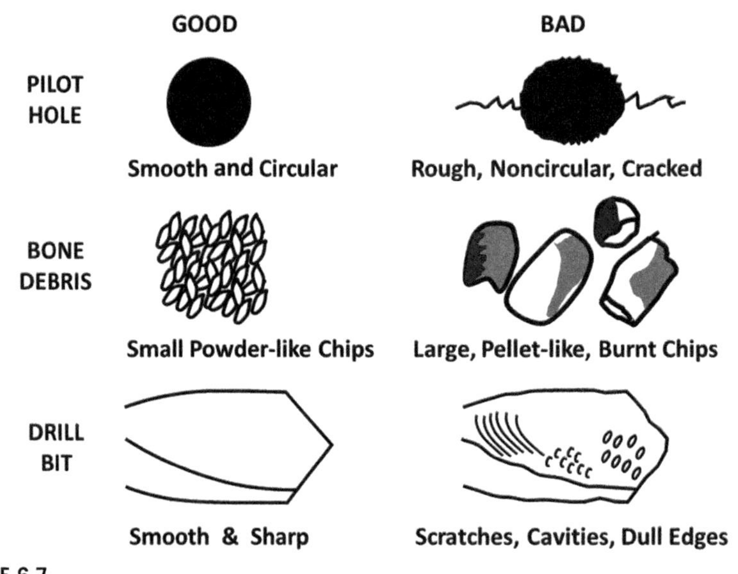

FIGURE 6.7

Qualitative drilling test outcomes for the pilot hole, bone debris, and drill bit.

ALTERNATIVES AND ADAPTATIONS

✓ **Sawbone whole bones.** Artificial whole bones, rather than biological bones, can be used as substitutes, since many commercially available sawbones mimic human bone geometry, they have been biomechanically validated, and they are increasingly used for research. 3-D density (i.e., g/cm^3), rather than 2-D BMD (i.e., g/cm^2), needs to be measured or obtained from the manufacturer, since DEXA scans for BMD are only suitable for biological bone. Note that sawbones have isotropic and homogeneous properties, unlike biological bone.

✓ **Sawbone shapes.** Artificial bones with standardized geometries and material properties may be useful when comparing different drill bit designs since bone parameters are fixed and known. These shapes are available as sheets, blocks, and cylinders. 3-D density (i.e., g/cm^3), rather than 2-D BMD (i.e., g/cm^2), needs to be measured or obtained from the manufacturer, since DEXA scans for BMD are only suitable for biological bone. Remember that sawbones have isotropic and homogeneous properties, unlike biological bone.

5. DISCUSSION

After completing all surgical drilling tests, data collection, data analysis, and data presentation, then final results can be considered and interpreted in the broader context of some important clinical, biomechanical, and/or technological considerations, as follows.

Drilling force and torque are not monitored during orthopaedic surgery. Instead, surgeons use an electrically-powered surgical hand drill to set the spindle speed, which depends on the surgical procedure being done. Moreover, the feed rate of the drill bit into the bone is controlled purely by "subjective feel." Alternately, pre-operative patient BMD scans could be done to permit engineering predictions of appropriate drilling force and torque, followed by the use of a surgical hand drill equipped with a digital force and torque display.

Drilling force and torque in bone have a wide range of values since they can be affected by feed rate, spindle speed, drill bit tip angle, drill bit diameter, and bone type. Drilling force studies have yielded average peaks of 5–300 N (human cortical bone),[9,10,13,14] 1–1.5 N (human cancellous bone),[13] 2–50 N (porcine cortical bone),[2,11,12] 4–32 N (porcine cancellous bone),[12] and 0–275 N (bovine cortical bone).[1,3,5,8] Similarly, drilling torque studies have reported average peaks of 0–186 N·mm (human cortical bone),[9,11,14] 2–3 N·mm (human cancellous bone),[13] 55 N·mm (porcine cortical bone),[2] and 0–400 N·mm (bovine cortical bone).[1,3,5,8]

Drilling force and torque can be estimated using engineering formulas, which are based on empirical research and mechanics theory.[2,17] Assume there are no effects from bone isotropy, bone homogeneity, bone viscoelasticity, bone chip clogging of the drill bit cutting flutes, bone heating, and drill bit wear. Thus, for each revolution of the drill bit around its long axis, the force $F = 2.5\sigma_U(f/\omega)D\sin(\theta/2)$ and the torque $T = 0.625\sigma_U(f/\omega)D^2$, where F is force [N], T is torque [N·mm], σ_U is ultimate tensile stress of bone [N/mm^2], f is drill bit feed rate [mm/min], ω is drill bit spindle speed [rpm], D is drill bit outer diameter [mm], and θ is drill bit tip angle [°].

Thermal generation during bone drilling is mainly caused by shear deformation of the bone, as well as friction between the drill bit, the bone chip being removed, and the underlying host bone.[18,19] About 60% of heat is dissipated into bone chips, while 40% is absorbed by host bone.[19] Heating could lead to bone necrosis, particularly if it is exposed to temperatures of 45°C for 5 h or more, 55°C for 30 s or more, or 70°C for even a moment.[3,18] This can be minimized during surgery by irrigation using a saline solution, as well as the presence of blood and other biofluids. There is no consensus on the influence of drill bit force, feed rate, and spindle speed on bone peak temperature, perhaps due to the large variety of drill bit cutting flute geometries.[18,19]

Clogging of bone debris in the drill bit cutting flutes substantially raises drilling force and torque with increasing drilling depth.[14,20] This occurs because bone chips lodged in the cutting flutes exert pressure against the inside surface of the hole that is being drilled. This, in turn, raises the friction at the interface between the bone chips and the host bone, thereby raising the temperature and creating greater axial and torsional resistance with each revolution of the cutting flutes. Bone chips that are generated during clogging will often be large, pellet-like, and burnt in appearance.

Wear of the drill bit has a negative effect on the efficiency of the drilling process and the quality of the pilot hole.[18,19] The main mechanisms of surface damage to the drill bit are abrasive wear, plastic deformation, cavity formation, and thermal load. Applying protective coatings to the drill bits has been tried, but this has not been particularly successful. Another consequence of drill bit wear is elevated bone temperatures, which can potentially cause bone necrosis.

6. SUMMARY

- Clinically, orthopaedic surgeons use electric hand drills for whole bone.
- Clinically, surgeons do not monitor drilling force or torque for whole bone.
- Drilling force and torque tests are done using a computer-controlled drill press.
- Drilling force and torque can be estimated using engineering formulas.
- Feed rate, spindle speed, drill bit design, and bone type affect force and torque.
- Orthopaedic surgeons could monitor drilling force and torque during surgery.

7. QUIZ QUESTIONS

1. What is the definition of surgical drilling force and torque?
2. Does a surgeon's experience level affect their use of an electric-powered hand drill?
3. How will the anisotropic nonhomogenous nature of biological bone affect force and torque?
4. How could orthopaedic surgeons control or monitor drilling force or torque during surgery?

5. Predict average force and torque for a unicortical drilling test. Assume ultimate tensile stress of cortical bone = 107 MPa, feed rate = 120 mm/min, spindle speed = 1000 rpm, drill bit outer diameter = 3.2 mm, and drill bit tip angle = 90° (answer: 73 N and 82 N·mm).

REFERENCES

1. Alam K, Mitrofanov AV, Silberschmidt VV. Experimental investigations of forces and torque in conventional and ultrasonically-assisted drilling of cortical bone. *Medical Engineering and Physics* 2011;**33**(2):234–9.
2. Allotta B, Giacalone G, Rinaldi L. A hand-held drilling tool for orthopedic surgery. *IEEE Transactions on Mechatronics* 1997;**2**(4):218–29.
3. Hillery MT, Shuaib I. Temperature effects in the drilling of human and bovine bone. *Journal of Materials Processing Technology* 1999;**92-93**:302–8.
4. Hobkirk JA, Rusiniak K. Investigation of variable factors in drilling bone. *Journal of Oral Surgery* 1977;**35**(12):968–73.
5. Jacob CH, Berry JT, Pope MH, Hoaglund FT. A study of the bone machining process – drilling. *Journal of Biomechanics* 1976;**9**(5):343–9.
6. Karalis T, Galanos P. Research on the mechanical impedance of human bone by a drilling test. *Journal of Biomechanics* 1982;**15**(8):561–81.
7. Karmani S, Lam F. The design and function of surgical drills and K-wires. *Current Orthopaedics* 2004;**18**(6):484–90.
8. Lee J, Gozen BA, Ozdoganlar OB. Modeling and experimentation of bone drilling forces. *Journal of Biomechanics* 2012;**45**(6):1076–83.
9. MacAvelia T, Salahi M, Olsen M, Crookshank M, Schemitsch EH, Ghasempoor A, et al. Biomechanical measurements of surgical drilling force and torque in human versus artificial femurs. *Journal of Biomechanical Engineering* 2012;**134**(12):124503-1-9.
10. Natali C, Ingle P, Dowell J. Orthopaedic bone drills: can they be improved? *Journal of Bone and Joint Surgery (British)* 1996;**78**(3):357–62.
11. Ong FR, Bouazza-Marouf K. Drilling of bone: a robust automatic method for the detection of drill bit break-through. *Proceedings of the Institution of Mechanical Engineers (Part H): Journal of Engineering in Medicine* 1998;**212**(3):209–21.
12. Ong FR, Bouazza-Marouf K. Evaluation of bone strength: correlation between measurements of bone mineral density and drilling force. *Proceedings of the Institution of Mechanical Engineers (Part H): Journal of Engineering in Medicine* 2000;**214**(4):385–99.
13. Tsai M-D, Hsieh M-S, Tsai C-H. Bone drilling haptic interaction for orthopedic surgical simulator. *Computers in Biology and Medicine* 2007;**37**(12):1709–18.
14. Wiggins KL, Malkin S. Drilling of bone. *Journal of Biomechanics* 1976;**9**(9):553–9.
15. Lewiecki EM, Borges JLC. Bone density testing in clinical practice. *Arquivos Brasileiros de Endocrinologia and Metabologia* 2006;**50**(4):586–95.
16. Pettersson U, Nordström P, Lorentzon R. A comparison of bone mineral density and muscle strength in young male adults with different exercise level. *Calcified Tissue International* 1999;**64**(6):490–8.

17. Allotta B, Belmonte F, Bosio L, Dario P. Study on a mechatronic tool for drilling in the osteosynthesis of long bones: tool/bone interaction, modeling, and experiments. *Mechatronics* 1996;**6**(4):447–59.
18. Pandey RK, Panda SS. Drilling of bone: a comprehensive review. *Journal of Clinical Orthopaedics and Trauma* 2013;**4**(1):15–30.
19. Bertollo N, Walsh WR. Drilling of bone: practicality, limitations, and complications associated with surgical drill-bits. In: Klika V, editor. *Biomechanics in applications*. Rijeka (Croatia): InTech; 2011 (Chapter 3). Available free online at: http://cdn.intechopen.com/pdfs-wm/19652.pdf.
20. MacAvelia T, Ghasempoor A, Janabi-Sharifi F. Force and torque modelling of drilling simulation for orthopaedic surgery. *Computer Methods in Biomechanics and Biomedical Engineering* 2014;**17**(12):1285–94.

CHAPTER

Insertion Torque Testing of Cortical and Cancellous Screws in Whole Bone

7

Radovan Zdero[1], Matthew R.S. Tsuji[2], Meghan C. Crookshank[2]

Western University, London, ON, Canada[1]; University of Toronto, Toronto, ON, Canada[2]

1. BACKGROUND

For whole bone fracture repair, orthopaedic surgeons typically use "subjective feel" to insert cortical or cancellous screws through a fixation plate's holes into bone or directly into bone (Fig. 7.1). Surgeons attempt to provide maximal screw tightening (i.e., stopping torque), while at the same time being careful not to accidentally overtighten the screw, which causes screw–bone interface failure (i.e., stripping torque). This is widely recognized to be a critical factor in the biomechanical stability of the resulting screw–bone interface immediately after surgery, although this will be augmented over time as bone fracture healing happens and

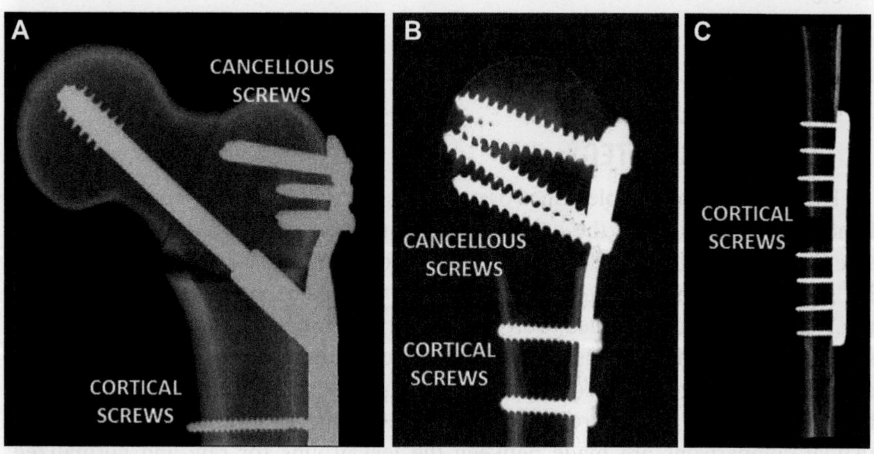

FIGURE 7.1

Radiographs showing cortical and cancellous screws used for fracture fixation. (A) Proximal femur, (B) proximal humerus, (C) humeral diaphysis.

as bony ongrowth occurs around screw threads. However, the few studies that have quantified this phenomenon show that surgeons achieve a stopping/stripping torque ratio of 66–92%, thus perhaps undertightening or overtightening screws on a routine basis.[1–4] Future orthopaedic surgeons will want more direct control for optimal screw insertion, making this topic clinically important. Therefore, this chapter explains how to effectively perform screw insertion experiments to measure stopping and stripping torque in whole bone, as well as how to analyze, present, and interpret the results.

2. RESEARCH QUESTIONS

Typical research questions might include one or more of the following:

- What are the stopping and stripping torques achieved by "feel" by orthopaedic surgeons?
- Do screw insertion results vary between orthopaedic surgeons with varying experience?
- Do human or animal bone characteristics influence screw insertion measurements?
- Do bone storage, bone-thawing time, bone cement, screw type, etc., affect results?
- Do tests on biological bone vs. artificial sawbone provide different data?
- etc.

3. METHODOLOGY

3.1 GENERAL STRATEGY

Human or animal whole bones or segments are obtained and grouped based on bone type depending on the exact research question. Cortical and/or cancellous screws are manually inserted through pilot holes until adequate screw tightness is "subjectively" achieved, thus representing adequate clinical fracture fixation (i.e., stopping torque). Then, screws are further tightened until failure of the screw–bone interface, thus representing accidental screw overtightening that can happen during surgery (i.e., stripping torque). Raw data are then normalized by the surface area of the screw–bone interface to eliminate geometric effects. Statistical comparisons are made between the test groups for raw and normalized screw insertion measurements. Finally, correlation coefficients are computed for raw and normalized data vs. biological bone demographics (i.e., age, sex, limb side, bone mineral density (BMD), and clinical T-score) to determine which factors are important.

GLOSSARY

✓ **BMD**. Bone mineral density for human or animal bone.
✓ **Cortical bone**. Exterior hard bone shell found in all biological bone.
✓ **Cancellous bone**. Internal soft bone most often found at the ends of long bones.
✓ **Homogeneous**. Material properties are evenly distributed throughout the object.
✓ **Isotropic**. Material properties are the same in all directions throughout the object.
✓ **Sawbone**. Artificial bone materials available in anatomical or geometric shapes.

SAFETY FIRST

✓ Remember to always wear goggles and gloves for protection.
✓ Secure the vice and bone before using the drill press or power hand drill.
✓ Clean the work area and all tools with bleach or disinfectant after testing.

3.2 MATERIALS AND TOOLS LIST

- cortical and/or cancellous screws
- digital torque screwdriver
- drill bits and drill press (or power hand drill)
- guide wire or pin
- human or animal bone
- manual surgical screwdriver
- metal washers
- screwdriver bits that match the screw heads
- tape measure, thickness gage, and calipers
- vice or clamp

3.3 SPECIMEN PREPARATION

Step 1. Store the bones. Fresh or fresh–frozen bone specimens need to be wrapped in plastic strips or vacuum-sealed in plastic bags for proper storage in a freezer prior to the study. The freezer should be held at −20°C or colder. Embalmed or dried/dehydrated bones do not need to be frozen and may simply be placed in a plastic box for storage at room temperature prior to tests.

Step 2. Thaw the bones. Bone specimens should be removed from the freezer, left in their plastic wrappings or vacuum-sealed bags, and placed on a suitable surface at ambient room temperature or in a warm water bath to thaw for at least 12 h.

Bones are then removed from their plastic wrappings or bags and soaked or sprayed with saline water solution.

Step 3. Identify screw insertion sites. Mount bone specimens into a vice or clamp them onto a table, make measurements for the screw insertion site, and mark the location with a pen. Because of geometric variability among biological bone, the relative location of screw insertion sites with respect to a common reference dimension needs to be computed and used for all specimens (e.g., % = relative dimension/reference dimension × 100 = distance from one end of the bone to the screw insertion site/total length of bone × 100).

Step 4. Choose drill bits. Choose the appropriate drill bit size, which depends on the screw's size, since the typical clinical and manufacturer recommendations are different for cortical screws (e.g., screw = 3.5 mm outer diameter and 2.4 mm core diameter; drill bit = 2.5 mm outer diameter) and cancellous screws (e.g., screw = 6.5 mm outer diameter and 3 mm core diameter; drill bit = 3.2 mm outer diameter) (Fig. 7.2A). The appropriate drilling depth for bone specimens will also be different due to the different clinical uses of cortical screws (i.e., partial engagement of the screw's threaded length by one or two cortical bone walls) and cancellous screws (i.e., full engagement of the screw's threaded length by cancellous bone). (Note: Some bone specimens may be awkward to mount in a vice in the drill press; thus, an electrically-powered hand drill can be used preferably with a drill bit guide to ensure pilot holes are drilled properly.)

Step 5. Drill pilot holes. Mount bone specimens in a drill press with the screw insertion site positioned directly under the drill bit (Fig. 7.2B and C). Lower the

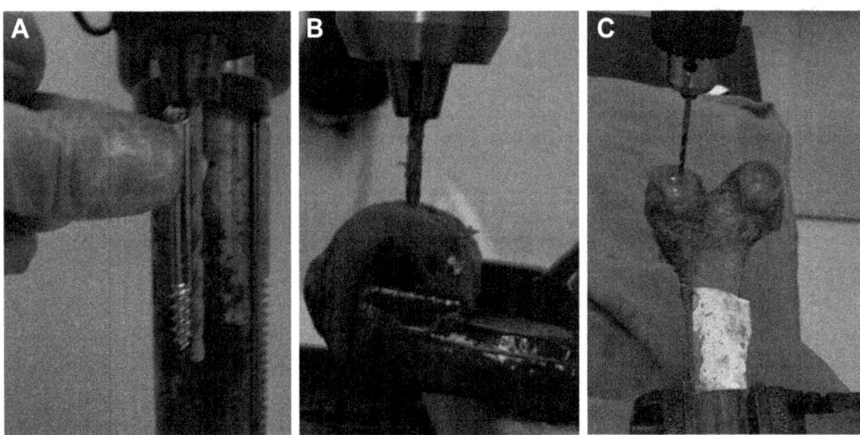

FIGURE 7.2

Creation of pilot holes for screw insertion tests. (A) Drill bit depth measurement corresponding to bone screw length, (B) drilling into cortical bone, (C) drilling into cancellous bone.

spinning drill bit very slowly and steadily until the tip of the drill bit superficially punctures the bone surface at the designated screw insertion site. Lowering the drill bit too quickly could cause the drill bit tip to slip overtop of the slippery bone surface so that the wrong location is punctured, as well as potentially bending or breaking the drill bit. Then raise the drill bit again in order to visually ensure the screw insertion site was located accurately. Then lower the drill bit and continue drilling slowly and steadily until completion. Raise the drill bit fully out of the pilot hole.

Step 6. Measure pilot hole geometry. Remove the bone from the drill press, and hold it firmly on a work surface, whereas large specimens may need to be secured using a vice or clamp on a work surface. To measure cortical wall thickness at the screw insertion site, insert a clinical or mechanical depth gage that has a hooked end, pull the hooked end until it engages the intramedullary underside of the cortical wall, and then measure this length using a ruler or Vernier caliper (Fig. 7.3A–C). If there is a need to double-check cancellous pilot hole depth, insert a thin rigid guide wire until it feels like it has reached the bottom of the hole, use a pen to mark the guide wire's surface that is at the level of the pilot hole, remove the guide wire, and then measure the length of the guide wire's tip-to-pen mark using a ruler or Vernier caliper (Fig. 7.3D–F). Guide wires can also be used to clear away any unwanted bone debris from the pilot hole.

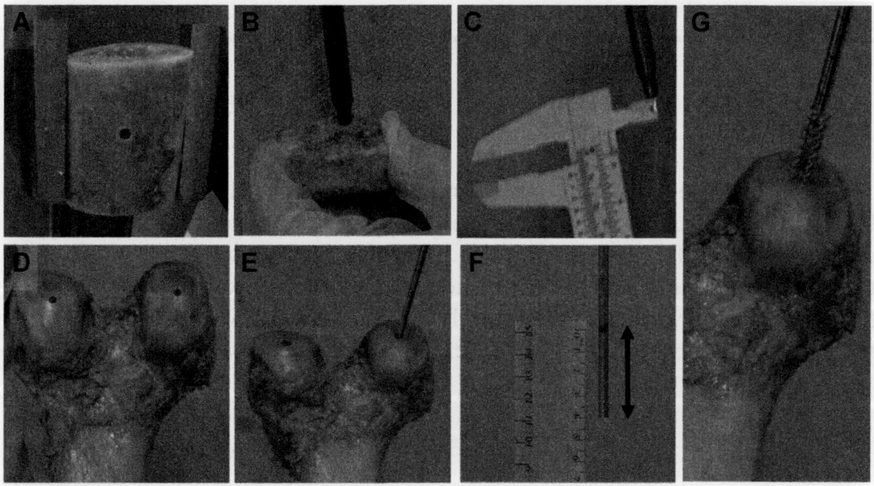

FIGURE 7.3

Measurement of pilot hole geometry. (A) Cortical bone specimen with a pilot hole, (B) depth gage insertion into a cortical pilot hole, (C) depth measurement of a cortical pilot hole using a depth gage and Vernier calipers, (D) cancellous bone specimen with a pilot hole, (E) guide wire insertion into a cancellous pilot hole, (F) depth measurement of a cancellous pilot hole using a guide wire and ruler, (G) tapping or threading the pilot hole for cancellous screw insertion tests.

Step 7. Create screw threads. Create internal threads in the pilot holes by "tapping" bone specimens using the manufacturer's recommended instrumentation to ensure proper screw–bone engagement and prevention of bone cracking (Fig. 7.3G). Make a pen mark on the tap shaft to indicate the correct depth is reached to tap the entire hole. Tapping is required for many types of clinically used nonself-tapping cancellous screws. However, many cortical and cancellous screws are self-tapping; thus, there is no need to tap the pilot hole.

TIPS AND TRICKS

- ✓ Long bones can be cut into segments with a band saw for easier handling.
- ✓ Use a rigid guide wire or screwdriver bit guide, so the screws go in straight.
- ✓ Insert a new screw for each new test to avoid biasing the results.
- ✓ The same orthopaedic surgeon should perform all tests for consistency.

THE "GOLD STANDARD"

No known international standard exists for testing "subjective" stopping torque, but the American Society for Testing and Materials (ASTM) provides guidelines in its document ASTM F543 (Standard specification and test methods for metallic medical bone screws) on medical bone screws for evaluating stripping torque.

3.4 SPECIMEN TESTING

Step 1. Choose the bone screw. Choose the right bone screw, since there is a large variety from medical manufacturers. Cortical screws are used mostly in the diaphysis (i.e., shaft) of long bones (e.g., humerus, femur, tibia, etc.) and inserted through one cortical bone wall (i.e., unicortical screw) or two cortical bone walls (i.e., bicortical screw) (Fig. 7.4A). Cancellous screws are used wherever there is substantial spongy or trabecular bone (e.g., femoral head, neck, and condyles)

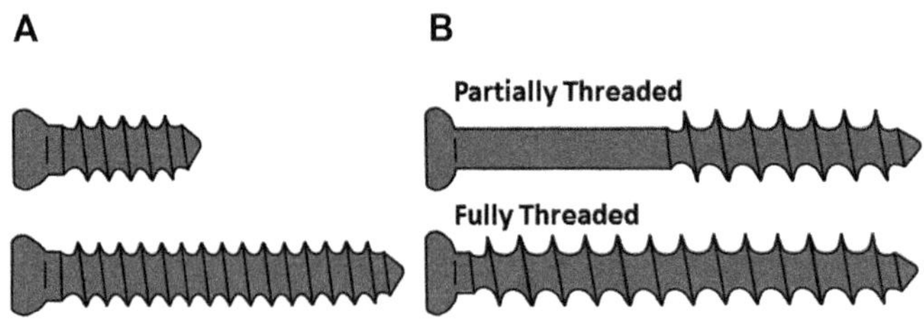

FIGURE 7.4

Common orthopaedic bone screws. (A) Cortical screws, (B) cancellous screws.

(Fig. 7.4B). Cortical screw threads are more closely spaced together (i.e., smaller thread pitch) and have shallower depth vs. cancellous screws since cortical bone is denser than cancellous bone.

Step 2. Obtain a screwdriver. Obtain a manual surgical screwdriver suggested by the screw manufacturer that matches the cortical and cancellous screw heads (Fig. 7.5A). For the digital torque screwdriver, attach the appropriate screwdriver bit that matches the heads of the cortical and cancellous screws (Fig. 7.5A). Secure cortical or cancellous bone specimens into a vice or clamp (Fig. 7.5B and C).

Step 3. Position the washer. Position a metal washer over the screw insertion site. This will prevent subsidence of the screw head into the bone surface at the end of the screw insertion test during stripping torque measurement. Thus, the washer eliminates any frictional effects between the screw head and bone surface (Fig. 7.5B and C).

Step 4. Position the screw. Use the manual surgical screwdriver to place cortical or cancellous screw tips into the pilot hole, turn the screw several times slowly so the screw threads begin to engage the bone, and then continue to drive the screws until the screw head approaches the washer to within 1–2 mm (Fig. 7.5B and C). This replicates the actual clinical scenario, since orthopaedic surgeons apply screws in this manner using the same kind of surgical screwdriver as shown here.

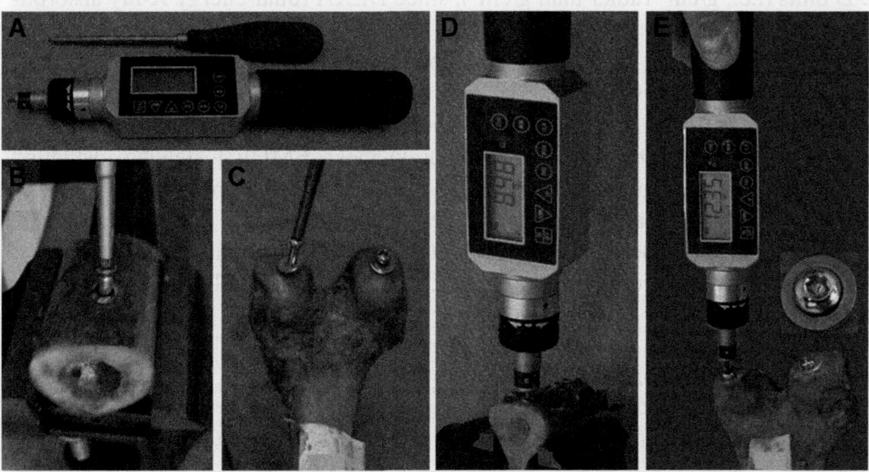

FIGURE 7.5

Biological bone screw insertion tests. (A) Manual surgical screwdriver and digital torque screwdriver, (B) superficial insertion of cortical screw tip through a metal washer using a manual surgical screwdriver, (C) superficial insertion of cancellous screw tip through a metal washer using a manual surgical screwdriver, (D) cortical screw insertion test using a digital torque screwdriver, (E) cancellous screw insertion test using a digital torque screwdriver.

Step 5. Measure stopping torque. Now use the digital torque screwdriver to tighten the screw head against the washer until there is a "subjective feeling" that a good tight purchase of the screw into the bone has been achieved (Fig. 7.5D and E). Record the peak torque reached as being the "stopping torque."

Step 6. Measure stripping torque. Continue using the digital screwdriver to tighten the screw against the washer, until there is a "subjective feeling" that the screw threads have fully stripped the bone material (Fig. 7.5D and E). This will allow the screw to be turned without much effort. Record the peak torque reached as being the "stripping torque." Then slowly remove the screw from the hole using the surgical screwdriver so that the hole or bone surface are not damaged.

Step 7. Inspect the specimen. Visually inspect the condition of the screw and hole, take a sufficient number of photos, and make a note of any damage to the screw (i.e., bending or breaking), any evulsed bone material still lodged in the screw threads, or any surface damage to the hole or surrounding bone.

3.5 RAW DATA COLLECTION

Step 1. Record bone characteristics. Enter human or animal demographic data (i.e., age, sex, and left or right limb) and quantitative bone properties (i.e., BMD and clinical T-score), which will help determine the factors that are statistically correlated with screw insertion torque measurements (Table 7.1). Note that BMD is in 2-D units (i.e., g/cm^2 rather than g/cm^3), since DEXA (dual-energy X-ray absorptiometry) bone density scans only provide 2-D scans and values. DEXA scan reports often provide BMD values and corresponding clinical T-scores (i.e., normal T-score ≥ -1, $-1 >$ osteopenic T-score > -2.5, and osteoporotic T-score ≤ -2.5). When such analysis is overlooked or missing, computations can be made as T-score = (bone BMD − reference population mean BMD)/(1 standard deviation of reference population BMD), where T-score reference populations are young healthy adults of the same sex that are aged 20–40 years.[5,6]

Step 2. Record bone–screw geometry. Record prior measurements of the bone length that engages the screw threads. This will vary greatly among specimens

Table 7.1 Raw data for biological bone screw insertion tests.

	Age [years]	Sex [M, F]	Limb side [L, R]	BMD [g/cm^2]	T-score	Engaged bone length [mm]	Stopping torque [N·mm]	Stripping torque [N·mm]	Failure mode
1									
2									
3									
etc.									
Avg		–	–						–
SD		–	–						–

Avg, *average;* BMD, *bone mineral density;* SD, *standard deviation.*

and should be known. Specifically, this is the cortical wall thickness at the screw insertion site or the cancellous screw threaded length engaged at the screw insertion site.

Step 3. Record screw insertion torques. Raw stopping and stripping torques should have been noted during the screw insertion tests and can now be entered into this summary data table.

Step 4. Record failure modes. Some final remarks may also be entered about the nature of failure at the screw–bone interface (i.e., Any visible fractures around the screw hole? Any nonsymmetry of the screw hole? Any bone material remaining lodged in the screw threads? Any bending or breaking of the screw itself? etc.). Obtain good photos of test specimens.

3.6 RAW DATA ANALYSIS

Step 1. Normalize raw data. Raw stopping and stripping torques should then be normalized by geometry by taking into account the approximate surface area engaged at the screw–bone interface. This can be computed as $T_{NORM} = T_{RAW}/A = T_{RAW}/(\pi D_o L)$, where T is torque, A is surface area, D_o is screw outer diameter, and L is threaded portion of the screw length engaged by bone.

Step 2. Compute torque ratio. Torque ratio = (stopping torque/stripping torque), which should be computed as an absolute value or %. Torque ratio numerically expresses how close the final achieved screw–bone "tightness" is to the allowable peak torque just prior to screw–bone interface failure. Since this is a ratio, it is automatically normalized (i.e., raw torque ratio = normalized torque ratio).

Step 3. Calculate correlation coefficients. Plot the measured result for each individual specimen (i.e., stopping torque 1, 2, 3, etc.) vs. its corresponding specimen characteristic (i.e., age 1, 2, 3, etc.) (Table 7.1) to help visualize the interrelationship between measurements vs. characteristics. Then, calculate the correlation coefficient R for each measurement–characteristic pair to determine which characteristic has an influence on measurements. $R > 0.8$ is a typical value considered to indicate a strong correlation. Do this for stopping torque, stripping torque, and stopping/stripping torque ratio.

Step 4. Perform statistical comparisons. The criterion for statistical difference needs to be chosen (e.g., $P < 0.01$ or < 0.05). Then data can be used to compare different patient groups (e.g., men ≤ 60 years old vs. men > 60 years old, "normal" women vs. "osteoporotic" women, etc.), bone types (e.g., femur vs. tibia, left humerus vs. right humerus, etc.), screw insertion sites (e.g., anterior vs. lateral surface), screw designs (e.g., 2.5-mm vs. 3.5-mm cortical screws), etc. Various software programs exist for comparing two test groups (e.g., paired t-test) or two or more test groups influenced by multiple factors (e.g., analysis of variance, ANOVA).

Step 5. Compute statistical power. Power analysis can be done after the study to ensure there were enough specimens per group to detect all statistical differences that were actually present (i.e., was type II statistical error avoided?). Statistical power >80% is usually considered to indicate there were enough specimens per test group. Note that if good predictions of averages and standard deviations are available from prior studies, then the number of specimens and/or tests can be chosen before the study begins to ensure a power >80%.

ENGINEER'S TOOLBOX

Stripping torque T can be approximated using a mathematical expression developed for a generic screw whose threads are perfectly engaged in isotropic homogeneous material. This engineering formula can be applied to scenarios relevant to orthopaedic surgery, as shown. Consider that σ_U is ultimate tensile strength of bone, L is engaged length of screw threads, D is screw pitch diameter $= D_o - 0.6495p$, D_o is screw outer diameter, p is screw thread pitch, and f is coefficient of friction between screw and bone. Therefore, stripping torque $T = 0.9069\sigma_U LD^2(p + fD)/(D - fp)$.

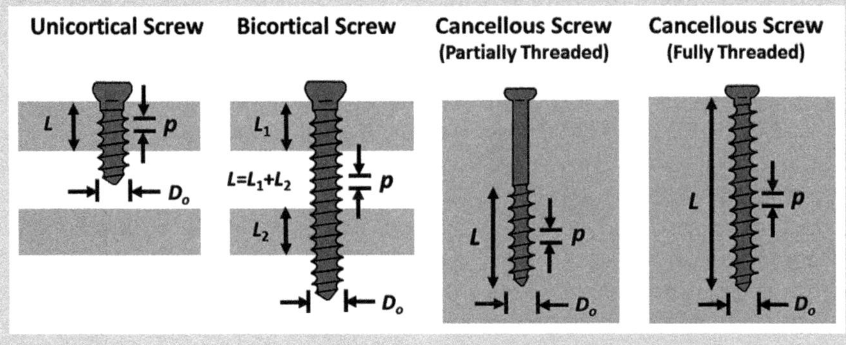

4. RESULTS

Once all screw insertion data collection and analysis have been performed, it is then important to communicate and present the primary results in an understandable and concise manner to the reader of a journal article, conference paper, technical report, or book chapter.

Step 1. Show raw torque profile. Begin with a typical raw torque profile, if available, which is recorded by many digital torque screwdrivers (Fig. 7.6).[7] Initially, torque is zero as the screw tip enters the pilot hole. Torque then grows relatively linearly as the screw turns to overcome thread–bone friction. "Seating torque" is then reached as the screw head contacts the washer (or fracture plate or bare bone). As the screw is tightened more, the torque rises very rapidly with

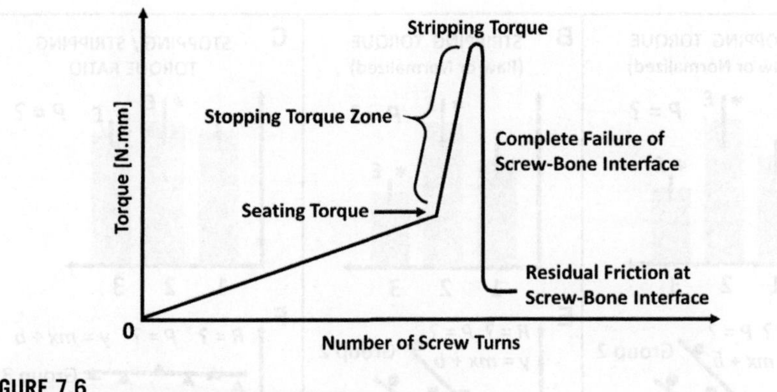

FIGURE 7.6

Torque raw data profile vs. the number of screw turns.[7]

very few turns as bone is stressed. It is within this zone where orthopaedic surgeons will subjectively feel that adequate screw tightness has been achieved (i.e., "stopping torque"). Additional screw turning will lead to peak torque such that the screw–bone interface experiences maximum possible stress (i.e., "stripping torque"), after which there is a rapid drop of torque as the screw–bone interface fails. Continued screw turning will show negligible (but not zero) torque, since there is still some residual friction between screw threads and bone. Torque magnitudes and the ratio of torque/number of screw turns (i.e., slopes of the two initial linear phases of the graph) depend on lubrication from marrow, blood, or synovial fluid, bone material properties, screw–thread design, and screw–bone friction factor.

Step 2. Show main torque data. The main numerical findings of the study can be presented (Fig. 7.7). This includes raw and normalized stopping torque, raw and normalized stripping torque, and stopping/stripping torque ratio. For each of these parameters, all statistical pairwise comparisons should be made (i.e., test group 1 vs. 2 vs. 3, etc.) to generate statistical P values (Fig. 7.7A–C). Also, any linear (or nonlinear) trends can be illustrated for torque vs. BMD (Fig. 7.7D–F). For each line of best fit, several items can be generated: an equation $y = mx + b$ showing slope m and intercept b, a linear correlation coefficient R, and its own statistical P value to ensure that the slope of each line of best fit (i.e., slope $= m$) is statistically different than a horizontal line (i.e., slope $= 0$).

Step 3. Show failure modes. Finally, photos of typical screws and/or screw holes after testing help visualize any kind of failure modes, such as bent or broken screws, residual bone in the screw threads, and any surface damage around the screw holes (Fig. 7.8).

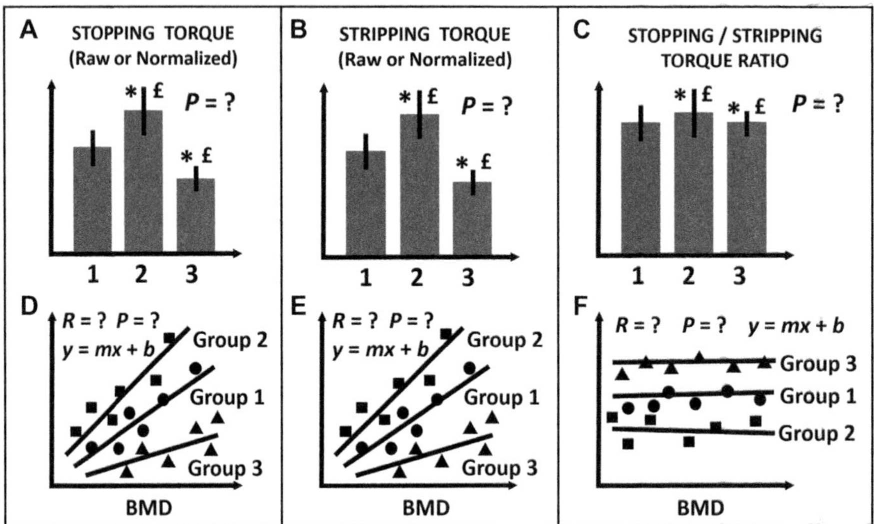

FIGURE 7.7

Screw insertion test results. (A) Stopping torque vs. test group, (B) stripping torque vs. test group, (C) stopping/stripping torque ratio vs. test group, where bar graphs show average $\pm$ 1 standard deviation and *P* is the statistical difference value for each pairwise comparison between each group which can be indicated by symbols like asterisks (*), pounds (£), etc., (D) stopping torque vs. BMD, (E) stripping torque vs. BMD, (F) stopping/stripping torque ratio vs. BMD, where BMD is bone mineral density, *R* is the linear correlation coefficient for each line of best fit, *P* is the statistical difference value to ensure that the slope of each line of best fit (i.e., slope $= m$) is statistically different than a horizontal line (i.e., slope $= 0$), and $y = mx + b$ is the equation of each line.

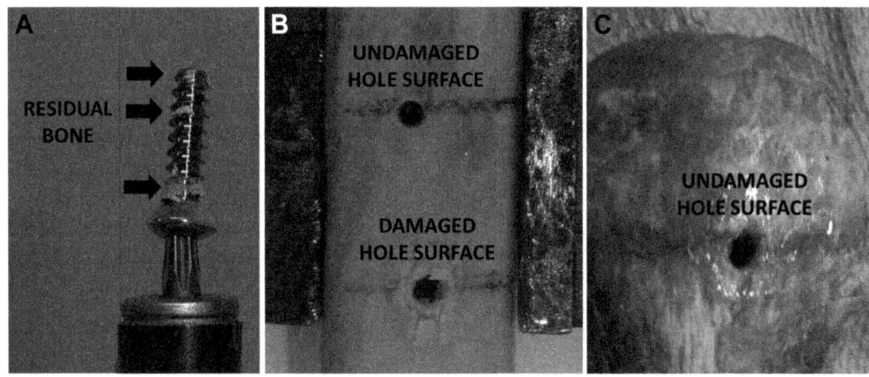

FIGURE 7.8

Failure photos after screw insertion tests. (A) Cortical screw that was undamaged but had residual bone lodged in its threads, (B) cortical bone specimen from humerus shaft with undamaged and damaged hole surface, (C) cancellous bone specimen from humerus head with undamaged hole surface.

ALTERNATIVES AND ADAPTATIONS

✓ **Torque testing machines.** Special mechanical testers designed for torque application can be used for research rather than manual screw insertion by an orthopaedic surgeon when fixed angulation rate is an important parameter in an investigation.

✓ **Metal fracture plates.** These devices, rather than metal washers, can prevent troublesome screw head subsidence into bone. This replicates numerous real-life clinical fracture repair techniques with a variety of metal fracture plates that are used by orthopaedic surgeons.

✓ **Sawbone models.** Artificial whole bones, or standardized blocks, cylinders, and plates, can be used for screw insertion tests, but 3-D density (i.e., g/cm^3), rather than 2-D BMD (i.e., g/cm^2), needs to be measured or obtained from the manufacturer since DEXA scans for BMD are only suitable for biological bone.

5. DISCUSSION

After completing all screw insertion tests, data collection, data analysis, and data presentation, then final results can be considered and interpreted in the broader context of some important clinical, biomechanical, and/or technological considerations, as follows.

Stopping torque is not measured during orthopaedic surgery. Instead, surgeons use a manual surgical screwdriver with "subjective feel" for adequate tightening. Surgeons achieve stopping/stripping torque ratios from 66–92%; thus, screws may be undertightened or overtightened.[1–4] Alternatively, preoperative patient BMD scans could be done to permit engineering predictions of stripping torque, followed by a digital torque screwdriver to achieve proper stopping torque.

Stripping torque can be predicted for a generic screw whose threads are perfectly engaged in an isotropic homogeneous material.[7] Assume σ_U is ultimate tensile strength of the host material, L is engaged length of screw threads, D is screw pitch diameter $= D_o - 0.6495p$, D_o is screw outer diameter, p is screw thread pitch, and f is coefficient of friction between the screw and host material. Thus, stripping torque $T = 0.9069\sigma_U LD^2(p + fD)/(D - fp)$.

Stripping torque is related to the force required to pull out or push out screws in bone.[7–10] Theoretically, this can be computed as stripping torque $T = FD_{AVG}(1/t + fD_{AVG})/(2D_{AVG} - 2f/t)$, where F is pullout force, D_{AVG} is average of outer and inner screw diameter, f is screw–bone friction factor, and t is number of screw threads per unit length.[7] Experimentally, for instance, sawbone tests with a 4.5-mm diameter cortical screw can yield a ratio of F/T = raw pullout force/raw stripping torque $= 0.498\ mm^{-1}$ over a range of densities.[8]

Stripping torques have a wide range of values.[1,3,4,9,11] Cortical screw tests have yielded 1089–2173 N·mm (human humeral shaft), 1126–2179 N·mm (human femoral shaft), 3265 N·mm (human tibial shaft), 1612–2331 N·mm (sawbone humeral shaft), and 2012 N·mm (sawbone femoral shaft). Cancellous screw tests have achieved 308–1176 N·mm (human humeral head), 554–2710 N·mm (human femoral condyles), and 1594–1675 N·mm (sawbone humeral head).

Stopping/stripping torque ratio remains constant over a wide density range for a given bone type when surgery is done by the same orthopaedic surgeon. For example, in tests on sawbone femurs and human femurs, one study reports ratios of $80.6 \pm 6.6\%$ (cortical screws in sawbone femurs), $76.8 \pm 6.4\%$ (cancellous screws in sawbone femurs), $66.6 \pm 10.4\%$ (cortical screws in human femurs), and $84.5 \pm 9.7\%$ (cancellous screws in human femurs).[3]

Stopping/stripping torque ratio remains constant for different bone types when assessed by the same orthopaedic surgeon. Cortical screw tests in five humerus groups have shown no statistical difference for torque ratio between fresh–frozen human, embalmed human, dried/dehydrated human, "normal" sawbone, and "osteoporotic" sawbone ($P = 0.1$), while cancellous tests also show no statistical difference for 9 of 10 group-to-group comparisons ($P > 0.05$).[4]

Screw design may alter screw–bone interface quality and screw insertion torque and, thus, may affect an orthopaedic surgeon's performance. Screw design can be quantified as thread shape factor $TSF = 0.5 + 0.57735d/p$, where d is screw thread depth and p is screw thread pitch.[12] Thread depth and pitch influence the overall angle of the thread. This means that deeper threads, which engage bone with better "bite," may "feel" tighter to an orthopaedic surgeon.

A torsion testing machine could improve intertest consistency by inserting screws at a fixed angulation rate, as described by ASTM standardized methodology.[2,13–15] This approach is useful to determine bone material properties (i.e., stripping torque). However, this does not replicate clinical practice in which surgeons tighten screws by "subjective feel" with a surgical screwdriver. This latter method is more helpful when mimicking clinical conditions (i.e., stopping torque).

6. SUMMARY

- "Subjective feel" is typically used in the clinic by orthopaedic surgeons to reach stopping torque.
- "Subjective feel" may cause under- or overtightening of screws, leading to suboptimal surgery.
- Stopping torque can be measured by an orthopaedic surgeon using a digital torque screwdriver.
- Stripping torque can be predicted using established engineering formulas.
- Screw design and bone type will substantially influence screw insertion results.
- Future orthopaedic surgeons could monitor insertion torque during surgery.

7. QUIZ QUESTIONS

1. What are stopping torque, stripping torque, and stopping/stripping torque ratio?
2. Will a surgeon's experience level affect stopping/stripping torque ratio?
3. How will anisotropy and nonhomogeneity of real bone affect stripping torque predictions?

4. How can orthopaedic surgeons more closely control or monitor stopping torque during surgery?
5. Predict stripping torque for a unicortical screw insertion test. Assume ultimate tensile strength = 107 MPa, screw outer diameter = 3.5 mm, screw thread pitch = 1.25 mm, average unicortical wall thickness = 4 mm, and steel-on-bone friction factor = 0.1 (answer: 1662 N·mm).

REFERENCES

1. Cordey J, Rahn BA, Perren SM. Human torque control in the use of bone screws. In: Uhthoff HK, editor. *Current concepts of internal fixation of fractures*. New York (NY): Springer-Verlag; 1980. p. 235–43.
2. Cleek TM, Reynolds KJ, Hearn TC. Effect of screw torque level on cortical bone pullout strength. *Journal of Orthopaedic Trauma* 2007;**21**(2):117–23.
3. Tsuji M, Crookshank M, Olsen M, Schemitsch EH, Zdero R. The biomechanical effect of artificial and human bone density on stopping and stripping torque during screw insertion. *Journal of the Mechanical Behavior of Biomedical Materials* 2013;**22**: 146–56.
4. Aziz MSR, Tsuji MRS, Nicayenzi B, Crookshank MC, Bougherara H, Schemitsch EH, et al. Biomechanical measurements of stopping and stripping torques during screw insertion in five types of human and artificial humeri. *Proceedings of the Institution of Mechanical Engineers (Part H): Journal of Engineering in Medicine* 2014;**228**(5): 446–55.
5. Lewiecki EM, Borges JLC. Bone density testing in clinical practice. *Arquivos Brasileiros de Endocrinologia and Metabologia* 2006;**50**(4):586–95.
6. Pettersson U, Nordström P, Lorentzon R. A comparison of bone mineral density and muscle strength in young male adults with different exercise level. *Calcified Tissue International* 1999;**64**(6):490–8.
7. Troughton MJ, editor. *Handbook of plastics joining: a practical guide*. Norwich (NY): Plastics Design Library; 2008. p. 180.
8. Edwards TR, Tevelen G, English H, Crawford R. Stripping torque as a predictor of successful internal fracture fixation. *ANZ Journal of Surgery* 2005;**75**(12):1096–9.
9. Lawson KJ, Brems J. Effect of insertion torque on bone screw pullout strength. *Orthopedics* 2001;**24**(5):451–4.
10. Lin J, Lin SJ, Chiang H, Hou SM. Bending strength and holding power of tibial locking screws. *Clinical Orthopaedics and Related Research* 2001;**385**:199–206.
11. Nicayenzi B, Crookshank M, Olsen M, Schemitsch EH, Bougherara H, Zdero R. Biomechanical measurements of cortical screw stripping torque in human versus artificial femurs. *Proceedings of the Institution of Mechanical Engineers (Part H): Journal of Engineering in Medicine* 2012;**226**(8):645–51.
12. Chapman JR, Harrington RM, Lee KM, Anderson PA, Tencer AF, Kowalski D. Factors affecting the pullout strength of cancellous bone screws. *Journal of Biomechanical Engineering* 1996;**118**(3):391–8.
13. Bahr W. Comparison of torque measurements between cortical screws and emergency replacement screws in the cadaver mandible. *Journal of Oral and Maxillofacial Surgery* 1992;**50**(1):46–9.

14. Thomas RL, Bouazza-Marouf K, Taylor GJ. Automated surgical screwdriver: automated screw placement. *Proceedings of the Institution of Mechanical Engineers (Part H): Journal of Engineering in Medicine* 2008;**222**(5):817–27.
15. ASTM F543. Standard specification and test methods for metallic medical bone screws. West Conshohocken (PA, USA): American Society for Testing and Materials (ASTM). Available at: http://www.astm.org.

CHAPTER

Pullout Force Testing of Cortical and Cancellous Screws in Whole Bone

8

Radovan Zdero[1], Mina S.R. Aziz[2], Bruce Nicayenzi[3]

Western University, London, ON, Canada[1]; University of British Columbia, Vancouver, BC, Canada[2]; Ryerson University, Toronto, ON, Canada[3]

1. BACKGROUND

To repair broken whole bones, orthopaedic surgeons apply cortical or cancellous screws through a fixation plate's holes into bone or directly into bone (Fig. 8.1). Surgeons try to use maximal screw tightening but avoid accidental screw overtightening, which could cause screw–bone interface failure.[1–4] This is a crucial factor in the biomechanical stability of repaired fractures. However, normal human activities and accidents that occur at home, in the workplace, during sports activities, or in the public square may cause screws to experience back-and-forth "toggling" and/or sudden impact forces.[5,6] This can damage the screw–bone interface and cause screw

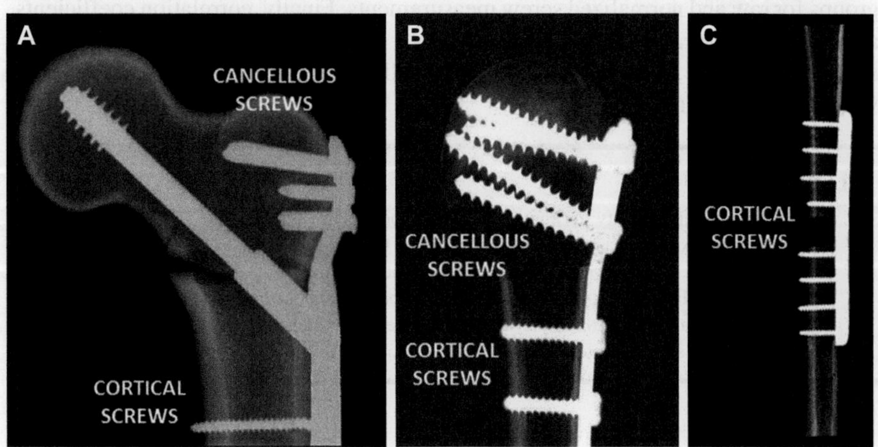

FIGURE 8.1

Radiographs showing cortical and cancellous screws for fracture fixation. (A) Proximal femur, (B) proximal humerus, (C) diaphysis of the humerus.

pullout. Therefore, this chapter explains how to effectively perform bone screw pullout tests to measure screw—bone interface quality, as well as how to analyze, present, and interpret results.

2. RESEARCH QUESTIONS

Typical research questions could include one or more of the following:

- What are the pullout strengths of cortical and cancellous bone screws?
- Do human or animal bone characteristics influence screw pullout?
- What is the effect of bone storage, bone-thawing time, screw type, etc.?
- Do tests on biological bone vs. artificial sawbone yield different results?
- etc.

3. METHODOLOGY

3.1 GENERAL STRATEGY

Human or animal whole bones or segments are obtained and grouped based on bone type depending on the particular research question. Cortical and/or cancellous screws are manually inserted using a surgical screwdriver through appropriately sized pilot holes. Bone specimens are rigidly mounted onto a test jig that is attached to a mechanical tester, a novel gripping mechanism pulls the head of the screw until it is completely evulsed from the bone, and raw force and displacement are recorded. Raw data are then normalized by the surface area of the screw—bone interface to eliminate geometric effects. Statistical comparisons are made between the test groups for raw and normalized screw measurements. Finally, correlation coefficients are computed for raw and normalized data vs. biological bone demographics (i.e., age, sex, limb side, bone mineral density (BMD), and clinical T-score) to identify which factors are important.

GLOSSARY

- ✓ **BMD**. Bone mineral density for human or animal bone.
- ✓ **Cortical bone**. Exterior hard bone shell found in all biological bone.
- ✓ **Cancellous bone**. Internal soft bone most often found at the ends of long bones.
- ✓ **Homogeneous**. Material properties are evenly distributed throughout the object.
- ✓ **Isotropic**. Material properties are the same in all directions throughout the object.
- ✓ **Sawbone**. Artificial bone materials available in anatomical or geometric shapes.

SAFETY FIRST

- ✓ Remember to always wear goggles and gloves for protection.
- ✓ Secure the vice and bone before using the drill press or power hand drill.
- ✓ Be careful not to get fingers caught in the test jig during specimen mounting.
- ✓ Place a plastic protective shield around the screw—bone assembly during testing.
- ✓ Clean the work area and all tools with bleach or disinfectant after testing.

3.2 MATERIALS AND TOOLS LIST

- cortical and/or cancellous screws
- drill bits and drill press (or power hand drill)
- guide wire or thin long rod
- human or animal bone
- manual surgical screwdriver
- mechanical tester
- screw pullout test jig
- screwdriver bits that match screw heads
- tape measure, thickness gage, and/or calipers
- vice or clamp

3.3 SPECIMEN PREPARATION

Step 1. Store the bones. Fresh or fresh–frozen bone specimens initially need to be wrapped in plastic strips or vacuum-sealed in plastic bags for proper storage in a freezer prior to the study. The freezer should be held at −20°C or colder. Embalmed or dried/dehydrated bones do not need to be frozen and may simply be placed in a plastic box for storage at room temperature.

Step 2. Thaw the bones. Bones should be removed from the freezer, left in their plastic wrappings or vacuum-sealed bags, and placed on a suitable surface at ambient room temperature or in a warm water bath to thaw for at least 12 h. Bone specimens are then removed from their plastic wrappings or bags and soaked or sprayed with saline water solution to prevent dehydration.

Step 3. Identify screw pullout sites. Mount bone specimens into a vice or clamp them onto a table, make measurements for the screw pullout site, and mark the location with a pen. Because of geometric variability among biological bones, the relative location of screw pullout sites with respect to a common reference dimension needs to be computed and used for all specimens (e.g., % = relative dimension/reference dimension × 100 = distance from one end of the bone to the screw pullout site/total length of bone × 100).

Step 4. Choose drill bits. Choose the appropriate drill bit size, which depends on the screw's size, since the typical clinical and manufacturer recommendations are different for cortical screws (e.g., screw = 3.5 mm outer diameter and 2.4 mm core diameter; drill bit = 2.5 mm outer diameter) and cancellous screws (e.g., screw = 6.5 mm outer diameter and 3 mm core diameter; drill bit = 3.2 mm outer diameter) (Fig. 8.2A). The appropriate drilling depth for bone specimens will also be different due to the different clinical uses of cortical screws (i.e., partial engagement of the screw's threaded length by one or two cortical bone walls) and cancellous screws (i.e., full engagement of the screw's threaded length by cancellous bone). (Note: Some bone specimens may be awkward to mount in a vice in the drill press; thus, an electrically-powered hand drill can be used preferably with a drill bit guide to ensure pilot holes are drilled properly.)

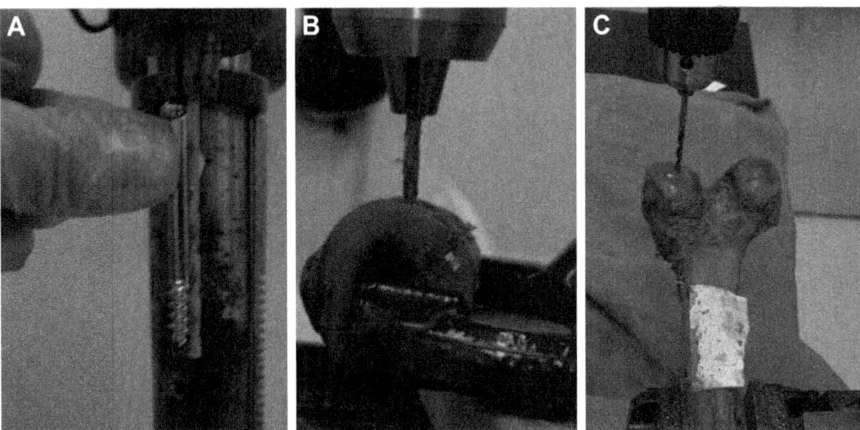

FIGURE 8.2

Creation of pilot holes for screw pullout tests. (A) Drill bit depth measurement corresponding to bone screw length, (B) drilling into cortical bone, (C) drilling into cancellous bone.

Step 5. Drill pilot holes. Mount bone specimens in a drill press with the screw pullout site positioned directly under the drill bit (Fig. 8.2B and C). Lower the spinning drill bit very slowly and steadily until the tip of the drill bit superficially punctures the bone surface at the designated screw pullout site. Lowering the drill bit too quickly could cause the drill bit tip to slip overtop of the slippery bone surface so that the wrong location is punctured, as well as potentially bending or breaking the drill bit. Then raise the drill bit again in order to visually ensure the screw pullout site was located accurately. Then lower the drill bit and continue drilling slowly and steadily until completion. Raise the drill bit fully out of the pilot hole.

Step 6. Measure pilot hole geometry. Remove the bone from the drill press and hold it firmly on a work surface, whereas large specimens may need to be secured using a vice or clamp on a work surface. To measure cortical wall thickness at the screw pullout site, insert a clinical or mechanical depth gage that has a hooked end, pull the hooked end until it engages the intramedullary underside of the cortical wall, and then measure this length using a ruler or Vernier caliper (Fig. 8.3A–C). If there is a need to double-check cancellous pilot hole depth, insert a thin rigid guide wire until it feels like it has reached the bottom of the hole, use a pen to mark the guide wire's surface that is at the level of the pilot hole, remove the guide wire, and then measure the length of the guide wire's tip-to-pen mark using a ruler or Vernier caliper (Fig. 8.3D–F). Guide wires can also be used to clear away any unwanted bone debris from the pilot hole.

Step 7. Create screw threads. Create internal threads in the pilot holes by "tapping" bone specimens using the manufacturer's recommended instrumentation to ensure proper screw–bone engagement and prevention of bone cracking

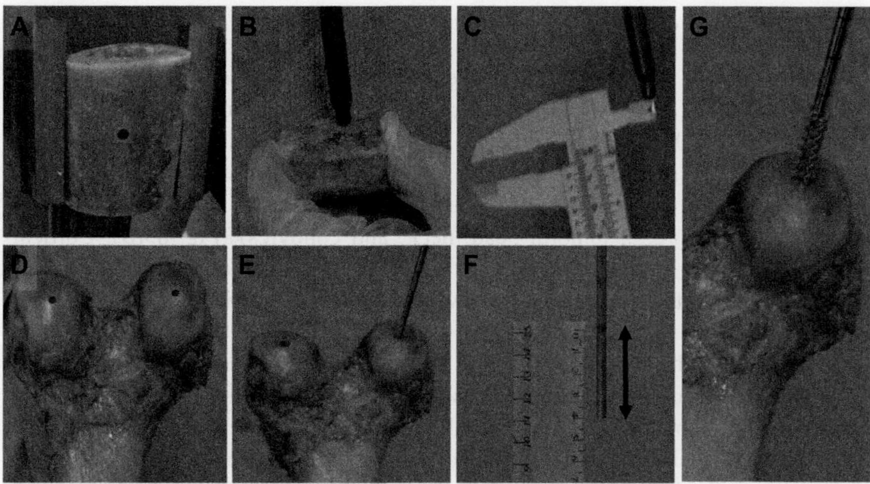

FIGURE 8.3

Measurement of pilot hole geometry. (A) Cortical bone specimen with a pilot hole, (B) depth gage insertion into a cortical pilot hole, (C) depth measurement of a cortical pilot hole using a depth gage and Vernier calipers, (D) cancellous bone specimen with a pilot hole, (E) guide wire insertion into a cancellous pilot hole, (F) depth measurement of a cancellous pilot hole using a guide wire and ruler, (G) tapping or threading the pilot hole for cancellous screw pullout tests.

(Fig. 8.3G). Make a pen mark on the tap shaft to indicate the correct depth is reached to tap the entire hole. Tapping is required for many types of clinically used nonself-tapping cancellous screws. However, many cortical and cancellous screws are self-tapping; thus, there is no need to tap the pilot hole.

TIPS AND TRICKS

- ✓ Bone shaft segments, heads, and condyles can be removed with a band saw for easier handling.
- ✓ Use a rigid guide wire or screwdriver bit guide, so the screws go in straight.
- ✓ Insert a new screw for each test to avoid biasing the results.
- ✓ The same researcher should perform all screw insertions for consistency.
- ✓ Two researchers will make specimen preparation and testing more efficient.

THE "GOLD STANDARD"

The American Society for Testing and Materials (ASTM) provides guidelines in its document ASTM F543 (Standard specification and test methods for metallic medical bone screws) on medical bone screws for evaluating screw pullout from standardized test blocks. However, this does not fully duplicate realistic clinical bone geometries or properties.

3.4 SPECIMEN TESTING

Step 1. Choose the screw. Choose the right bone screw since there is a large variety from medical manufacturers. Cortical screws are used mostly in the diaphysis (i.e., shaft) of long bones (e.g., humerus, femur, tibia, etc.) and inserted through one cortical bone wall (i.e., unicortical screw) or two cortical bone walls (i.e., bicortical screw) (Fig. 8.4A). Cancellous screws are used wherever there is substantial spongy or trabecular bone (e.g., femoral head, neck, and condyles) (Fig. 8.4B). Cortical screw threads are more closely spaced together (i.e., smaller thread pitch) and have shallower depth vs. cancellous screws since cortical bone is denser than cancellous bone.

Step 2. Assemble mounting plates. Mount the bone specimen into a vice or clamp with the pilot hole facing up. Insert a metal guide wire to ensure the pilot hole is clear of debris and to get a visual idea about the direction of the pilot hole (Fig. 8.5A). The middle hole of the metal support plate is then positioned directly over the bone pilot hole. This metal support plate prevents bending of the bone specimen on either side of the pilot hole during the screw pullout test. Then, the middle hole of the screw head gripper is also positioned over the pilot hole (Fig. 8.5B–D).

Step 3. Insert the screw. Obtain a manual surgical screwdriver suggested by the screw manufacturer that matches the cortical and cancellous screw heads. Use the surgical screwdriver to insert the screw through the holes of the gripper and support plate and then into the bone's pilot hole. Turn the screw several times slowly so the screw threads engage the bone, and then continue to drive the screw until the screw head approaches the gripper to within about 5 mm (Fig. 8.5C and D). The screw head is not tightened against the gripper, so that residual axial stresses are not introduced along the screw–bone interface. (Note: There are two reasons for this, namely, screw pullout tests will be a true measure of the material properties of the bone itself, and the screw head needs to have some physical clearance around it for the gripper to work properly.)

Step 4. Assemble the test jig. Position the specimen by hand into the test jig that is fixed to the base of the mechanical tester (Fig. 8.6). Lower the universal joint so that its eye (i.e., hole) is lined up with the holes in the side of the gripper. Place a metal screw or pointed metal pin through the entry hole, the eye, and the exit

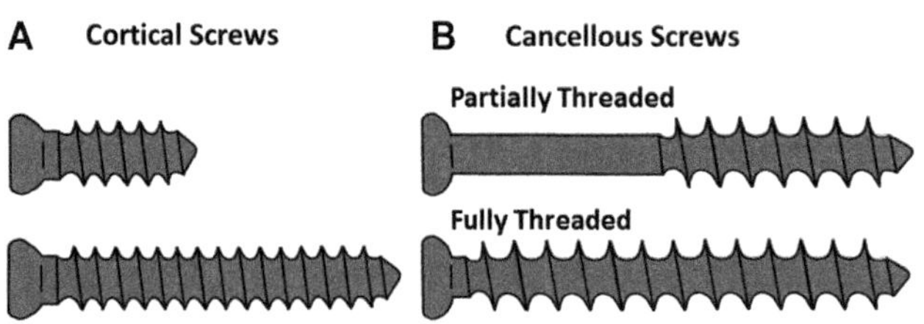

FIGURE 8.4

Common orthopaedic bone screws. (A) Cortical screws, (B) cancellous screws.

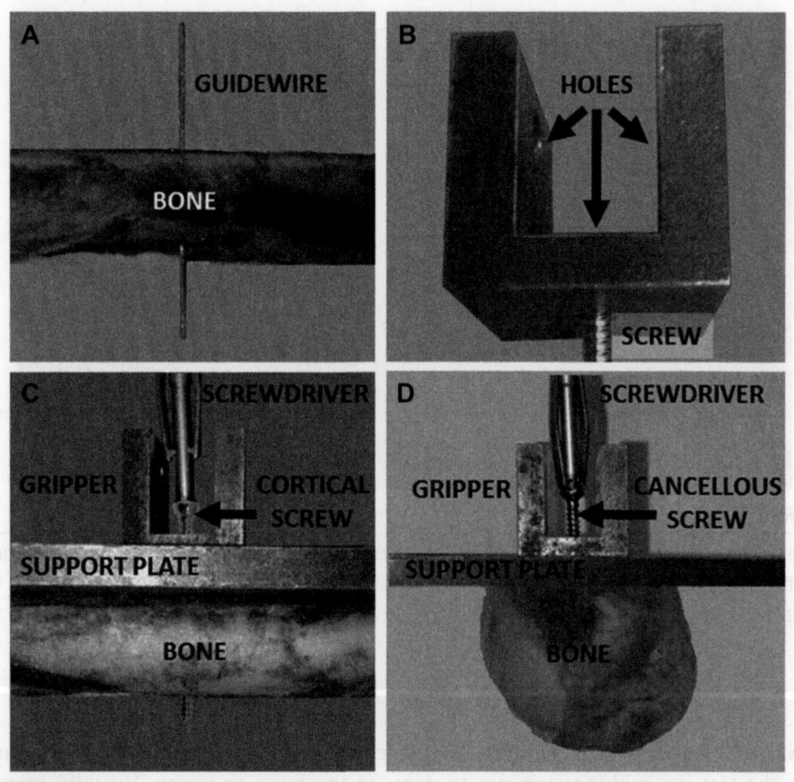

FIGURE 8.5

Screw insertion into bone specimens. (A) Guide wire insertion to clear away debris and to visualize hole direction, (B) screw head gripper, (C) cortical screw insertion through the midshaft cortical bone segment of a human long bone, (D) cancellous screw insertion into the cancellous bone of a head or condyle that was removed from a human long bone.

hole of the gripper. Jiggle the specimen by hand to make sure it is not impinging against any surface. Then, very slowly raise the actuator of the mechanical testing system so that the top side of the metal support plate engages the test jig arms, thus creating a low axial preload of about 50 N in the screw–bone assembly.[7–10]

Step 5. Conduct the pullout test. Place a plastic protective shield around the test assembly. Execute the manual or computer-controlled test sequence at a low fixed screw pullout rate (e.g., 10 mm/min) until the screw tip is completely removed from the bone's pilot hole. Prior studies have shown there may be a moderate influence of speed over a wide range of pullout speeds (i.e., 1–60 mm/min); thus, a lower pullout rate is recommended to minimize viscoelastic effects.[10,11]

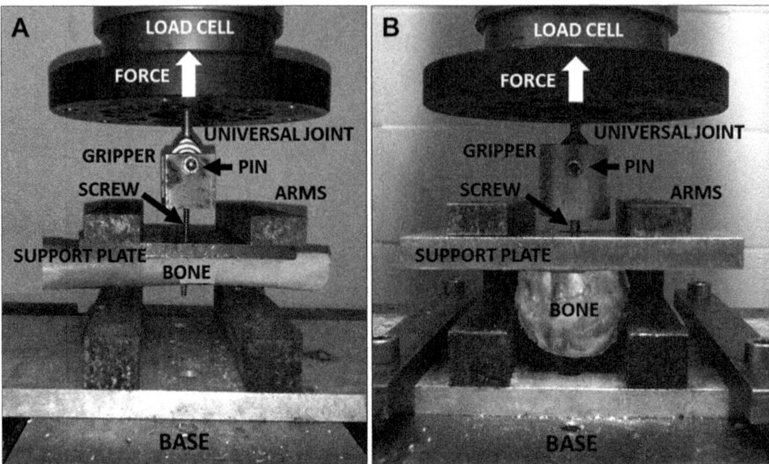

FIGURE 8.6

Bone screw pullout tests. (A) Cortical screw pullout test using the cortical midshaft segment of a human long bone, (B) cancellous screw pullout test from the cancellous bone of the proximal head or distal condyle that has been removed from a human long bone.

Step 6. Inspect the specimen. Stop the test sequence. Remove the specimen from the test jig, making sure not to damage the bone or screw. Visually inspect the screw and hole, take a sufficient number of photos, and make a note of any damage to the screw (i.e., bending or breaking), any evulsed bone still lodged in the screw threads, or any surface damage to hole or surrounding bone.

3.5 RAW DATA COLLECTION

Step 1. Record bone characteristics. Enter human or animal demographic data (i.e., age, sex, and left or right limb) and quantitative bone properties (i.e., BMD and clinical T-score), which will help determine the factors that are statistically correlated with screw pullout measurements (Table 8.1). Note that BMD is in 2-D units (i.e., g/cm^2 rather than g/cm^3) since DEXA (dual-energy X-ray absorptiometry) bone density scans only provide 2-D scans and values. DEXA scan reports often provide BMD values and corresponding clinical T-scores (i.e., normal T-score ≥ -1, $-1 >$ osteopenic T-score > -2.5, and osteoporotic T-score ≤ -2.5). When such analysis is overlooked or missing, computations can be made as T-score = (bone BMD − reference population mean BMD)/(1 standard deviation of reference population BMD), where T-score reference populations are young healthy adults of the same sex that are aged 20–40 years.[12,13]

Table 8.1 Raw data collected for biological bone screw pullout tests.

Test	Age [years]	Sex [M, F]	Limb side [L, R]	BMD [g/cm^2]	T-score	Engaged bone length [mm]	Pullout force [N]	Pullout displacement [mm]	Failure mode
1									
2									
3									
etc.									
Avg		–	–						–
SD		–	–						–

Avg, *Average;* BMD, *bone mineral density;* SD, *standard deviation.*

Step 2. Record bone–screw geometry. Measure and record the bone length that engages the screw threads. This will vary greatly among specimens and should be known (i.e., this is the cortical wall thickness at the screw insertion site or the cancellous screw threaded length engaged at the screw insertion site).

Step 3. Record pullout parameters. Raw pullout force and displacement at screw pullout should be noted during tests and can now be entered into this table. Raw pullout force is the maximum force achieved during the test, while the corresponding displacement is the pullout displacement.

Step 4. Record failure modes. Final remarks may be entered about the nature of failure at the screw–bone interface (i.e., Any visible fractures around the screw hole? Any nonsymmetry of the screw hole? Any bone material remaining lodged in the screw threads? Any bending or breaking of the screw itself? etc.). Good photos of removed screws and screw holes allow for proper review.

3.6 RAW DATA ANALYSIS

Step 1. Compute pullout energy. Raw pullout force vs. displacement graphs can then be used to compute the raw pullout energy (i.e., $1\ \mathrm{N \cdot m} = 1\ \mathrm{J}$), which is the total area under the graph from the start of the test, up to and including the maximum pullout force which occurs at screw pullout.

Step 2. Normalize raw data. Raw pullout parameters like force, displacement, and energy can now be normalized by geometry by considering the approximate surface area engaged at the screw–bone interface. This can be computed as pullout parameter (normalized) = pullout parameter (raw)/surface area = pullout parameter (raw)/($\pi D_o L$), where D_o is screw outer diameter and L is threaded portion of the screw length engaged by bone. Note that pullout force (normalized) is also sometimes called "shear strength" or "shear stress."

Step 3. Calculate correlation coefficients. Plot the measured result for each individual specimen (i.e., pullout force 1, 2, 3, etc.) vs. its corresponding specimen characteristic (i.e., age 1, 2, 3, etc.) (Table 8.1) to help visualize the interrelationship of measurements vs. characteristics. Then, calculate the correlation coefficient R for each measurement–characteristic pair to determine which characteristic has an influence on measurements. $R > 0.8$ is a typical value considered to indicate a strong correlation. Do this for pullout force, displacement, and energy.

Step 4. Perform statistical comparisons. The criterion for statistical difference first needs to be chosen (e.g., $P < 0.01$ or < 0.05). Then data can be used to compare different patient groups (e.g., men ≤ 60 years old vs. men > 60 years old, "normal" women vs. "osteoporotic" women, etc.), bone types (e.g., femur vs. tibia, left humerus vs. right humerus, etc.), screw insertion sites (e.g., anterior vs. lateral surface), screw designs (e.g., 2.5-mm vs. 3.5-mm cortical screws), etc. Various software programs exist for comparing two test groups (e.g., paired t-test) or two or more test groups influenced by multiple factors (e.g., analysis of variance, ANOVA).

Step 5. Compute statistical power. Power analysis can be done after the study to ensure there were enough specimens per group to detect all statistical differences that were actually present (i.e., was type II statistical error avoided?). Statistical power $> 80\%$ is usually considered to indicate there were enough specimens per test group. Note that if good predictions of averages and standard deviations are available from prior studies, then the number of specimens and/or tests can be chosen before the study begins to ensure a power $> 80\%$.

ENGINEER'S TOOLBOX

Screw pullout force F can be approximated using a mathematical expression developed for a generic screw whose threads are perfectly engaged in isotropic homogeneous material. The formula can be applied to scenarios relevant to orthopaedic surgery, as shown. Consider that σ_U is ultimate tensile stress of bone, D is screw pitch diameter $= D_o - 0.6495p$, D_o is screw outer diameter, p is screw thread pitch, and L is engaged length of screw threads. Therefore, screw pullout force $F =$ (ultimate shear strength of bone) $\times$ (shear surface area) $= (\sigma_U/\sqrt{3}) \times (\pi DL)$.

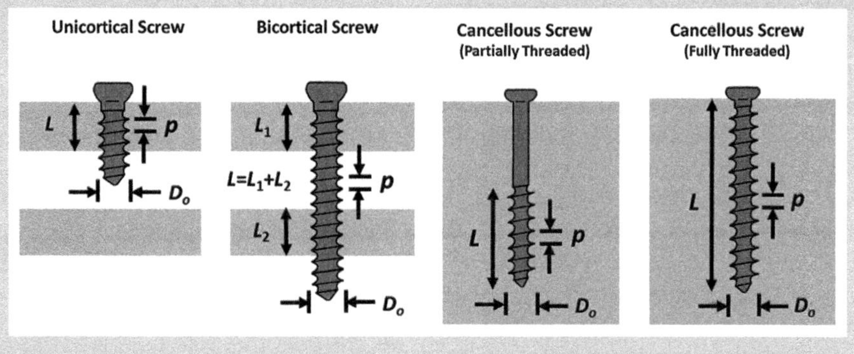

4. RESULTS

Once all screw pullout data collection and analysis have been performed, it is then important to communicate and present the primary results in an understandable and concise manner to the reader of a journal article, conference paper, technical report, or book chapter.

Step 1. Show raw pullout profile. Begin with a typical profile of raw pullout force vs. displacement (Fig. 8.7). Initially, force is at the low preload level, or perhaps at zero if no preload was applied. Then, as the screw begins to be extracted out of the bone, there is a sudden rise in the force, which is usually quite linear; this indicates the bone is still within the linear elastic region and has not sustained any permanent damage yet. Upon continued screw extraction, the linear elastic yield point is surpassed and the graph becomes nonlinear; this indicates some initial permanent bone damage. A peak force value is eventually reached, which is the maximum force the screw–bone interface can withstand. This is followed by a rapid drop in the force as the screw–bone interface completely fails. The screw threads are no longer engaged with the bone; however, there is still some frictional contact between the outer edges of the screw threads and the bone hole until the screw is fully removed from the hole. This graph will be influenced by lubrication from marrow, blood, or synovial fluid, bone material properties, screw thread design, screw–bone friction factor, and screw pullout rate.

Step 2. Show main pullout data. The main numerical findings of the study can be presented for cortical and/or cancellous screw pullout tests (Fig. 8.8). This includes raw or normalized pullout force, raw or normalized pullout displacement,

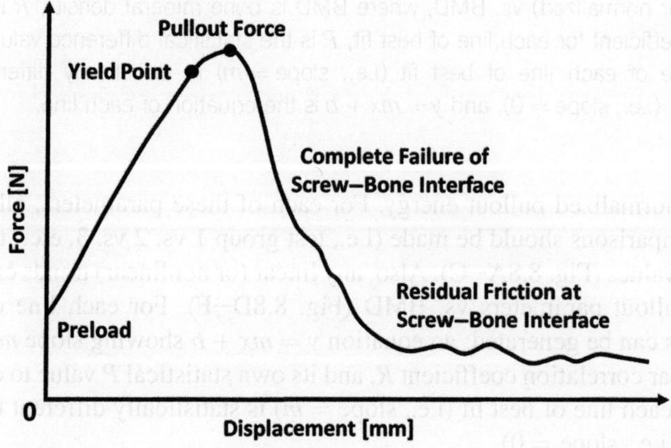

FIGURE 8.7

Raw screw pullout force vs. displacement.

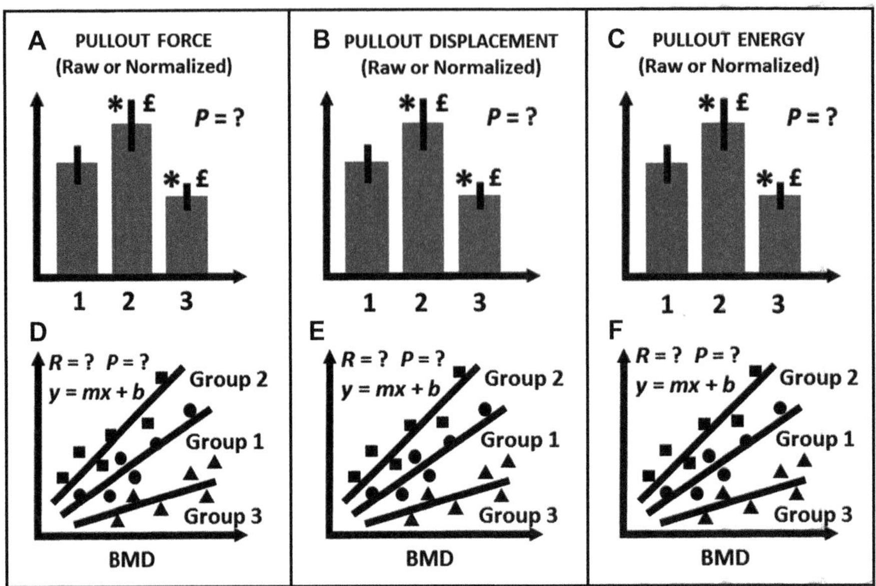

FIGURE 8.8

Screw pullout results. (A) Pullout force (raw or normalized) vs. test group, (B) pullout displacement (raw or normalized) vs. test group, (C) pullout energy (raw or normalized) vs. test group, where bar graphs show average ± 1 standard deviation and P is the statistical difference value for each pairwise comparisons between each group which can be indicated by symbols like asterisks (*), pounds (£), etc., (D) pullout force (raw or normalized) vs. BMD, (E) pullout displacement (raw or normalized) vs. BMD, (F) pullout energy (raw or normalized) vs. BMD, where BMD is bone mineral density, R is the linear correlation coefficient for each line of best fit, P is the statistical difference value to ensure that the slope of each line of best fit (i.e., slope $= m$) is statistically different than a horizontal line (i.e., slope $= 0$), and $y = mx + b$ is the equation of each line.

and raw or normalized pullout energy. For each of these parameters, all statistical pairwise comparisons should be made (i.e., test group 1 vs. 2 vs. 3, etc.) to generate statistical P values (Fig. 8.8A–C). Also, any linear (or nonlinear) trends can be illustrated for pullout parameters vs. BMD (Fig. 8.8D–F). For each line of best fit, several items can be generated: an equation $y = mx + b$ showing slope m and intercept b, a linear correlation coefficient R, and its own statistical P value to ensure that the slope of each line of best fit (i.e., slope $= m$) is statistically different than a horizontal line (i.e., slope $= 0$).

Step 3. Show failure modes. Finally, photos of typical screws and/or screw holes after testing help visualize any kind of failure modes, such as bent or broken screws, residual bone in the screw threads, and any surface damage around the screw holes (Fig. 8.9).

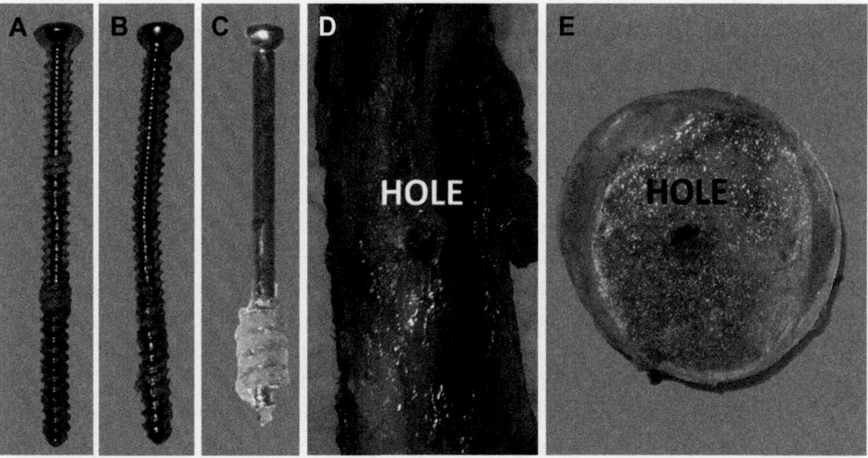

FIGURE 8.9

Failure photos after screw pullout tests. (A) Undamaged cortical screw with residual bone in its threads, (B) bent cortical screw with residual bone in its threads, (C) undamaged cancellous screw with residual bone in its threads, (D) cortical bone from a humerus shaft with undamaged hole surface, (E) cancellous bone from a femur head with undamaged hole surface.

ALTERNATIVES AND ADAPTATIONS

- ✓ **Metal fracture plates**. These devices can be fixed to biological bone with multiple screws, as routinely done clinically. Then, a "pull off" force can be applied to the plate to measure screw purchase strength.
- ✓ **Sawbone models**. Artificial anatomically correct whole bones, or standardized blocks, cylinders, and plates, can be used for screw pullout tests. 3-D density (i.e., g/cm^3), rather than 2-D BMD (i.e., g/cm^2), needs to be measured or obtained from the manufacturer, since DEXA scans for BMD are only suitable for biological bone. Remember that sawbones have isotropic and homogeneous properties.
- ✓ **Slotted gripper**. This would allow screw heads to be inserted and grabbed quickly and easily, like a hook. But, there may be an asymmetric stress distribution on screw heads during pullout tests, causing screws to be unevenly pulled out of pilot holes.

5. DISCUSSION

After completing all screw pullout tests, data collection, data analysis, and data presentation, then final results can be considered and interpreted in the broader context of some important clinical, biomechanical, and/or technological considerations, as follows.

Pullout force can be predicted for a generic screw whose threads are perfectly engaged in an isotropic homogeneous material.[14] Consider that σ_U is ultimate tensile stress of the host bone, D is screw pitch diameter $= D_o - 0.6495p$, D_o is screw outer diameter, p is screw thread pitch, and L is engaged length of screw threads. Consequently, screw pullout force $F =$ (ultimate shear stress) $\times$ (shear surface area) $= \left(\sigma_U/\sqrt{3}\right) \times (\pi DL)$.

Pullout forces can have a wide range of values.[7–10,15–17] Cortical screw tests have yielded 1100–4460 N (human humeral shaft), 1646–3085 N (human femoral shaft), 1100–1700 N (human tibial shaft), 2218–3112 N (sawbone humeral shaft), and 2949–5286 N (sawbone femoral shaft). Cancellous screw tests have achieved 1523 N (human femoral head), 56–1230 N (human tibial plateau), 927–1410 N (sawbone femoral head), and 915–1093 N (sawbone blocks).

Pullout force is related to the stripping torque at the screw–bone interface because of accidental screw overtightening.[14] Theoretically, this is computed as pullout force $F = T(2D_{AVG} - 2f/t)/[D_{AVG}(1/t + fD_{AVG})]$, where T is stripping torque, D_{AVG} is average of outer and inner screw diameter, f is screw–bone friction factor, and t is number of screw threads per unit length.[14] Experimentally, for instance, sawbone tests with a 4.5-mm diameter cortical screw can yield a ratio of $F/T =$ raw pullout force/raw stripping torque $= 0.498\ \text{mm}^{-1}$ over a range of densities.[18]

Screw design may alter screw–bone interface quality and screw pullout parameters and, thus, may affect the success of fracture repair surgery. Screw design can be quantified as thread shape factor $TSF = 0.5 + 0.57735d/p$, where d is screw thread depth and p is screw thread pitch.[19] Thread depth and pitch influence the overall angle of the thread. This means that deeper threads, which engage bone with better "bite," may "feel" tighter to orthopaedic surgeons.

The American Society for Testing and Materials (ASTM) provides guidelines in its document ASTM F543 on medical bone screws for evaluating screw pullout from test blocks with standardized geometry and properties.[20] However, this does not fully duplicate realistic clinical conditions involving complex cortical geometry, cancellous architecture, or bone properties.

6. SUMMARY

- Bone screws are used to repair bone fractures with or without plates.
- Bone screws can eventually be pulled out due to "toggling" or impact events.
- Bone screw pullout force, displacement, and energy can be measured experimentally.
- Bone screws can be bent or broken during screw pullout events.
- Pullout force can be predicted using established engineering formulas.
- Screw design and bone type will substantially influence pullout results.

7. QUIZ QUESTIONS

1. What is the definition of pullout force, pullout displacement, and pullout energy?
2. Are raw or normalized data more representative of the material properties of biological bone?
3. How do the anisotropy and nonhomogeneity of real bone affect pullout force predictions?
4. What is the relationship between pullout force and stripping torque?
5. Compute pullout force for a unicortical screw test. Assume ultimate tensile strength for bone = 107 MPa, screw outer diameter = 3.5 mm, screw thread pitch = 1.25 mm, and average unicortical wall thickness = 4 mm (answer: 2087 N).

REFERENCES

1. Cordey J, Rahn BA, Perren SM. Human torque control in the use of bone screws. In: Uhthoff HK, editor. *Current concepts of internal fixation of fractures*. New York (NY): Springer-Verlag; 1980. pp. 235–43.
2. Cleek TM, Reynolds KJ, Hearn TC. Effect of screw torque level on cortical bone pullout strength. *Journal of Orthopaedic Trauma* 2007;**21**(2):117–23.
3. Tsuji M, Crookshank M, Olsen M, Schemitsch EH, Zdero R. The biomechanical effect of artificial and human bone density on stopping and stripping torque during screw insertion. *Journal of the Mechanical Behavior of Biomedical Materials* 2013;**22**: 146–56.
4. Aziz MSR, Tsuji MRS, Nicayenzi B, Crookshank MC, Bougherara H, Schemitsch EH, et al. Biomechanical measurements of stopping and stripping torques during screw insertion in five types of human and artificial humeri. *Proceedings of the Institution of Mechanical Engineers (Part H): Journal of Engineering in Medicine* 2014;**228**(5): 446–55.
5. Fulkerson E, Koval K, Preston CF, Iesaka K, Kummer FJ, Egol KA. Fixation of periprosthetic femoral shaft fractures associated with cemented femoral stems: a biomechanical comparison of locked plating and conventional cable plates. *Journal of Orthopaedic Trauma* 2006;**20**(2):89–93.
6. Law M, Tencer AF, Anderson PA. Caudo-cephalad loading of pedicle screws: mechanisms of loosening and methods of augmentation. *Spine* 1993;**18**(16):2438–43.
7. Aziz MSR, Nicayenzi B, Crookshank MC, Bougherara H, Schemitsch EH, Zdero R. Biomechanical measurements of cortical screw purchase in five types of human and artificial humeri. *Journal of the Mechanical Behavior of Biomedical Materials* 2014;**30**: 159–67.
8. Zdero R, Elfallah K, Olsen M, Schemitsch EH. Cortical screw purchase in synthetic and human femurs. *Journal of Biomechanical Engineering* 2009;**131**(9):094503-1-7.

9. Zdero R, Olsen M, Bougherara H, Schemitsch EH. Cancellous bone screw purchase: a comparison of synthetic femurs, human femurs, and finite element analysis. *Proceedings of the Institution of Mechanical Engineers (Part H): Journal of Engineering in Medicine* 2008;**222**(8):1175–83.
10. Zdero R, Schemitsch EH. The effect of screw pullout rate on screw purchase in synthetic cancellous bone. *Journal of Biomechanical Engineering* 2009;**131**(2):024501-1-5.
11. Zdero R, Shah S, Mosli M, Bougherara H, Schemitsch EH. The effect of the screw pullout rate on cortical screw purchase in unreamed and reamed synthetic long bones. *Proceedings of the Institution of Mechanical Engineers (Part H): Journal of Engineering in Medicine* 2010;**224**(3):503–13.
12. Lewiecki EM, Borges JLC. Bone density testing in clinical practice. *Arquivos Brasileiros de Endocrinologia and Metabologia* 2006;**50**(4):586–95.
13. Pettersson U, Nordström P, Lorentzon R. A comparison of bone mineral density and muscle strength in young male adults with different exercise level. *Calcified Tissue International* 1999;**64**(6):490–8.
14. Troughton MJ, editor. *Handbook of plastics joining: a practical guide*. Norwich (NY, USA): Plastics Design Library; 2008. pp. 180.
15. Tankard SE, Mears SC, Marsland D, Langdale ER, Belkoff SM. Does maximum torque mean optimal pullout strength of screws? *Journal of Orthopaedic Trauma* 2013;**27**(4): 232–5.
16. Lyon WF, Cochran JR, Smith L. Actual holding power of various screws in bone. *Annals of Surgery* 1941;**114**(3):376–84.
17. Westmoreland GL, McLaurin TM, Hutton WC. Screw pullout strength: a biomechanical comparison of large fragment and small fragment fixation in the tibial plateau. *Journal of Orthopaedic Trauma* 2002;**16**(3):178–81.
18. Edwards TR, Tevelen G, English H, Crawford R. Stripping torque as a predictor of successful internal fracture fixation. *ANZ Journal of Surgery* 2005;**75**(12):1096–9.
19. Chapman JR, Harrington RM, Lee KM, Anderson PA, Tencer AF, Kowalski D. Factors affecting the pullout strength of cancellous bone screws. *Journal of Biomechanical Engineering* 1996;**118**(3):391–8.
20. ASTM F543. Standard specification and test methods for metallic medical bone screws. West Conshohocken (PA, USA): American Society for Testing and Materials (ASTM); Available at: http://www.astm.org.

CHAPTER

Biomechanical Testing of the Intact and Surgically Treated Spine

9

Alejandro A. Espinoza Orías[1], Jade He[1], Mei Wang[2]
Rush University Medical Center, Chicago, IL, United States[1]; Medical College of Wisconsin, Milwaukee, WI, United States[2]

1. BACKGROUND

The spine is a complex anatomic structure comprising a long and articulated column that exhibits substantial range of motion (*ROM*) and distinct differences in its three regions: the cervical, thoracic, and lumbar spines (Fig. 9.1). The primary biomechanical functions of the spinal column are to facilitate physiologic movement of the head and trunk, and to bear the loads from activities of daily living and protect the neural tissues. As such, the largest spinal motion occurs in the cervical region, while the axial compressive strength progressively increases in a cranial to caudal direction. Structural components that contribute to the biomechanical functions include vertebrae, intervertebral discs, facet joints, and spinal ligaments. There are six degrees of freedom at an intervertebral joint: rotations in the sagittal, frontal, and axial planes, as well as translations along the axis of these planes.[1]

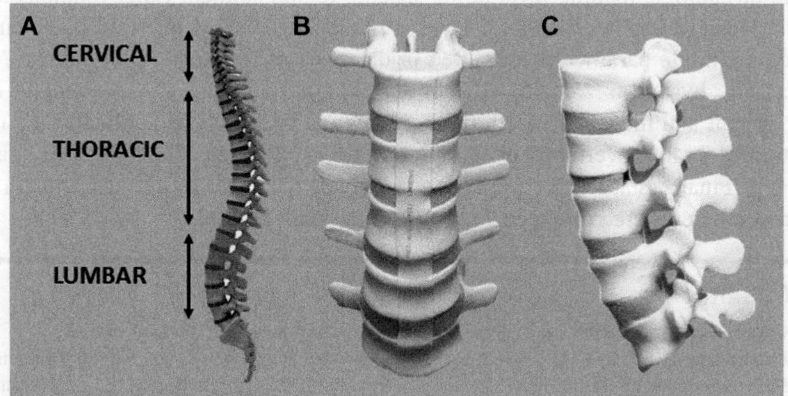

FIGURE 9.1

Spine anatomy. (A) Full spine showing the three main regions, (B) close-up (front view), (C) close-up (side view).

Understanding biomechanics of the spine is an important aspect in managing diseases and injuries to the spine and restoring its normal function. Therefore, this chapter explains how to perform biomechanical testing of the spine, as well as how to analyze, report, and interpret results.

2. RESEARCH QUESTIONS

Typical research questions might include one or more of the following:

- What are the effects of spinal fusion on the *ROM* of adjacent segments?
- How does disc degeneration change the kinematic behavior of a spinal segment?
- Does artificial disc replacement replicate the kinematic patterns of the native spine?
- Can anterior lumbar fusion with an interbody cage alone provide adequate stability?
- What is the best surgical approach in stabilizing a thoracolumbar burst fracture?
- Does spinal instrumentation alter the loading of the facet joints?
- etc.

3. METHODOLOGY

3.1 GENERAL STRATEGY

The flexibility test is a common test protocol in spine biomechanics. Test specimens include fresh–frozen spine specimens from human cadavers or animal models with similar mechanical properties (i.e., calf, pig, sheep).[2] In this protocol, pure moments in flexion, extension, lateral bending, and axial rotation are applied to the spinal segment one direction at a time, while the resultant 3-D motion of the spinal segment is recorded using a motion tracking system. The peak moment is based on the physiologic loading level at the specific region studied. Euler rotations and translations at each intervertebral level are calculated based on rigid body kinematics. Common parameters measured in the analysis are *ROM*, neutral zone *NZ*, stiffness *K*, hysteresis *H*, coupled motion, axis of rotation, etc. To simulate the body weight and musculature effects, compressive preloads are applied to the unconstrained cranial vertebra of the tested segment using the "follower load" concept.[3] Finally, statistical comparisons are made between test groups, and correlation coefficients are computed to identify important factors.

GLOSSARY

✓ **Coupled motion**. The motion in directions other than the direction of the applied load.
✓ **Euler angle**. Three sequence-dependent rotations about a moving body's local coordinate system.
✓ **FSU (functional spinal unit)**. The smallest motion unit of the spine, consisted of two adjacent vertebrae, the intervertebral disc, facet joints, and surrounding ligaments.
✓ **NZ (neutral zone)**. Movement from the neutral position using minimal applied load.
✓ **ROM (range of motion)**. The path traversed in space under a given direction of motion.

SAFETY FIRST

✓ Use proper safeguards and precautions when handling cadaveric tissue.
✓ When procuring tissue, make sure the serology report shows no blood-borne pathogens.
✓ Cadaveric tissue must be handled under the assumption that it might be contaminated.
✓ Do not use bare hands to install or remove scalpel blades from their holders.
✓ All sharp instruments used in the laboratory must be handled with extreme care.
✓ Any sharps used in the lab need to be discarded in the biohazard sharps container.

3.2 MATERIALS AND TOOLS LIST

- cadaveric lumbar spine (L1–S1)
- dynamometer (i.e., 6-degree-of-freedom load cell)
- electric drill or wire driver
- infrared motion tracking markers
- mechanical tester
- potting mix (e.g., bone cement, dental cement, etc.)
- potting mold (e.g., metallic mold or PVC pipe)
- spine implants and surgical tools
- support stand to hold the spine while potting
- wood screws or Kirschner wires

3.3 SPECIMEN PREPARATION

Step 1. Thaw the specimen. A lumbar spine specimen can be thawed either in a refrigerator or in air at room temperature. For thawing in a refrigerator at 4°C, perform the following tasks: (1) transfer the specimen from a −20°C freezer to a 4°C refrigerator; (2) record the time to keep track of how long the specimen was not refrigerated; and (3) keep the specimen in the refrigerator for ≤48 h. For thawing in air at room temperature at about 22°C, perform the following tasks: (1) record the room temperature; (2) transfer the specimen from a −20°C freezer to the laboratory work table, such as a cadaver dissection table; (3) record the time to keep track of how long the specimen was not refrigerated; and (4) keep the specimen on the dissection table for no longer than 24 h. In both cases, only open the bag containing the specimen after the thawing period is completed. This will protect the specimen from drying out.

Step 2. Dissect the specimen. Make careful notes during dissection. Decide on whether experiments will be done on the whole spine or a single motion segment, which is comprised of two adjacent vertebrae and the intervening intervertebral disc. Use blunt dissection techniques to remove muscles. Then, use a scalpel to strip the specimen of musculature, but keep ligaments and other important soft tissue. Be careful not to damage the facet capsule and any of the anterior, posterior, and interspinous ligaments.

Step 3. Image the specimen. Image the specimen with clinical imaging modalities [e.g., computed tomography (CT) scans, DEXA (dual energy X-ray absorptiometry) scans, MRI, X-rays] to determine infrared marker location, specimen morphology, and specimen quality. Specifically, CT scans provide very detailed geometry of the vertebral bodies and posterior components of the spine. Also, intervertebral disc height is a key parameter to measure using X-rays or MRI in combination with the Pfirrmann scale[4] to evaluate motion segment quality. Moreover, DEXA scans can be used to quantify the bone mineral density of the vertebrae.

Step 4. Pot the specimen. Obtain potting mix, such as anchoring cement, body filler putty, bone cement, or dental cement for potting the specimen (Fig. 9.2). Then, pot the most caudal vertebrae in a mold (e.g., metal cube or PVC pipe), which will later be attached to a base dynamometer, but make sure to align screw holes or place thread inserts in the base so machine screws can be inserted from the dynamometer side. Also, if the top vertebra will not be attached directly to the mechanical tester's load actuator during tests, perform a second potting of the cranial vertebrae to mechanically attach the mechanical tester piston or cross-bar to the spines. Then, in either case, do the following tasks. Insert one or two wood screws (or threaded rods) halfway into the potting segment to enhance anchoring, but be careful to keep a reference level at a consistent segment. Use the top vertebrae at L3 because it is parallel to the horizontal plane. Next, use a vertical support stand to clamp the specimen from the spinous process in order to set the specimen in position inside the

FIGURE 9.2

Potting the spine specimen. The specimen should be properly aligned with the horizontal plane. The L3 level should be parallel with the horizontal plane, while S1 should have an angle of approximately 40° with respect to the horizontal. The support stand at the right helps to hold the specimen above the potting mold.

potting frame. The angle between S1 and the side of the metal frame should be approximately 40°, thereby rendering L2/L3 endplates parallel to the horizontal plane (i.e., base of the potting mold). To avoid specimen drying, wrap the spine in a gauze soaked with phosphate buffered saline. Finally, once the specimen is in the desired position, the potting mix should be poured carefully and allowed to cure.

Step 5. Mount motion analysis markers. Place three motion analysis markers (or triad) on each vertebral body to track the spatial location of the spinal segment. Use screws, threaded rods, or the exposed ends of Kirschner wires driven into the bone using a power drill to obtain rigid placement of markers. This should result in a vertebral body with markers that will behave as a rigid body (i.e., the spatial relationship between the vertebral body and the motion tracking markers is constant). Note that depending on the motion tracking system, markers can be passive or active. In both cases, ensure that markers will be visible by the camera. For active markers, cables should be arranged to prevent blocking the camera's view or interfering with spine segment movement.

Step 6. Position the dynamometer. Place a 6-degree-of-freedom dynamometer under the whole spine specimen or the single motion segment, depending on the specific research question. This instrument allows for recording of moments and guarantees that pure moment is applied.

TIPS AND TRICKS

- ✓ To improve anchoring, add screws or other protruding segments to the potted vertebra.
- ✓ To preserve tissue quality, put the spine in a plastic bag and thaw in a refrigerator for 2 days.
- ✓ Overnight thawing outside a refrigerator is possible, but make sure to avoid tissue drying.
- ✓ Maintain spine quality by wrapping it in saline-soaked gauze and storing it in a bag.
- ✓ For specimen potting, a PVC pipe or metal tubing mold are the best options.

THE "GOLD STANDARD"

No known international standards exist specifically for the flexibility test for spine biomechanics. However, the American Society for Testing and Materials (ASTM) has a document that provides some guidelines, namely, ASTM F2423 (Standard guide for functional, kinematic and wear assessment of total disc prostheses) that recommends loads for pure moment testing of the lumbar spine in flexion–extension, lateral bending, and axial rotation, with or without axial compression. Also, researchers should consult peer-reviewed journal articles on spine biomechanics that can serve as precedents.

3.4 SPECIMEN TESTING

Step 1. Calibrate the motion tracking system. Before testing, the motion tracking system must be calibrated. Different manufacturers provide different methods, but the main idea is to recognize the markers on the moving bodies within a region of interest in the testing space. Care must be taken so that no extraneous signals are picked up by the motion tracking system.

Step 2. Define test requirements. Test requirements must be defined (i.e., is this test meant to evaluate motion preserving or fusion "motion-restricting" devices?). For motion-preserving devices (e.g., artificial discs), the kinematic parameters, such as the Euler angles, the helical axis of motion, and total *ROM*, will be of interest. For motion-restricting devices (e.g., cages, rod constructs, screw constructs) that restrict the motion of the spine, these will show noticeable changes in *ROM* and *K*, which the tests seek to quantify.

Step 3. Choose the loading method. Choose from among the two types of commonly used loading methods, depending on what experimental equipment is available.

The "floating ring method" provides continuous motion to test the spine in pure moment.[5,6] This can be achieved using commercially available mechanical testers that provide sinusoidal motion in a vertical direction through their piston actuators. This vertical motion can be converted into cyclic pure moment bending motion that mimics the flexion–extension, lateral bending, and axial rotation motions on a whole lumbar spine or a single motion segment (Fig. 9.3). Since the input parameter to drive the test is the position of the testing frame piston, this is called a "displacement control" experiment.

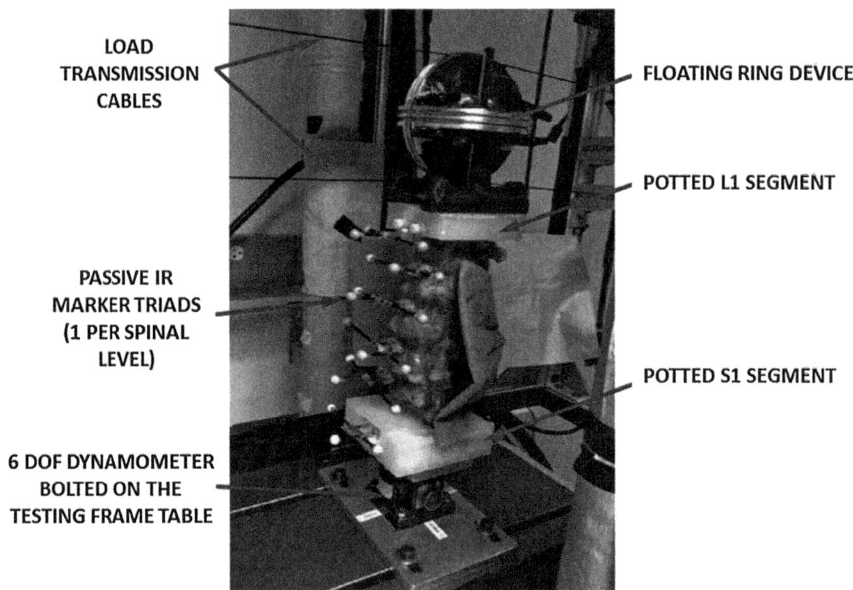

FIGURE 9.3

The floating ring method. A whole lumbar spine placed in the testing jig. Passive infrared marker triads are attached to each spinal level.

The "cross bar method" is an alternative when a mechanical tester is not available.[7] A cross bar of known length is placed atop the most cranial vertebra, and two counterweights are attached at both ends (Fig. 9.4). To produce motion, a load differential must exist between the two arms. Both ends must have the same length to provide equal moment arms. The arms can be actuated by adding known incremental weights or actuated with a system of pumps transferring water from one end to the other. In either case, a complete flexion-to-extension cycle can be completed, which allows the recording of a complete loop of a load vs. displacement curve. In contrast to the previous method, since a known load is added to produce motion, this is a "load control" experiment.

Step 4. Test the native intact spine. Place the whole spine or spine segment in either full extension or full flexion, although the middle or neutral position is hard to determine consistently. Then, determine the pure moment loading mode, loading level, and testing order, such as flexion–extension (e.g., ±7.5 N·m), lateral bending (e.g., ±6 N·m), and axial rotation (e.g., ±3 N·m) without or with axial compression (e.g., 1.2 kN).[10] Next, apply 3–5 preconditioning sinusoidal cycles slower than 1 Hz (typically 0.1 Hz), as these are considered quasi-static tests to minimize viscoelastic effects, using the chosen loading mode and level in order to remove any mechanical slack in the setup and to allow the specimen to reach mechanical steady state. However, only the last loading cycle from the load vs. displacement curve will be used later for data analysis. Remember to simultaneously record the movement of the motion tracking markers. After finishing the test, remove the spine.

Step 5. Mount spinal implants and repeat tests. If a spine implant will also be tested to simulate surgical injury repair or disease treatment, the implant manufacturer's surgical toolkit is needed to attach the device to the spine. Typical surgical

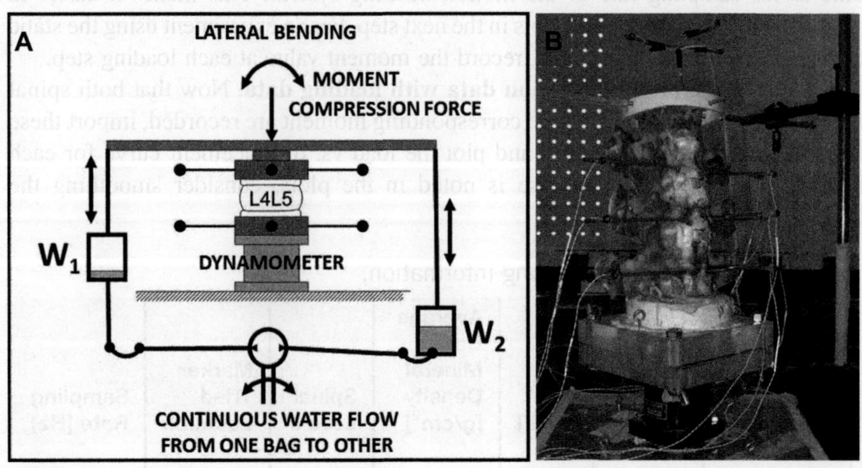

FIGURE 9.4

The cross bar method. (A) Schematic for a water pump system, (B) active infrared markers and a 6-degree-of-freedom dynamometer at the base are employed.

procedures include discectomy, distraction, and insertion of the implant, making sure to always following implant manufacturer's guidelines. This will allow for biomechanical comparison of different configurations of the spine, such as intact specimen vs. fusion cage (e.g., PEEK/titanium) vs. screw fixation vs. rod fixation, etc. In this case, the intact specimen serves as its own control to compare data. Then, remount the new specimen into the loading jig and repeat mechanical tests as in step 4.

3.5 RAW DATA COLLECTION

Step 1. Record specimen information. Enter demographic information, bone density measurements, and spinal levels included in the experiment (Table 9.1).

Step 2. Record marker triad locations. To track the spatial location of each spinal segment, a marker triad is mounted onto the segment and their coordinates are recorded with the motion tracking system. The three markers on the triad are used to establish a local coordinate system of the spinal segment. The movement of one spinal segment with respect to another (i.e., the relative spinal motion) is derived from the transformation matrix between the two local coordinate systems that are set up by the triads. This spinal motion is presented in the form of three sequential rotations (i.e., Euler angles). This process is often carried out automatically by the built-in software of the motion tracking system. However, it is still important to record the triad location or its three markers with respect to the segment it represents.

Step 3. Collect moment data. For an experiment using the continuous loading configurations either with the floating ring method (Fig. 9.3) or cross bar method (Fig. 9.4A), record the sampling rate of the moment data and confirm that it is the same as the sampling rate of the motion tracking system. This makes it easier to generate load vs. displacement plots in the next step. For an experiment using the static loading configuration (Fig. 9.4B), record the moment value at each loading step.

Step 4. Combine spine motion data with loading data. Now that both spinal motion values and the value of the corresponding moment are recorded, import these data onto the same spreadsheet, and plot the load vs. displacement curve for each loading plane. If excessive noise is noted in the plots, consider smoothing the data with a low-pass filter.

Table 9.1 Spine specimen testing information.

Specimen ID	Age [years]	Sex [M, F]	Average Bone Mineral Density [g/cm^2]	Spinal Levels	Marker Triad Location	Sampling Rate [Hz]
1						
2						
3						
etc.						

3.6 RAW DATA ANALYSIS

Step 1. Plot the load vs. displacement curve. This important curve has a characteristic sigmoidal shape like an "S," which is analyzed to obtain all other spine parameters (Fig. 9.5).

Step 2. Identify *ROM* at specified load levels. Record the *ROM*, which is the extent of the angular displacement for the applied loading (Fig. 9.5, Table 9.2).

Step 3. Identify *NZ* at specified load levels. Record the *NZ*, which describes the instability of the construct or the laxity of the joint (Fig. 9.5, Table 9.2). In the *NZ*, the specimen moves essentially free of applied loading. The more the *NZ* is "flat," the more unstable is the construct. *NZ* is reported numerically through its *ROM* (or length) in degrees.

Step 4. Compute stiffness *K*. Record the slope of the curve, which is the stiffness *K* of the construct or spinal motion segment (Fig. 9.5, Table 9.2). This represents the ratio of moment to angular deformation. The higher the *K* value, the more resistance to the moment is exhibited.

Step 5. Compute sustained moment *M*. If the *ROM* is the extent of the plot on the abscissa, the limits of the plot on the coordinates represents the total sustained moment *M* (i.e., how much moment the spine can withstand) (Fig. 9.5, Table 9.2). Note that if these values are measured only from the origin, they would represent the load or motion in each direction (e.g., flexion–extension or left–right lateral bending).

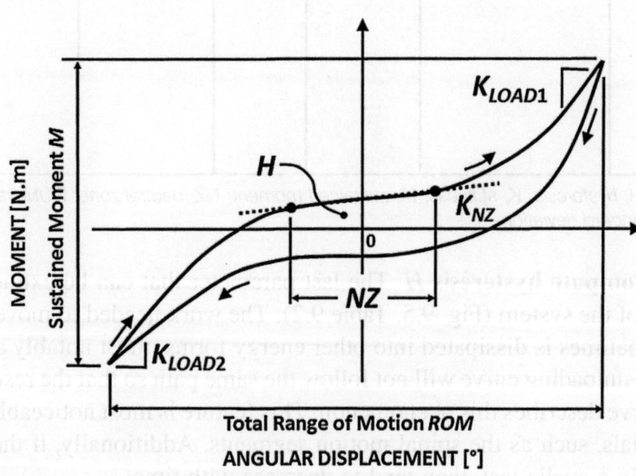

FIGURE 9.5

The spine load vs. displacement curve and its parameters. The hysteresis *H* of the curve is denoted by the area inside the loop. The two circles show the inflexion points that act as limits for the neutral zone (*NZ*). The arrows show the loading direction. The stiffness *K* of the loading sections $Load_1$ and $Load_2$ are not necessarily the same (i.e., if $Load_1$ corresponds to flexion, the stiffness in the countermotion of extension will be different). The same happens with the range of motion in each direction.

Table 9.2 Spine biomechanics data from a typical spinal segment flexibility test.

Specimen ID	Moment [N·m]	*ROM* [°]	*NZ* [°]	*K* [N·m/°]	*M* [N·m]	*H* [J]
Testing Plane: Flexion–Extension						
1						
2						
3						
etc.						
Avg						
SD						
Testing Plane: L–R Axial Rotation						
1						
2						
3						
etc.						
Avg						
SD						
Testing Plane: L–R Lateral Bending						
1						
2						
3						
etc.						
Avg						
SD						

Avg, *average;* H, *hysteresis;* K, *stiffness;* M, *sustained moment;* NZ, *neutral zone;* ROM, *range of motion;* SD, *standard deviation.*

Step 6. Compute hysteresis *H*. The last parameter that can be extracted is the hysteresis *H* of the system (Fig. 9.5, Table 9.2). The work needed to move the spine construct sometimes is dissipated into other energy forms, most notably as friction. The loading–unloading curve will not follow the same path so that the resulting area within the curve describes this phenomenon. This feature is most noticeable in viscoelastic materials, such as the spinal motion segments. Additionally, if these curves are repeated in a cyclic test, they tend to decrease with time.

Step 7. Normalize to controls. Normalize all the parameter values of the surgical or experimental group that were collected (Table 9.2) with respect to the values of intact (i.e., control) in order to get a sense of the percentage increase or decrease in the parameters.

Step 8. Perform statistical analysis. Prior to conducting statistical comparisons between the experimental and control groups, the group average value and the 95% confidence intervals need to be calculated for each parameter, and a normality check should also be conducted using the Shapiro–Wilk test. Parametric analysis of variance can be used if data are normally distributed, otherwise nonparametric tests, such as Wilcoxon paired comparison, should be used.

Step 9. Compute statistical power. Power analysis can be done after the study to ensure there were enough specimens per test group to detect all statistical differences that were actually present (i.e., was type II statistical error avoided?). Statistical power >80% is usually considered to indicate there were enough specimens per test group. Note that if good predictions of averages and standard deviations are available from prior studies, then the number of specimens and/or tests can be chosen before the study begins to ensure a power >80%.

ENGINEER'S TOOLBOX

The spine can be modeled as a straight cantilever beam with a load applied perpendicular to it at one end, producing a reaction moment at the base. If this moment is constrained to a single plane, without shear load reactions, then it is a pure moment loading scenario. The moment of the spine on the base (and the equal but opposite reaction moment of the base on the spine) is calculated as $M = FL$, where F is applied force and L is spine segment length. Also, the compressive or tensile stress acting along the spine segment length is $\sigma = Mc/I$, where c is the distance from the neutral axis (NA) of the beam and I is the area moment of inertia of the beam cross-section.

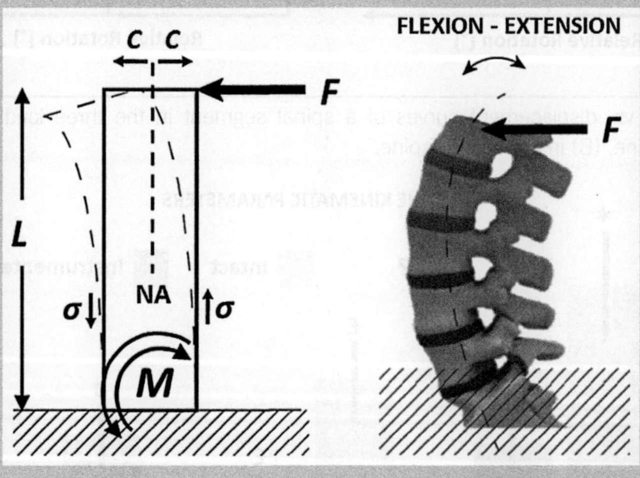

4. RESULTS

Once all spine test data collection and analysis have been performed, it is then important to communicate and present the primary results in an understandable and concise manner to the reader of a journal article, conference paper, technical report, or book chapter.

Step 1. Show load vs. displacement curves. Present the load vs. displacement graphs of the spinal segment in the three loading planes in its native intact condition and in the surgically treated condition, such as after instrumented with fusion implant (Fig. 9.6). This provides a quick overview of the 3-D kinematic responses of spine in the instrumented condition over the control.

Step 2. Show statistical comparisons. Present a descriptive analysis of the reduced data, such as the group averages, standard deviations, etc., of the *ROM*, *NZ*, and *K* derived from experimental data (Fig. 9.7). Use averages and standard deviations for normally distributed data; otherwise use medians and quartiles. A bar

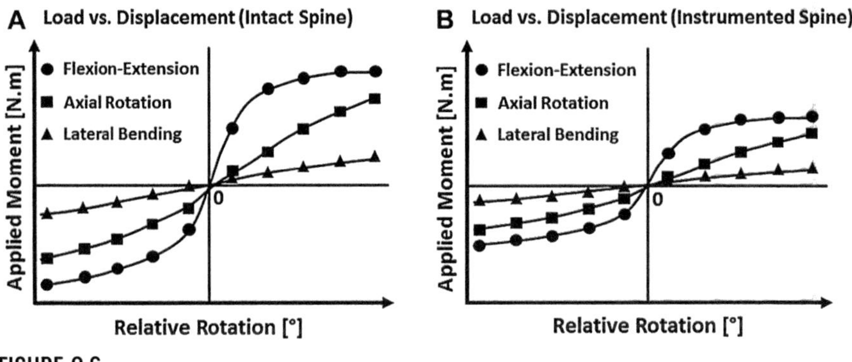

FIGURE 9.6

Typical load vs. displacement curves of a spinal segment in the three loading planes. (A) Intact spine, (B) instrumented spine.

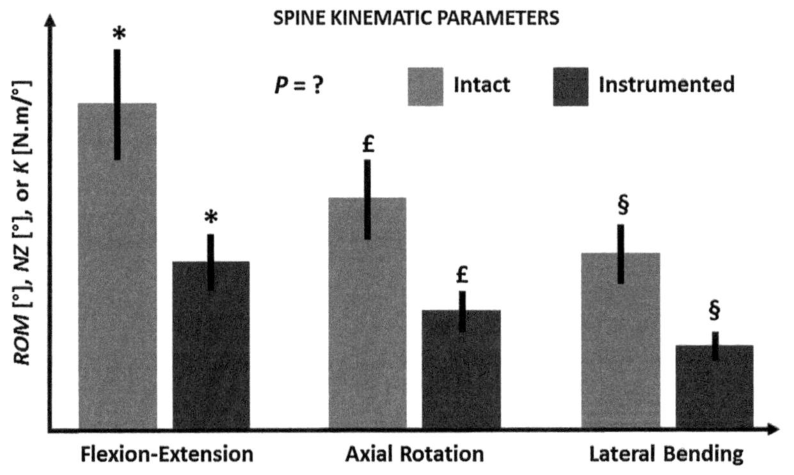

FIGURE 9.7

Typical spine kinematic parameters of instrumented vs. intact control groups in the three loading planes. Each result for range of motion *ROM*, neutral zone *NZ*, or stiffness *K* is an average ± 1 standard deviation. *P* indicates the statistical probability of not having intergroup differences obtained for each pairwise comparison between test groups, which can be indicated by symbols like asterisks (*), pounds (£), etc.

chart is typically the graph of choice for this purpose. If necessary, the sustained torsional load and hysteresis can also be compared statistically, but these are not critical.

ALTERNATIVES AND ADAPTATIONS

✓ **Passive vs. active tracking**. 3-D motion capture methods are the norm and are available in two varieties. For passive tracking, cameras emit and record in the infrared spectrum of light. For active tracking, markers themselves are infrared light emitting diodes seen only by the camera. Both systems are comparable in cost and accuracy. The decision to use one or the other usually is due to equipment availability and/or available space to place markers on the moving bodies.

✓ **Sawbones models**. Artificial foam models of the spine are good for practicing the use of the testing system and learn about analysis methodologies for spine kinematics. These are economical, widely used, and allow for practice with, or prototyping of, spinal instrumentation.

5. DISCUSSION

After completing all spine biomechanics tests, data collection, data analysis, and data presentation, then final results can be considered and interpreted in the broader context of some important clinical, biomechanical, and/or technological considerations, as follows.

Most knowledge on spine biomechanics has been gained through experiments using human cadaveric specimens or validated animal models. Quantitative information on the 3-D movement of each motion segment or FSU under loading at physiologic levels is obtained through motion tracking systems. This facilitates depiction of biomechanical load vs. displacement behavior of the spine for conditions of interest from which parameters such as *ROM*, *NZ*, *K*, etc., can be derived.

Biomechanical parameters such as *ROM*, *NZ*, and *K* of the natural intact motion segment can be used to quantify regional differences between cervical, thoracic, and lumbar spines, or the effects of aging and disc degeneration.[8] When studying the performance of various spinal fusion and stabilization procedures, *ROM* and *K* of a motion segment have been the primary variables to measure spine stability.[9] However, the concept of *NZ* is unique to spine studies, it was introduced to describe behavior near its neutral position, and it is a good indicator of instability or injuries of the motion segment.[10] Typical numerical *ROM* values for the *NZ* in human cadaveric intact lumbar spine segment L4–L5 are $8.9 \pm 1.8°$ in flexion, $5.8 \pm 1.8°$ in extension, $4.4 \pm 0.7°$ in combined axial rotation, and $11.4 \pm 3.2°$ in combined lateral bending.[11] Following the application of spinal implants used to stabilize the segment, the *ROM* may be reduced by 82% in flexion, 55% in extension, 61% in axial rotation, and 70% in lateral bending.[12]

Facet joints bear up to one-third of spinal loads and, thus, are also a motive for investigating spinal biomechanics. Since facet joints are the only synovial joint in the spine that work in a sliding/compressive fashion, contact pressure maps can be investigated using static or dynamic pressure films and sensors.[13–15] Typical facet contact forces measured for human intact cadaveric spine can range from 2–15 N (flexion), 13–95 N (extension), 11–42 N (lateral bending), and 55–165 N (rotation).[13,15,16]

Spines from animal models are an alternative to human specimens, which may be costly and difficult to obtain. Good biomechanical comparability with human spines has been shown from calf, goat, pig, and sheep models. However, distinctive anatomic differences are noted with animal models, such as flatter sagittal alignment, disc curvature, and longer transverse and spinous processes.

Degenerative changes of the spine, such as disc degeneration and facet arthrosis, are associated with aging. Degeneration alters the mechanical behavior of the motion segment of the spine. Experimentally, it is beneficial to grade the degenerative condition of donor spines prior to tests using MRI imaging, the Pfirrmann scale, or some other imaging system.[17] Clinically, surgical treatments for symptomatic disc degeneration include removal of the diseased or collapsed intervertebral disc, implanting an interbody bone graft and/or fusion device to restore the disc space, and immobilizing the motion segment with a fusion procedure to eliminate pain.[18]

6. SUMMARY

- The flexibility test is a commonly performed experiment in spine biomechanics.
- Spinal segment movement is 3-D in nature due to the complex anatomical geometry.
- Spine biomechanics is often characterized in terms of *ROM*, *NZ*, and *K*.
- Spine biomechanics can be applied to study the performance of spinal implants.
- Degenerative conditions of the intervertebral disc and other structures affect the spine.

7. QUIZ QUESTIONS

1. What is the definition of the FSU?
2. What are the definitions of *ROM*, *NZ*, and *K*?
3. What is the definition of coupled motion for a spinal segment?
4. Why does rigid spinal fusion of one or more motion segments affect adjacent segments?
5. Assume that a spine segment can be described as a solid circular cylinder. If length $L = 21$ cm, diameter $D = 7$ cm, and applied load $F = 36$ N to produce flexion, calculate the bending moment M at the base and the maximum tensile (or compressive) stress σ on the spine's outer surface (answer: $M = 7.56\ \text{N}\cdot\text{m}$, $\sigma = 224.5$ kPa).

REFERENCES

1. Panjabi MM, Brand Jr RA, White 3rd AA. Three-dimensional flexibility and stiffness properties of the human thoracic spine. *Journal of Biomechanics* 1976;**9**:185–92.
2. Kettler A, Liakos L, Haegele B, Wilke HJ. Are the spines of calf, pig and sheep suitable models for pre-clinical implant tests? *European Spine Journal* 2007;**16**:2186–92.

3. Patwardhan AG, Havey RM, Meade KP, Lee B, Dunlap B. A follower load increases the load-carrying capacity of the lumbar spine in compression. *Spine* 1999;**24**:1003–9.
4. Pfirrmann CW, Metzdorf A, Zanetti M, Hodler J, Boos N. Magnetic resonance classification of lumbar intervertebral disc degeneration. *Spine* 2001;**26**:1873–8.
5. Crawford NR, Brantley AG, Dickman CA, Koeneman EJ. An apparatus for applying pure nonconstraining moments to spine segments in vitro. *Spine (Phila, PA, 1976)* 1995; **20**(19):2097–100.
6. Lysack JT, Dickey JP, Dumas GA, Yen D. A continuous pure moment loading apparatus for biomechanical testing of multi-segment spine specimens. *Journal of Biomechanics* 2000;**33**(6):765–70.
7. Wang M, Tang S-J, McGrady LM, Rao RD. Biomechanical comparison of supplemental posterior fixations for two-level anterior lumbar interbody fusion. *Proceedings of the Institution of Mechanical Engineers (Part H): Journal of Engineering in Medicine* 2012;**227**(3):245–50.
8. Mimura M, Panjabi MM, Oxland TR, Crisco JJ, Yamamoto I, Vasavada A. Disc degeneration affects the multidirectional flexibility of the lumbar spine. *Spine* 1994;**19**(12): 1371–80.
9. Toth JM, Estes BT, Wang M, Scifert JL, Turner AS, Cornwall GB. Evaluation of 70/30 poly (L-lactide-co-D,L-lactide) for use as a resorbable interbody fusion cage. *Journal of Neurosurgery* 2002;**97**:423–32.
10. Panjabi M, Abumi K, Duranceau J, Oxland T. Spinal stability and intersegmental muscle forces. A biomechanical model. *Spine* 1989;**14**:194–200.
11. Yamamoto I, Panjabi MM, Crisco T, Oxland T. Three-dimensional movements of the whole lumbar spine and lumbosacral joint. *Spine* 1989;**14**(11):1256–60.
12. Voor MJ, Mehta S, Wang M, Zhang Y-M, Mahan J, Johnson J. Biomechanical evaluation of posterior and anterior lumbar interbody fusion techniques. *Journal of Spinal Disorders* 1998;**11**(4):328–34.
13. Takigawa T, Espinoza Orías AA, An HS, Gohgi S, Udayakumar RK, Sugisaki K, et al. Spinal kinematics and facet load transmission after total disc replacement. *Spine (Phila, PA, 1976)* 2010;**35**(22):E1160–6.
14. Wilson DC, Niosi CA, Zhu QA, Oxland TR, Wilson DR. Accuracy and repeatability of a new method for measuring facet loads in the lumbar spine. *Journal of Biomechanics* 2006;**39**:348–53.
15. Niosi CA, Wilson DC, Zhu Q, Keynan O, Wilson DR, Oxland TR. The effect of dynamic posterior stabilization on facet joint contact forces: an in vitro investigation. *Spine* 2008; **33**:19–26.
16. Rundell SA, Auerbach JD, Balderston RA, Kurtz SM. Total disc replacement positioning affects facet contact forces and vertebral body strains. *Spine (Phila, PA, 1976)* 2008; **33**(23):2510–7.
17. McNally DS, Adams MA. Internal intervertebral disc mechanics as revealed by stress profilometry. *Spine (Phila, PA, 1976)* 1992;**17**(1):66–73.
18. Okada E, Matsumoto M, Ichihara D, Chiba K, Toyama Y, Fujiwara H, et al. Aging of the cervical spine in healthy volunteers: a 10-year longitudinal magnetic resonance imaging study. *Spine* 2009;**34**(7):706–12.

CHAPTER

Biomechanical Testing of the Intact and Surgically Treated Pelvis

10

Mina S.R. Aziz[1], Emil H. Schemitsch[2], Radovan Zdero[3]

University of British Columbia, Vancouver, BC, Canada[1]; London Health Sciences Centre, London, ON, Canada[2]; Western University, London, ON, Canada[3]

1. BACKGROUND

Pelvic fractures account for over one-tenth of all human bone fractures, but acetabular fractures account for about half of all pelvic fractures.[1] Acetabular fractures can be simple (or elementary) and complex (or associated), being caused by a low-energy fall in the elderly and high-energy impact in the young (Fig. 10.1).[2,3] No "gold standard" exists for surgical repair of these injuries; however, surgeons most often use plate-and-screw fixation, although cable fixation may be used when osteoporosis prevents good screw fixation into bone. Acetabular fracture repair is often accompanied by simultaneous total hip arthroplasty for elderly patients who eventually require a hip prosthesis due to osteoarthritis.[4,5] The goal of surgical repair is perfect anatomical reduction of the hip to minimize pain, improve strength, and restore

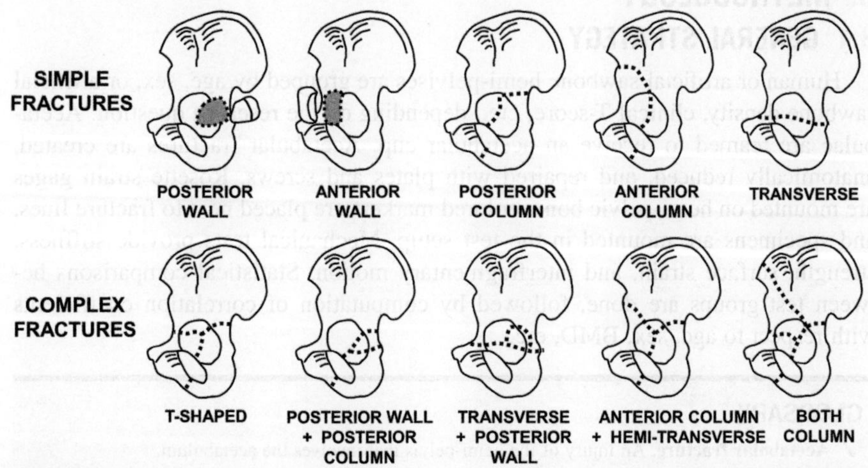

FIGURE 10.1

Acetabular fracture patterns in a hemi-pelvis. Dotted lines indicate fractures, while gray areas show sections of displaced bone.

Experimental Methods in Orthopaedic Biomechanics.

function.[3–5] A key element is the biomechanical stability provided by various acetabular fracture fixation methods. Therefore, this chapter shows how to surgically repair a pelvic acetabular fracture and perform biomechanical testing, as well as how to analyze, report, and interpret data.

2. RESEARCH QUESTIONS

Typical research questions might include one or more of the following:

- What are the biomechanical properties of human vs. artificial sawbone hemi-pelvises?
- Which of the 10 most common acetabular fractures is the most unstable biomechanically?
- What are the stiffnesses and strengths of various acetabular fracture fixation constructs?
- What is the effect of an acetabular cup on the biomechanics of an acetabular fixation construct?
- Do age, sex, or bone mineral density (BMD) influence the best way to repair an acetabular fracture?
- Does the experience level of an orthopaedic surgeon alter acetabular fixation construct stability?
- etc.

3. METHODOLOGY

3.1 GENERAL STRATEGY

Human or artificial sawbone hemi-pelvises are grouped by age, sex, or artificial sawbone density, clinical T-score, etc., depending on the research question. Acetabulae are reamed to receive an acetabular cup. Acetabular fractures are created, anatomically reduced, and repaired with plates and screws. Rosette strain gages are mounted on hemi-pelvic bone, colored markers are placed next to fracture lines, and specimens are mounted in the test setup. Mechanical tests provide stiffness, strength, surface strain, and interfragmentary motion. Statistical comparisons between test groups are done, followed by computation of correlation coefficients with respect to age, sex, BMD, etc.

GLOSSARY

✓ **Acetabular fracture**. An injury of the hemi-pelvis that involves the acetabulum.
✓ **Acetabular fracture fixation**. Surgical repair of a cracked acetabulum with metal implants.
✓ **Acetabulum**. The cartilage-covered cup-like hip socket that is part of the hemi-pelvis.
✓ **Hemi-pelvis**. The left or right half of the pelvis, which has a symmetric geometry.
✓ **Total hip arthroplasty**. Surgical replacement of the hip joint with artificial materials.

SAFETY FIRST

✓ Remember to always wear goggles and gloves for protection.
✓ Clamp the hemi-pelvis before using the drill press, power hand drill, or jig saw.
✓ Place a plastic protective shield around the test setup during experiments.
✓ Clean the work area and all tools with bleach or disinfectant if testing human bone.

3.2 MATERIALS AND TOOLS LIST

- acetabular cup and surgical kit
- C-clamps and heavy duty vices
- drill press and/or power hand drill
- fixation screws, plates, and surgical kit
- hemi-pelvis test jig
- human or artificial sawbone hemi-pelvises
- mechanical tester
- reciprocating jig saw
- rosette strain gages and accessories
- video camera and colored markers

3.3 SPECIMEN PREPARATION

Step 1. Store the hemi-pelvis. Fresh or fresh–frozen human hemi-pelvises should be wrapped in plastic strips or vacuum-sealed in plastic bags for storage in a freezer. The freezer should be −20°C or colder. Human embalmed, human dried/dehydrated, or artificial sawbone hemi-pelvises do not need to be frozen and may be placed in a box for storage at room temperature.

Step 2. Thaw the hemi-pelvis. Fresh or fresh–frozen hemi-pelvises should be removed from the freezer, left in their plastic wrappings or vacuum-sealed bags, and placed on a suitable surface at ambient room temperature or in a warm water bath to thaw for at least 12 h. Fresh, fresh–frozen, or embalmed hemi-pelvises are then removed from their wrappings or bags and soaked or sprayed with saline water to prevent dehydration. Dried/dehydrated or sawbone hemi-pelvises, of course, do not need to be thawed or soaked in saline.

Step 3. Ream the acetabulum. An acetabular cup should be implanted to simulate total hip arthroplasty in the presence of an acetabular fracture, but the acetabulum must first be reamed. To do this, firmly mount each intact hemi-pelvis onto a table using a large C-clamp or vice (Fig. 10.2A). Then, attach a stainless steel acetabular reamer dome to a reamer handle, attach the reamer handle to a power drill, and then ream the acetabulum using progressively larger diameter domes until the acetabular cup can fit snugly into the acetabulum (Fig. 10.2B and C).

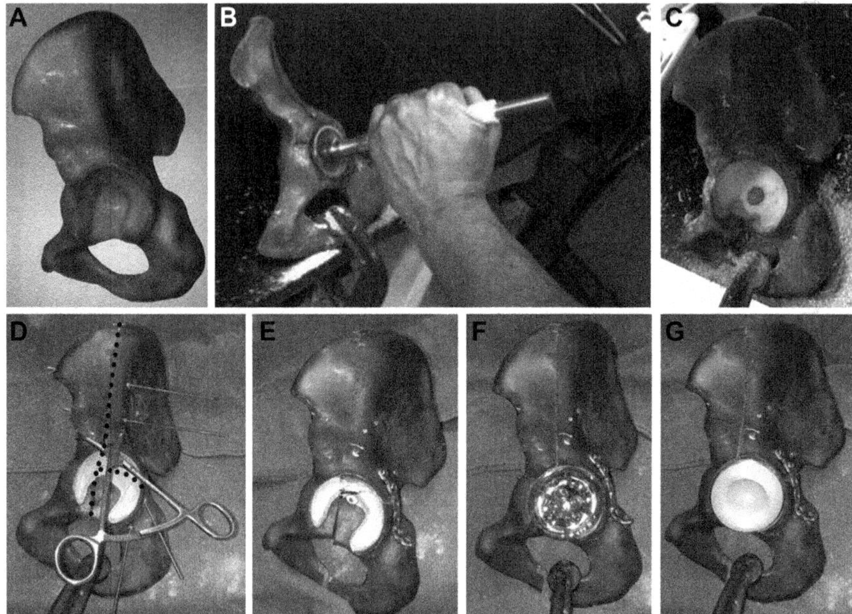

FIGURE 10.2

Hemi-pelvis preparation for acetabular fracture fixation and total hip arthroplasty. (A) Intact hemi-pelvis, (B) reaming the acetabulum, (C) fully reamed acetabulum, (D) reduction with a clamp and K-wires of the acetabular fracture created with a jig saw (dotted line represents an anterior column plus hemi-transverse fracture), (E) fixation of the acetabular fracture with a plate and screws, (F) insertion of the acetabular metal shell, (G) insertion of the acetabular polymer liner on top of the metal shell.

Step 4. Fracture the acetabulum. Draw a line along the planned fracture line with a pen. Then use a power hand drill with a small drill bit (e.g., 1/16 or 1/32 inch diameter) to drill multiple holes at equal distances (e.g., 3 or 5 mm) along the planned fracture line to weaken the bone. Finally, standardize the fracture line from specimen to specimen by using a reciprocating jig saw to cut along the planned fracture line. Alternatively, for a more clinically realistic irregular fracture line from specimen to specimen, apply a bending force by hand to the hemi-pelvis across the drilled holes, which will crack the acetabulum along the planned fracture line. (Note: In the real world, the following chronological steps occur: (1) injury causes acetabular fracture, (2) surgical alignment of fracture fragments, (3) surgical insertion of plates and screws to join the fracture fragments, (4) surgical reaming of the acetabulum, and (5) surgical implantation of the acetabular cup. However, to create geometrically standardized specimens for mechanical tests in the laboratory, acetabular reaming should be done first in an intact hemi-pelvis to avoid practical difficulties of aligning and clamping fracture fragments for proper reaming.)

Step 5. Reduce the fracture. Align the fracture fragments to reproduce the original anatomy of the intact uninjured hemi-pelvis. Then insert metal K-wires into the fracture fragments and apply surgical reduction clamps across the fracture line in order to secure this alignment (Fig. 10.2D). This process is called "anatomical reduction" or "surgical reduction."

Step 6. Perform fracture fixation. Apply a plate–screw fixation technique to provide accurate reduction and restoration of hemi-pelvis anatomy (Fig. 10.2E). To do so, choose the correct plate size, use a plate bender to bend the plate to perfectly conform to the bone anatomy where it will be applied, create pilot holes through the plate screw holes, and fix the plate with the appropriate number of stainless steel cortical screws above and below the fracture line. If needed, lag screw(s) can be applied across the fracture line after pilot hole(s) are drilled to allow easier insertion of the screw(s) without cracking the hemi-pelvis.

Step 7. Insert the acetabular cup. Insert the acetabular cup (i.e., metal shell and polymer liner) of a total hip arthroplasty in the correct position using an impactor and mallet provided in a surgical kit (Fig. 10.2F and G). Use two to five cancellous screws to stabilize the metal shell into position and the liner impactor to "snap" into place the polymer liner inside the metal shell. The final acetabular cup orientation angle is usually 45° superoinferiorly and 45° mediolaterally.

Step 8. Mount rosette strain gages. Keep several things in mind when mounting rosettes on the highly complex geometry of the hemi-pelvis (Fig. 10.3). First, rosette gages are more suitable for curved and complex surfaces since they provide 2-D data, whereas linear gages may only be useful for relatively flat surfaces since they provide only 1-D data. Second, mount rosettes on pelvic bone, rather than on fixation implants, since the bone has more surface area to accommodate rosettes.

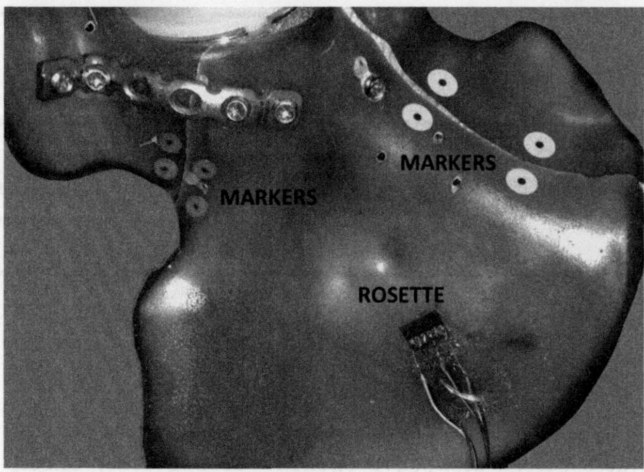

FIGURE 10.3

Rosette gage and colored marker positions on a surgically repaired hemi-pelvis.

Third, mount rosettes away from sharp bone ridges, fracture lines, or fixation plates and screws, since rosettes do not respond consistently to sudden changes in geometry or strain. Fourth, mount rosettes where high and low pelvic bone strains are expected in order to cover a wide range of potential strains. Finally, mount rosettes on several fracture fragments or on both sides of a fracture line to prevent any statistical biasing of the data.

Step 9. Mount colored markers. The relative parallel motion of fracture fragments along a fracture line (i.e., "sliding") and the relative perpendicular motion between fracture fragments (i.e., "gapping") can be monitored using a digital video camera and colored markers (Fig. 10.3).[6,7] To do this, visual reference points need to be created on each specimen along the fracture line of interest. Go to a stationery store to obtain small, circular, colored, adhesive markers used for color coding of paper documents or file folders. Then, choose markers whose color is easily contrasted against the color of the hemi-pelvis, mount matching pairs of markers at known distances on both sides of the fracture line of interest, draw a small black dot in the center of each marker with a black permanent ink pen so measurement of marker movement will be more precise, and place a strip of transparent adhesive tape over each marker to secure it in place.

TIPS AND TRICKS

- ✓ Several researchers will make the preparation and testing more efficient and safe.
- ✓ The same orthopaedic surgeon should perform all surgical procedures for consistency.
- ✓ Long lag screw insertion in tight areas may be easier using X-ray or fluoroscopy.
- ✓ Mount rosette strain gages after all surgeries to prevent damaging fragile strain gages.
- ✓ Rosette strain gages, rather than linear gages, are more suitable for complex geometries.
- ✓ Double-check femoral ball–acetabular cup alignment using pressure sensing Fujifilm.

THE "GOLD STANDARD"

No known international standards exist for surgical repair or biomechanical testing of acetabular fractures. Researchers should use surgery textbooks and peer-reviewed journal articles for precedents and guidelines.

3.4 SPECIMEN TESTING

Step 1. Determine testing order. Hemi-pelvises should be tested for nondestructive stiffness before destructive strength. Testing order effects should be avoided, but this requires changing the specimen and test jig between tests.

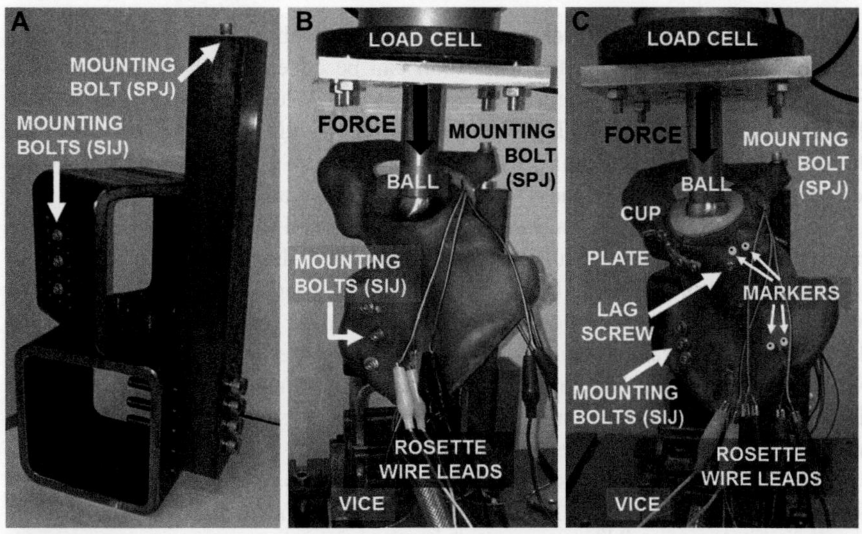

FIGURE 10.4

Biomechanical testing of hemi-pelvises. (A) Test jig composed of steel hollow cubes, a steel block, and steel bolts, (B) intact sawbone hemi-pelvis mounted in the test jig, (C) sawbone hemi-pelvis with an acetabular fracture (type: anterior column plus hemi-transverse) was repaired with a posterior column plate secured by four cortical screws plus one anterior column lag screw, implanted with an acetabular cup secured by four cancellous screws, and then mounted in the test jig. *SIJ*, sacroiliac joint simulated using three mounting bolts; *SPJ*, symphysis pubis joint simulated using one mounting bolt.

Alternatively, individual specimens can be randomly chosen (e.g., pull numbers from a box) for all stiffness tests and then for all strength tests.

Step 2. Mount the hemi-pelvis and test jig. A unique test jig needs to be designed suitable for the hemi-pelvis geometry, but which also simulates the sacroiliac and symphysis pubis joints (Fig 10.4A). To the load cell, attach a 75–100 mm long metal rod with a smooth metal ball at its end, which will then resemble the femoral head of a total hip replacement through which vertical load will be applied to the acetabulum. Next, secure an angle vice to the base of the mechanical tester, mount the bottom steel cube of the test jig into the angle vice, and then adjust the angle vice to orient the acetabulum at approximately 45° in the superomedial and 15° in the posterior direction relative to the sagittal plane. This orientation simulates the hip joint force vector during standing and other daily activities.[6–8] Mount the hemi-pelvis onto the test jig using four metal bolts: one 1/4 inch metal bolt inserted into an 11/64 inch pilot hole in the symphysis pubis region, and three #10 metal bolts inserted into 9/64 inch pilot holes in the sacroiliac region.

Step 3. Align the hemi-pelvis and test jig. To double-check metal ball alignment prior to actual testing, the metal ball can be covered with pressure sensing Fujifilm, inserted into the acetabular cup using a low load (e.g., 50–100 N), and then removed from the cup. The Fujifilm can be visually inspected to ensure that the discolored region indicating ball–cup contact is in the center of the Fujifilm for proper ball–cup congruency. Make final alignment adjustments.

Step 4. Connect strain gages and position colored markers. Connect rosette strain gages to the digital acquisition system, and then double-check that electrical signals are read properly by the computer. Position a video camera in front of the fracture line which has the colored markers, place a ruler (or 2-D object with known dimensions) in the camera's field of view, and record the image of the ruler, which will later be used as a 2-D reference for measurements of marker motion. Strain gage data acquisition and video camera recording need to be synchronized with the start of the stiffness and strength tests, as described below.

Step 5. Perform stiffness tests. Lower the metal ball into the acetabulum or acetabular cup to a preload of 50–100 N to overcome "mechanical slack" or hysteresis (Fig 10.4B and C).[9,10] Next, apply 1–25 load cycles at 0.5–2 Hz to precondition the pelvis so stiffness reaches "steady state," and then use the last one or two load cycles to compute stiffness.[11,12] For these load cycles, the specific research question(s) and the hemi-pelvis's physical robustness may influence the choice of displacement or load control, displacement or load rate (e.g., 8–300 mm/min and 37.5–150 N/s are common for hip joint tests), and max displacement or load (e.g., 2 mm and 200–2000 N are common for hip joint tests).[6–18] [Note: Low-speed quasi-static tests may not replicate acetabular impact injuries, but they are an acceptable standard approach for comparative studies (i.e., group 1 vs. 2 vs. 3, etc.), since all test groups undergo the same low-speed load regime. Also, most pelvic biomechanical studies do not report cyclic loading and/or cyclic preconditioning.][6–10,14,17,18]

Step 6. Perform strength tests. Apply a preload as described above. Then, increase the vertical displacement or load at the same rate as for the stiffness tests (unless a high-speed impact force is desired), until the intact or repaired hemi-pelvis completely fails structurally.

Step 7. Inspect failed constructs. Visually inspect the failed acetabular constructs, and take a sufficient number of good photos for later analysis of the failure mechanisms.

3.5 RAW DATA COLLECTION

Step 1. Record acetabular characteristics. Enter demographic information for the intact human hemi-pelvises (i.e., age, sex, left or right limb) and quantitative bone properties (i.e., BMD, clinical T-score), which will help determine which factors are statistically correlated with stiffness and strength (Table 10.1). If using sawbone hemi-pelvises, then only data for artificial cortical bone and/or artificial cancellous bone densities will be available.

Table 10.1 Raw characteristics for intact human or sawbone hemi-pelvises.

	Human Hemi-Pelvis					Sawbone Hemi-Pelvis	
Specimen	Age [years]	Sex [M, F]	Limb [L, R]	BMD [g/cm^2]	T-Score	Cortical Density [g/cm^3]	Cancellous Density [g/cm^3]
1							
2							
3							
etc.							
Avg		–	–				
SD		–	–				

Avg, *average;* BMD, *bone mineral density;* SD, *standard deviation.*

Step 2. Record raw stiffness. Stiffness is the slope of the average line of best fit of the load vs. displacement graph obtained from a stiffness test (Table 10.2). Any hysteresis (i.e., a gap between load and unload curve) may be due to the nonlinear material behavior of repaired acetabular constructs caused by too much load being applied, mechanical slack in the test jig or fracture fixation construct, slippage of the acetabular cup under the metal loading ball, etc. This is acceptable if the line of best fit has a linearity of $R^2 \geq 0.9$, indicating no permanent damage has been done to the specimen. However, if $R^2 < 0.9$, then these other factors should be checked.

Step 3. Record raw strength. Strength is the maximum load of the load vs. displacement graph obtained from a strength test and is considered "mechanical failure" (MF) (Table 10.2). Alternatively, some clinical criterion for allowable vertical load (e.g., 500 N), vertical displacement (e.g., 5 mm), or interfragmentary sliding or gapping (e.g., 2 mm) can be used to define "clinical failure" (CF). Also, if separate stiffness tests are not performed, stiffness can still be obtained from a strength graph (i.e., slope of the initial linear region of the load vs. displacement graph up to the linear yield point). However, this stiffness depends on how the linear region is defined in a strength graph since linear yield point is sometimes difficult to determine. One approach is to use the engineer's 0.2% offset displacement method. Another approach is to choose a load or displacement range based on a clinical criterion (e.g., 0–500 N or 0–5 mm).

Step 4. Record raw rosette data. Von Mises strain (i.e., overall equivalent strain) of each rosette strain gage is a dimensionless value (Table 10.2). It can be computed using established formulas for different possible geometric arrangements of the three individual linear gages that comprise each rosette. Of course, for each repaired acetabular construct, there may be one or more rosettes that have been used which need to be individually recorded in the data summary table.

Table 10.2 Raw biomechanical measurements on the hemi-pelvis specimen.

Repaired Acetabular Construct	Stiffness [N/mm]	Strength at CF [N]	Strength at MF [N]	Rosette Strain [μm/m]	Sliding [mm]	Gapping [mm]	Failure Pattern
1							
2							
3							
etc.							
Avg							–
SD							–

Avg, *average;* CF, *clinical failure;* MF, *mechanical failure;* SD, *standard deviation.*

Step 5. Record raw sliding and gapping. Examine the recorded video images of the colored markers placed along the fracture line of interest. Sliding and gapping can be computed by measuring appropriate distances between colored markers from images that represent the start (image 1) and end (image 2) of a desired event (Table 10.2). For example, the event may be represented by zero load (image 1) to CF (image 2), zero load (image 1) to MF (image 2), or zero load (image 1) to maximum load that covers the linear elastic "stiffness" region during nondestructive loading (image 2). For better precision, sliding and gapping distances should be initially measured in pixels and then converted to mm units, based on the mm/pixel conversion factor obtained from known marker-to-marker distances at initial conditions (image 1). Of course, for each pelvic construct, there may be one or more sliding or gapping results, which need to be individually recorded in the data summary table.

Step 6. Record failure patterns. Enter general remarks about the nature of the failure pattern, such as breakage of plates or screws, loosening of the acetabular cup, the shape of failure line(s) through pelvic bone, the creation of loose pelvis bone fragments, etc. (Table 10.2).

3.6 RAW DATA ANALYSIS

Step 1. Calculate correlation coefficients. Plot the measured result for each individual specimen (i.e., stiffness 1, 2, 3, etc.) (Table 10.2) vs. its corresponding pelvic repair construct characteristic (i.e., age 1, 2, 3, etc.) (Table 10.1) to help visualize the interrelationship between measurements vs. characteristics. Then, calculate the correlation coefficient R for each measurement−characteristic pair to determine which characteristic has an influence on measurements. $R > 0.8$ is a typical value considered to indicate a strong correlation. Do this for each measured parameter (i.e., stiffness, strength, strain, sliding, and gapping).

Step 2. Perform statistical comparisons. The criterion for statistical difference first needs to be chosen (e.g., $P < 0.01$ or < 0.05). Then, data can be used to compare different pelvic groups (e.g., men ≤ 60 years old vs. men > 60 years old, "normal" women vs. "osteoporotic" women, etc.), fracture fixation types (e.g., plate fixation 1 vs. 2 vs. 3, etc.), etc. Various software programs can compare two test groups (e.g., paired t-test) or two or more test groups influenced by multiple factors (e.g., analysis of variance, ANOVA).

Step 3. Compute statistical power. Power analysis can be done after the study to ensure there were enough specimens per group to detect all statistical differences that were actually present (i.e., was type II statistical error avoided?). Statistical power $> 80\%$ is usually considered to indicate there were enough specimens per test group. Note that if good predictions of averages and standard deviations are available from prior studies, then the number of specimens and/or tests can be chosen before the study begins to ensure a power $> 80\%$.

ENGINEER'S TOOLBOX

A rosette strain gage can help evaluate how close to MF a specimen is at that location. To avoid failure, Von Mises stress σ_{VM} should be less than the specimen's ultimate tensile stress σ_U, which is often a known material property. Thus, $\sigma_{VM} = \varepsilon_{VM}E$, where σ_{VM} is Von Mises stress, ε_{VM} is Von Mises strain $= \left(\varepsilon_{MAX}{}^2 + \varepsilon_{MIN}{}^2 - \varepsilon_{MAX}\varepsilon_{MIN}\right)^{1/2}$, $\varepsilon_{MAX,\ MIN}$ is maximum and minimum principal strain, and E is modulus of elasticity for the material. For a 45° rectangular rosette $\varepsilon_{MAX,\ MIN} = (\mathcal{E}_1 + \mathcal{E}_3)/2 \pm (1/\sqrt{2})[(\mathcal{E}_1 - \mathcal{E}_2)^2 + (\mathcal{E}_2 - \mathcal{E}_3)^2]^{1/2}$. For a 60° delta rosette $\varepsilon_{MAX,\ MIN} = (\mathcal{E}_1 + \mathcal{E}_2 + \mathcal{E}_3)/3 \pm (\sqrt{2}/3)[(\mathcal{E}_1 - \mathcal{E}_2)^2 + (\mathcal{E}_2 - \mathcal{E}_3)^2 + (\mathcal{E}_3 - \mathcal{E}_1)^2]^{1/2}$. Note that $\mathcal{E}_1$, $\mathcal{E}_2$, $\mathcal{E}_3$ are raw linear strain readings from the three individual linear gages that comprise each rosette.

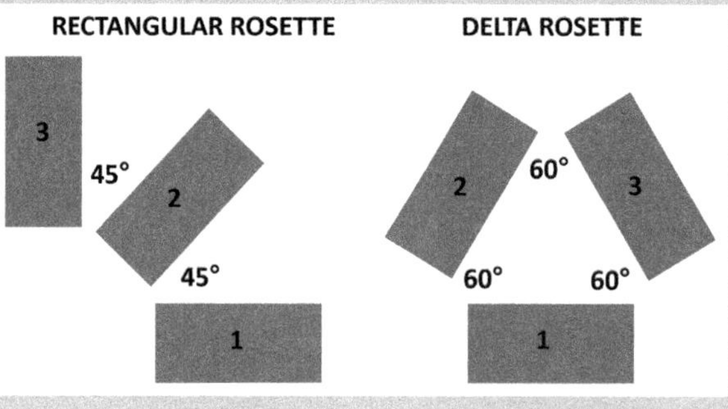

4. RESULTS

Once all pelvic test data collection and analysis have been performed, it is then important to communicate and present the primary results in an understandable and concise manner to the reader of a journal article, conference paper, technical report, or book chapter.

Step 1. Show raw loading profile. Begin with a typical profile of raw load vs. displacement (Fig. 10.5). Initially, load is at the low preload level, or perhaps at zero if no preload was applied. Then, load increases linearly within the elastic region up to the yield point, the slope of which is defined as stiffness. Soon after, pelvic bone and fixation implants begin to fail unpredictably because of the specimen's complex 3-D geometry. Finally, a peak load is reached where complete structural collapse occurs, which is defined as MF. CF can also be indicated, which is based on a clinical criterion the researcher chooses for allowable load (e.g., 500 N), displacement (e.g., 5 mm), or interfragmentary sliding or gapping (e.g., 2 mm), which may occur before or after the yield point.

Step 2. Show main results. Stiffness, strength at CF or MF, strain, sliding, and gapping should be presented (Fig. 10.6). All statistical pairwise comparisons should be made (i.e., test group 1 vs. 2 vs. 3, etc.) to generate statistical P values (Fig. 10.6A). Linear (or nonlinear) trends can be illustrated vs. age, BMD, or T-score for human hemi-pelvises (Fig. 10.6B). Alternatively, trends can be plotted vs. artificial cortical or cancellous bone density for sawbone hemi-pelvises (Fig. 10.6C).

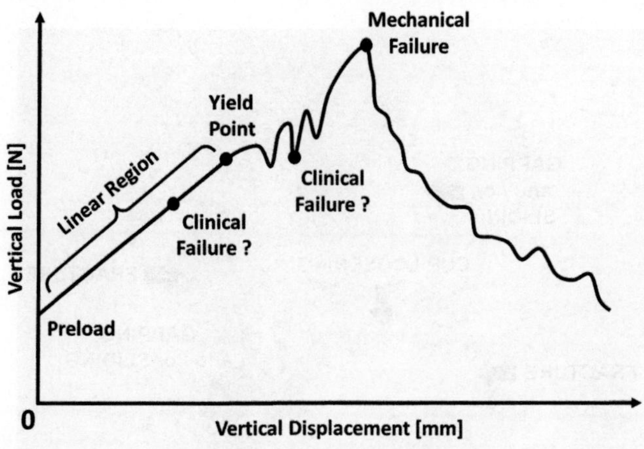

FIGURE 10.5

Repaired acetabular construct load vs. displacement profile.

Step 3. Show failure modes. A typical photo from each test group of the repaired acetabular constructs after strength tests helps visualize typical failure patterns at MF or CF and identifies weak points in the specimens (Fig. 10.7).

Step 4. Show failure distribution. The different failure patterns can also be classified to identify any trends. The test groups can be statistically compared to see if they yielded the same distribution of failure types.

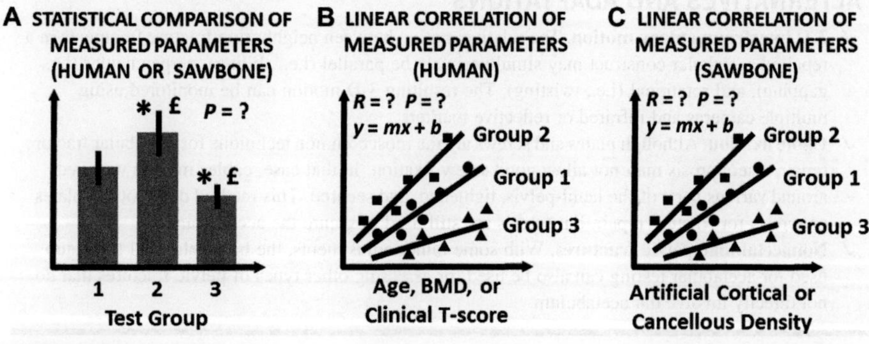

FIGURE 10.6

Repaired acetabular construct results. (A) Statistical comparison of measured parameters for human or sawbone hemi-pelvises, where bar graphs show average $\pm$ 1 standard deviation and P is the statistical difference value for each group-to-group comparison, which can be indicated by symbols like asterisks (*), pounds (£), etc., (B) linear correlation of measured parameters for human hemi-pelvises, (C) linear correlation of measured parameters for sawbone hemi-pelvises, where R is the linear correlation coefficient for each line of best fit, P is the statistical difference value to ensure that the slope of each line of best fit (i.e., slope $= m$) is statistically different than a horizontal line (i.e., slope $= 0$), and $y = mx + b$ is the equation of each line. Measured parameters to be statistically compared and correlated are stiffness, strength at CF and/or MF, rosette strain, sliding, and gapping. *BMD*, bone mineral density.

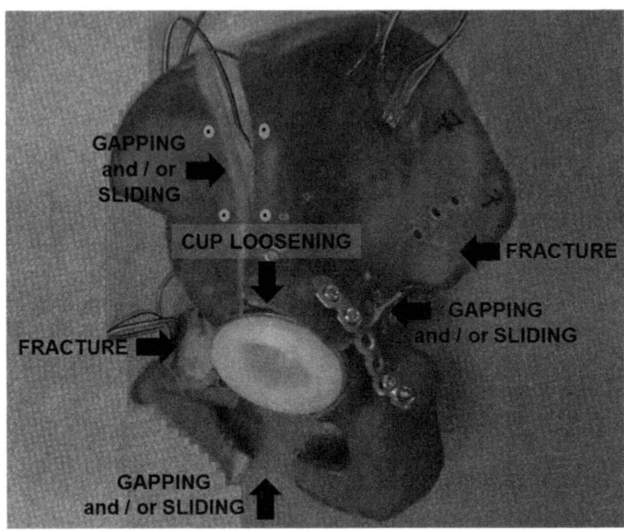

FIGURE 10.7

Typical failure modes for a surgically repaired hemi-pelvis with an acetabular cup that is mechanically loaded to failure.

ALTERNATIVES AND ADAPTATIONS

✓ **3-D interfragmentary motion.** The relative motion between neighboring fracture fragments in a repaired acetabular construct may simultaneously be parallel (i.e., sliding), perpendicular (i.e., gapping), and rotational (i.e., twisting). The resulting 3-D motion can be monitored using multiple cameras and infrared or reflective markers.

✓ **Cable fixation.** Although plates and screws are the most common technique for acetabular fracture repair, osteoporosis may not allow good screw fixation. In that case, cables may be wrapped around various parts of the hemi-pelvis, tightened, and secured. This method does not use plates or screws for fracture repair, but screws are still used to secure the acetabular cup.

✓ **Nonacetabular pelvic fractures.** With some minor adjustments, the biomechanical test setup used for acetabular testing can also be used for assessing other types of pelvic fractures that do not directly involve the acetabulum.

5. DISCUSSION

After completing all pelvic tests, data collection, data analysis, and data presentation, then final results can be considered and interpreted in the broader context of some important clinical, biomechanical, and/or technological considerations, as follows.

Acetabular fracture mechanisms are influenced by patient age. For those under 60 years of age, the most common acetabular injury mechanisms are motor vehicle accidents (66%), followed by falls (17.7%), and auto vs. pedestrian impact (9.1%).[19] In contrast, for those over 60, due to increased occurrence of osteoporosis in the elderly, the most frequent mechanisms are falls (49.8%), followed by motor vehicle accidents (37.4%), and auto vs. pedestrian impact (6.4%).[19]

Acetabular fracture subtypes vary greatly in frequency depending on age. For those under 60 years of age, the most frequent injury types are both column (27.9%), transverse plus posterior wall (18.2%), posterior wall (14.7%), and T-shaped (12.9%).[19] For those over 60 years of age, the most common injuries are both column (26.4%), anterior column (19.2%), anterior column plus hemi-transverse (14.9%), and posterior wall (13.2%).[19]

Acetabular fracture fixation methods are often determined by an orthopaedic surgeon's personal preference and clinical outcome studies, rather than biomechanical stability. Plate application using screws is the most common, although a "gold standard" for each of the 10 subtypes of acetabular fractures has not been established.[2–8,10,12–15] For instance, the number, type, and orientation of the plate(s), as well as the number, type, and orientation of the screw(s), is still a matter of disagreement. However, if osteoporotic bone does not allow for adequate screw purchase, then cables alone may be an alternate approach.[2–8,10,12–15] Moreover, some elderly patients receive an acetabular cup during the same operation as fracture plating or cabling since they may eventually need a total hip arthroplasty due to osteoarthritis.[4,5,10]

Experimentally, hemi-pelvis orientation and load vector direction will vary, depending on the research question. Hemi-pelvises can be oriented 45° superomedially and 15–25° posteriorly in the sagittal section to simulate the position of the femoral neck while a person is standing.[6–8] Loads can be applied perpendicular to the acetabulum to simplify experiments[13] or in other directions to duplicate more physiological activities like walking (6.5° flexion with 7.2° abduction) and descending stairs (9.2° extension with 9.5° abduction).[12]

Experimentally, parameters measured for repaired acetabular constructs differ widely. Average values can range from 386—3130 N/mm (stiffness),[6,7,10,12,13,15] 848–3210 N (strength at CF),[10,13] 1012–4753 N (strength at MF),[10,13] 0.04–13.58 mm (sliding),[7,10,14] and 0.07–4.72 mm (gapping).[10,13,14] Such variation is due to different types of pelvic bone (human vs. sawbone), pelvic bone quality (normal vs. osteoporotic), fracture type (simple vs. complex), fracture fixation method (plating vs. cabling), acetabular condition (intact acetabulum vs. acetabular cup), load type (quasi-static vs. cyclic), and load vector orientation (more vs. less posterior).

6. SUMMARY

- The human acetabulum and hemi-pelvis have very complicated geometries.
- Surgical repair of acetabular fractures is often guided by the surgeon's preference.

- Elderly patients with poor bone quality often also undergo a total hip arthroplasty.
- Biomechanical tests can objectively evaluate the various fracture fixation techniques.
- Biomechanical tests provide stiffness, strength, surface strain, and interfragmentary motion.
- Age, sex, and bone density influence the biomechanics of surgical repair techniques.

7. QUIZ QUESTIONS

1. How many types of acetabular fractures are there, and how are they categorized?
2. What are some common fixation techniques used to surgically repair acetabular fractures?
3. List the surgical steps for implanting an acetabular cup in an elderly fractured acetabulum.
4. Would surgical acetabular repair vary based on age, sex, limb side, or bone density?
5. Does a repaired acetabular construct fail under load at a particular location where a 45° rectangular rosette gage has been placed? The rosette's linear readings are ε_1, ε_2, $\varepsilon_3 = 30$, 300, 200 μm/m. Pelvic cortical bone's ultimate tensile stress $\sigma_U = 106$ MPa and modulus of elasticity $E = 15$ GPa (answer: no failure, since Von Mises stress $\sigma_{VM} = 5.7$ MPa $< \sigma_U$).

REFERENCES

1. Court-Brown CM, Aitken SA, Forward D, O'Toole RV. The epidemiology of fractures. In: Bucholz RW, Heckman JD, Court-Brown CM, Tornetta P, editors. *Rockwood and Green's fractures in adults*, vol. 1. Lippincott; 2010. pp. 53–83.
2. Letournel E, Judet R. *Fractures of the acetabulum*. 2nd ed. New York (NY, USA): Springer-Verlag; 1993.
3. Tile M, Helfet DL, editors. *Fractures of the pelvis and acetabulum*. 4th ed. (vol. 2: Acetabulum). New York (NY, USA): Thieme/AO; 2015.
4. Boraiah S, Ragsdale M, Achor T, Zelicof S, Asprinio DE. Open reduction internal fixation and primary total hip arthroplasty of selected acetabular fractures. *Journal of Orthopaedic Trauma* 2009;**23**(4):243–8.
5. Herscovici D, Lindvall E, Bolhofner B, Scaduto JM. The combined hip procedure: open reduction internal fixation combined with total hip arthroplasty for the management of acetabular fractures in the elderly. *Journal of Orthopaedic Trauma* 2010;**24**(5):291–6.
6. Shazar N, Brumback RJ, Novak VP, Belkoff SM. Biomechanical evaluation of transverse acetabular fracture fixation. *Clinical Orthopaedics and Related Research* 1998;**352**: 215–22.

7. Khajavi K, Lee AT, Lindsey DP, Leucht P, Bellino MJ, Giori NJ. Single column locking plate fixation is inadequate in two column acetabular fractures: a biomechanical analysis. *Journal of Orthopaedic Surgery and Research* 2010;**5**(30):1–6.
8. Chang JK, Gill SS, Zura RD, Krause WR, Wang GJ. Comparative strength of three methods of fixation of transverse acetabular fractures. *Clinical Orthopaedics and Related Research* 2001;**392**:433–41.
9. Papini M, Zdero R, Schemitsch EH, Zalzal P. The biomechanics of human femurs in axial and torsional loading: comparison of finite element analysis, human cadaveric femurs, and synthetic femurs. *Journal of Biomechanical Engineering* 2007;**129**(1):12–9.
10. Aziz MSR. *Biomechanics of acute total hip arthroplasty after acetabular fracture: plate vs. cable fixation* [M.Sc. thesis]. Toronto (Canada): Institute of Medical Science, University of Toronto; 2014.
11. Zdero R, Gallimore CH, McConnell AJ, Patel H, Nisenbaum R, Morshed G, et al. A preliminary biomechanical study of cyclic preconditioning effects on canine cadaveric whole femurs. *Journal of Biomechanical Engineering* 2012;**134**(9):094502-1-7.
12. Gilliland JM, Anderson LA, Henninger HB, Kubiak EN, Peters CL. Biomechanical analysis of acetabular revision constructs: is pelvic discontinuity best treated with bicolumnar or traditional unicolumnar fixation? *Journal of Arthroplasty* 2013;**28**(1): 178–86.
13. Mehin R, Jones B, Zhu Q, Broekhuyse H. A biomechanical study of conventional acetabular internal fracture fixation versus locking plate fixation. *Canadian Journal of Surgery* 2009;**52**(3):221–8.
14. Sawaguchi T, Brown TD, Rubash HE, Mears DC. Stability of acetabular fractures after internal fixation: a cadaveric study. *Acta Orthopaedica Scandinavica* 1984;**55**(6):601–5.
15. Wu YD, Cai XH, Liu XM, Zhang HX. Biomechanical analysis of the acetabular buttress-plate: are complex acetabular fractures in the quadrilateral area stable after treatment with anterior construct plate-1/3 tube buttress plate fixation? *Clinics* 2013;**68**(7): 1028–33.
16. Fulkerson E, Koval K, Preston CF, Iesaka K, Kummer FJ, Egol KA. Fixation of periprosthetic femoral shaft fractures associated with cemented femoral stems: a biomechanical comparison of locked plating and conventional cable plates. *Journal of Orthopaedic Trauma* 2006;**20**(2):89–93.
17. McConnell A, Zdero R, Syed K, Peskun C, Schemitsch EH. The biomechanics of ipsilateral intertrochanteric and femoral shaft fractures: a comparison of 5 fracture fixation techniques. *Journal of Orthopaedic Trauma* 2008;**22**(8):517–24.
18. Zdero R, Walker R, Waddell JP, Schemitsch EH. Biomechanical evaluation of periprosthetic femoral fracture fixation. *Journal of Bone and Joint Surgery* (American) 2008; **90**(5):1068–77.
19. Ferguson TA, Patel R, Bhandari M, Matta JM. Fractures of the acetabulum in patients aged 60 years and older: an epidemiological and radiological study. *Journal of Bone and Joint Surgery* (British) 2010;**92**(2):250–7.

CHAPTER

Tension and Compression Testing of Cortical Bone

11

Tarik Attia[1], Thomas L. Willett[2]
University of Toronto, Toronto, ON, Canada[1]; University of Waterloo, Waterloo, ON, Canada[2]

1. BACKGROUND

Cortical bone is a biological material, which acts as the external shell for all bones in the human body (Fig. 11.1). As a natural composite material with a hierarchical structure, cortical bone mainly consists of type I collagen with an embedded mineral phase (i.e., hydroxyapatite) and water.[1] It is anisotropic and heterogeneous since its composition, microstructure, and mechanical properties vary with direction and location. Cortical bone quality at a tissue level is critical for overall bone and joint health, but also for the mechanical stability of orthopaedic implants. Consequently, cortical bone's mechanical properties have been experimentally assessed using clinically relevant loads in tension,[2–5] compression,[2–4] shear,[6–9] torsion,[2,5,10] and cycling.[3,6,11] Therefore, this chapter explains how to effectively perform mechanical testing on geometrically standardized coupons of cortical bone in tension and compression, as well as how to analyze, present, and interpret results.

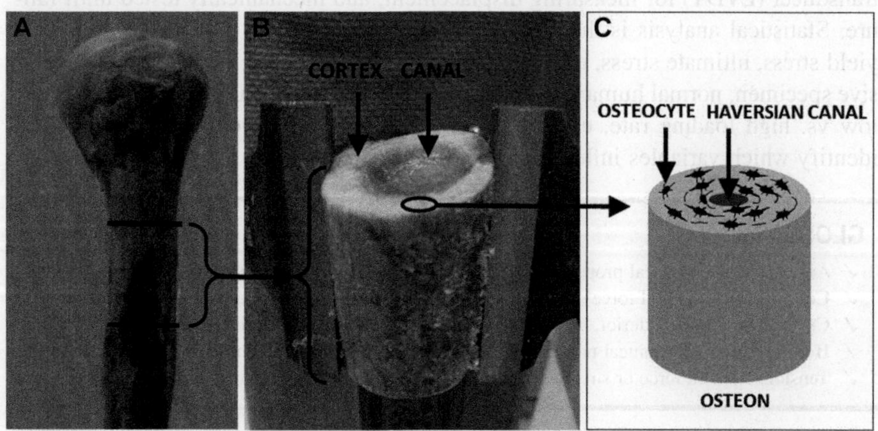

FIGURE 11.1

Cortical bone. (A) Typical human long bone, (B) long bone diaphysis showing the cortex surrounding the marrow-filled intramedullary canal, (C) tube-like osteons are cortical bone's basic micro-structural units. Osteocytes are bone cells that have become trapped within the hard bone tissue.

Experimental Methods in Orthopaedic Biomechanics.

2. RESEARCH QUESTIONS

Typical research questions could include one or more of the following:

- How do the mechanical properties of cortical bone differ under tension vs. compression?
- Are test results affected by test parameters, such as loading rate, specimen size, etc.?
- Do sterilization processes and chemical pretreatments alter the properties of cortical bone?
- How do biological features like age, sex, and disease influence cortical bone?
- etc.

3. METHODOLOGY

3.1 GENERAL STRATEGY

Cortical bone is an anisotropic material whose mechanical properties depend on loading direction; thus, bone is tested in the longitudinal direction to obtain tensile and compressive properties. The midshaft diaphysis of human or animal long bone is cut into longitudinal halves (for tension tests) or transverse slices (for compression tests). Then, bone is machined into coupons of standardized "dog bone" geometry (for tension tests) or cylindrical geometry (for compression tests), mounted into a mechanical tester, equipped with an extensometer or linear velocity displacement transducer (LVDT) for measuring displacement, and mechanically tested until failure. Statistical analysis is then done to compare properties (e.g., elastic modulus, yield stress, ultimate stress, etc.) between the test groups (e.g., tensile vs. compressive specimen, normal human vs. osteoporotic human bone, human vs. bovine bone, low vs. high loading rate, etc.). Finally, correlation coefficients are computed to identify which variables influence cortical bone properties.

GLOSSARY

- ✓ **Anisotropy.** Mechanical properties differ when measured along different directions.
- ✓ **Compression.** Applied force or stress causes the cortical bone specimen to be shortened.
- ✓ **Cortical bone.** The exterior, hard, and dense bone shell found in all biological bone.
- ✓ **Heterogeneity.** Mechanical properties are unevenly distributed through the cortical bone.
- ✓ **Tension.** Applied force or stress causes the cortical bone specimen to be lengthened.

SAFETY FIRST

- ✓ Secure the cortical bone specimen before cutting with power or machine tools.
- ✓ Wear goggles, lab coat, gloves, and N95 mask when machining cortical bone.
- ✓ Place a protective shield around the test area if high-velocity flying debris is generated.
- ✓ Clean the work area and tools with bleach or disinfectant after cortical bone testing.

3.2 MATERIALS AND TOOLS LIST

- band saw
- computer numerically controlled (CNC) mill
- LVDT
- diamond wire saw
- drill press and drill bit for coring
- extensometer
- human or animal long bone
- mechanical tester and fixtures
- micrometer
- sandpaper

3.3 SPECIMEN PREPARATION

Whole Bone Preparation

Step 1. Store whole bones. Fresh or fresh–frozen human or animal whole bone (e.g., femur, tibia, etc.) initially needs to be wrapped in plastic strips or vacuum-sealed in plastic bags for proper storage in a freezer prior to tests. The freezer should be held at −20°C or colder. (Note: Embalmed bone does not need to be frozen and may be stored in a plastic box at room temperature.)

Step 2. Thaw whole bones. Whole bones should be removed from the freezer, left in plastic wrappings or vacuum-sealed bags, and placed on a surface at room temperature or in a warm water bath to thaw for at least 12 h. Then, whole bones should be removed from plastic wrappings or bags and soaked or sprayed with phosphate buffered saline (PBS) to prevent dehydration.

Step 3. Remove soft tissue. All soft tissue should be removed from the area of the whole bone, which will be used for mechanical tests. This can be done using surgical tools, like a scalpel, and/or machine shop tools, like a precision knife or scraper.

Cortical Bone Preparation (Tension Tests)

Step 1. Make preliminary cuts. Use a band saw to remove proximal and distal parts of whole bone, leaving only the midshaft diaphysis, which is made of cortical bone. Then, use the band saw to cut the bone in half along the longitudinal axis to obtain posterior and anterior segments.

Step 2. Create dog bone shapes. Use a CNC mill to cut out a dog bone from each posterior and anterior segment of bone (Fig. 11.2A). The dog bone can be drawn with a computer-aided design program and converted into G-code, which directs the CNC mill.

Step 3. Slice the specimens. Use a diamond wire saw to slice the full cortical thickness dog bone specimens (Fig. 11.2B) into two or more thinner specimens to obtain the required thickness for mechanical testing (Fig. 11.2C). Recommended thicknesses can reach up to 4 mm.[12]

Step 4. Polish the specimens. To eliminate surface defects (and potential stress risers) caused by prior cutting, polish the specimens. To do so, polish all four sides

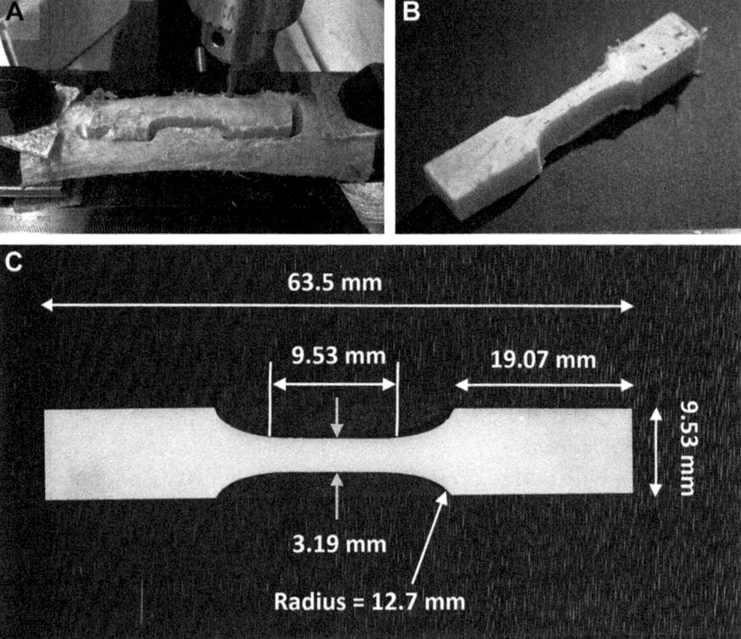

FIGURE 11.2

Fabrication of dog bone specimens from cortical bone for tensile tests. (A) CNC milling of a whole bone diaphysis, (B) intermediate specimen having full cortical thickness, (C) final specimen with required thickness and outer dimensions.

using sandpaper that has been moistened with water, move progressively from 400 grit to 600 grit to 6 μm to 1 μm sandpaper, use circular motion while evenly applying hand pressure to hold the specimen against a stationary sanding belt or wheel, and repeat the process for about 30–40 s per side.

Step 5. Rinse and soak. After polishing, rinse the specimens with PBS to remove residual sanding debris. Then, soak the specimens in PBS for 4 h at room temperature before testing.

Step 6. Measure dimensions. Measure the final width and thickness of the specimens at several locations using a micrometer with a minimum resolution of 1 ± 4 μm.

Step 7. Obtain BMD values. The bone mineral density (BMD) along with the corresponding clinical T-score can now be obtained using DEXA (dual-energy X-ray absorptiometry) scans.

Cortical Bone Preparation (Compression Tests)

Step 1. Make preliminary cuts. Use a band saw to remove proximal and distal parts of the whole bone, leaving the midshaft diaphysis, which is composed of cortical bone.

Step 2. Cut transverse segments. Use a band saw to cut transversely (i.e., perpendicular to the diaphysis long axis) across the diaphysis to obtain 15-mm-thick slices to have enough length to create cylindrical beams. The length of the cylindrical beam should be at least twice its diameter.[13]

Step 3. Core the segments. Clamp each slice into an industrial vice. Then, use a coring drill bit to create cylindrical beams, so their long axes are parallel to the diaphyseal long axis (Fig. 11.3A).

Step 4. Slice and sand. Use the diamond wire saw to slice off the far ends of the cylindrical sections to produce two perfectly parallel ends perpendicular to the cylinder long axis. Some light hand sanding using fine sandpaper may be necessary to remove any rough edges at the ends.

Step 5. Rinse and soak. Rinse the specimens with PBS to remove residual cutting debris. Then, soak the specimens in PBS for 4 h at room temperature prior to testing.

Step 6. Measure dimensions. Measure the final diameter and thickness of the specimens at several locations using a micrometer with a minimum resolution of 1 μm (±4 μm measurement uncertainty). A typical cylindrical beam has a 4.7-mm diameter and a 9.5-mm length (Fig. 11.3B).

Step 7. Obtain BMD values. The BMD, along with the corresponding clinical T-score, can now be obtained using DEXA scans.

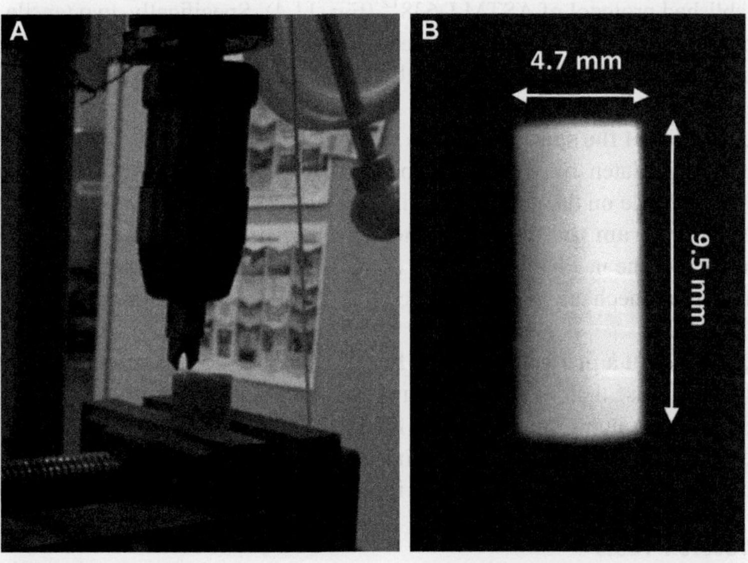

FIGURE 11.3

Fabrication of cylindrical specimens from cortical bone for compression tests. (A) Drill press with coring drill bit and clamped bone segment, (B) final cylindrical specimen with required diameter and length.

TIPS AND TRICKS

- ✓ Long bones can be initially cut into segments with a band saw for easier handling.
- ✓ Always use PBS to keep the whole bone and machined bone sufficiently hydrated.
- ✓ A tenoning jig ensures longitudinal alignment of microstructure during band saw cutting.
- ✓ An extensometer avoids strain measurement errors caused by mechanical tester stiffness.

THE "GOLD STANDARD"

No international standard exists for quasi-static mechanical testing of cortical bone in tension or compression. The American Society for Testing and Materials (ASTM) provides guidelines for testing plastics in ASTM D638 (Standard test method for tensile properties of plastics) and ASTM D695 (Standard test method for compressive properties of rigid plastics), which can be applied to cortical bone since it is also an anisotropic viscoelastic material. Researchers should also consult published peer-reviewed studies in the bone mechanics literature.

3.4 SPECIMEN TESTING

Tension Tests

Step 1. Install fixtures. Mount the grip fixtures in a mechanical tester based on the established protocol of ASTM D638[12] (Fig. 11.4). Specifically, two tensile grips are mounted tightly to the mechanical tester. One grip is attached directly to the load cell, and the other grip is mounted to the base. It is recommended to use stainless steel grips in order to avoid corrosion.

Step 2. Mount the specimen. Place the wide ends of the specimen into the grips of the loading platen fixtures. Then, place an extensometer with a sensitivity of 0.15% of full scale on the gage length (i.e., narrow middle section) of the specimen.

Step 3. Program the mechanical tester. Enter the appropriate loading regime parameters into the mechanical tester's computer software, such as preload to eliminate any initial mechanical hysteresis (e.g., 20 N) and linear loading rate (e.g., 0.5% strain rate).

Step 4. Mount a protective shield. Place a clear plastic protective shield around the test area in case there is any concern about flying debris from the bone specimen during loading or upon final specimen fracture.

Step 5. Run the test. Start the loading regime for the test until the cortical bone specimen experiences major structural failure. End the test.

Compression Tests

Step 1. Install platen fixtures. Mount the loading platen fixtures in a mechanical tester based on the established protocol of ASTM D695[13] (Fig. 11.5). Specifically, the LVDT is attached to the moving plate to measure displacement. Again, stainless steel plates are used to avoid corrosion.

Step 2. Mount the specimen. Place the specimen's flat ends between the metal compression plates' flat ends. Then, attach an LVDT (± 0.5 mm stroke) to

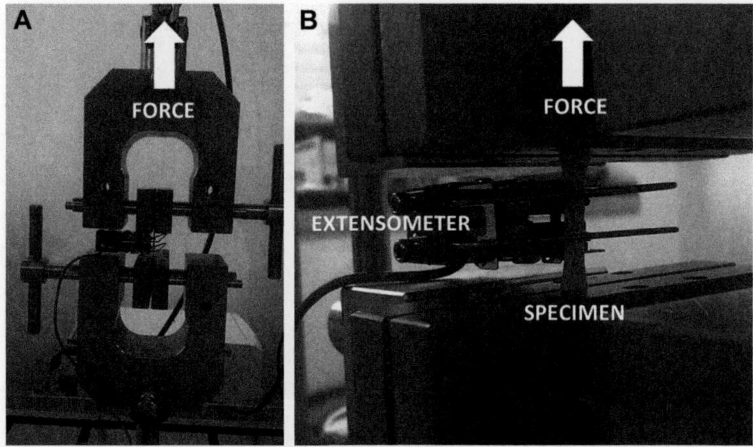

FIGURE 11.4

Cortical bone tension test setup. (A) Dog bone specimen mounted in the grips and equipped with an extensometer, (B) close-up of the test setup.

the compression plate that will move during the test. Use some type of lubricant (e.g., machine oil or grease) between the compression plates and specimen to minimize friction, thus allowing the specimen to expand transversely due to Poisson's effect.

Step 3. Program the mechanical tester. Enter the appropriate loading regime parameters into the mechanical tester's computer software, such as preload to eliminate any initial mechanical hysteresis (e.g., 100 N) and linear loading rate (e.g., 0.5% strain rate).

Step 4. Mount a protective shield. Place a clear plastic protective shield around the test area in case there is any concern about flying debris from the bone specimen during loading or upon final specimen fracture.

Step 5. Run the test. Start the loading regime for the test until the cortical bone specimen experiences major structural failure. End the test.

3.5 RAW DATA COLLECTION

Step 1. Record bone characteristics. Enter human or animal demographics (i.e., age, sex, and left or right limb) and quantitative bone properties (i.e., BMD and clinical T-score) (Table 11.1). BMD has 2-D units (i.e., g/cm^2 rather than g/cm^3), since DEXA scans only provide 2-D scans. DEXA scan reports often provide BMD values and clinical T-scores (i.e., normal T-score ≥ -1, $-1 >$ osteopenic T-score > -2.5, and osteoporotic T-score ≤ -2.5). When such analysis is overlooked or missing, computations can be made as T-score = (bone BMD − reference population mean BMD)/(1 standard deviation of reference population BMD), where T-score reference populations are young healthy adults of the same sex that are aged 20–40 years.[14,15]

FIGURE 11.5

Cortical bone compression test setup. The specimen is placed between the compression plates, while the linear velocity displacement transducer measures displacement of the upper plate.

Step 2. Record geometric dimensions. Enter the specimen length, width, and thickness of the middle narrow gage length section of the dog bone tension specimens, as well as the specimen length and diameter of the "cylindrical" compression specimens (Table 11.1).

Step 3. Record raw force vs. displacement data. Save the raw data files from mechanical tests, which can be used to generate raw force vs. displacement curves (Fig. 11.6). This includes time and force from the mechanical tester, as well as the displacement from the extensometer and LVDT.

Table 11.1 Cortical bone specimen characteristics.

Specimen	Age [Years]	Sex [M, F]	Limb [L, R]	BMD [g/cm^2]	Clinical T-Score	Dimensions [mm]
1						
2						
3						
etc.						
Avg						
SD						

Avg, *average;* SD, *standard deviation.*

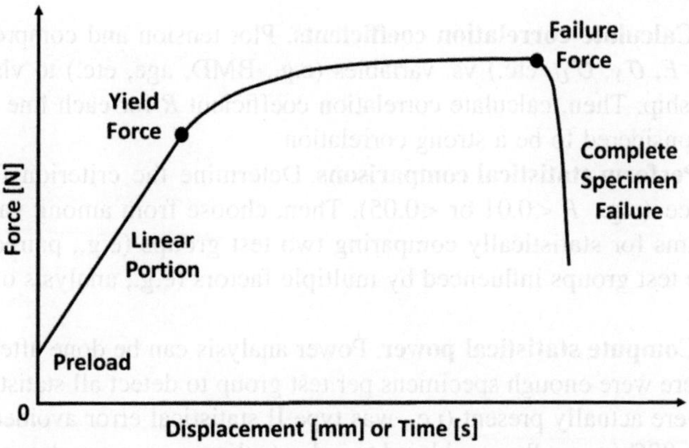

FIGURE 11.6

Typical force vs. displacement curve for tension or compression tests on cortical bone.

Step 4. Record fracture modes. Some final remarks should also be entered about the nature of specimen failure (i.e., Where on the specimen did the fracture occur? What was the orientation of the fracture line? etc.). Good photos of the failed specimens will allow for proper review later.

3.6 RAW DATA ANALYSIS

Step 1. Compute mechanical properties. Use the geometric dimensions, along with the raw force and displacement measurements to compute the properties of the cortical bone.

First, convert raw force vs. displacement data into stress vs. strain data. Formulas for tensile and compressive stress, respectively, are $\sigma = F/A = F/(wt)$ and $\sigma = F/A = F/(\pi d^2/4)$, where σ is average stress, F is raw force, A is cross-sectional area, w is width of the dog bone gage section, t is thickness of the dog bone gage section, and d is diameter of the cylindrical specimen. For tensile and compression tests, the formula for strain is $\varepsilon = \Delta L/L$, where ε is average strain, ΔL is measured length − original length, and L is original length.

Second, obtain elastic modulus E, which is computed as the slope (i.e., $\Delta\sigma/\Delta\varepsilon$) of the initial elastic linear portion of the stress vs. strain curve.

Third, obtain yield stress σ_Y and strain ε_Y points on the stress vs. strain curve that indicate the end of the elastic linear portion (i.e., no permanent damage) and the initiation of permanent damage.

Fourth, obtain the ultimate stress σ_U and strain ε_U, which occur at the point on the stress vs. strain curve that corresponds to the complete fracture of the specimen.

Finally, obtain the modulus of toughness U_T, which is the total area under the stress vs. strain curve all the way from zero stress (and strain) to the ultimate stress (and strain) point.

Step 2. Calculate correlation coefficients. Plot tension and compression outcomes (e.g., E, σ_Y, σ_U, etc.) vs. variables (e.g., BMD, age, etc.) to visualize the interrelationship. Then, calculate correlation coefficient R for each line of best-fit. $R > 0.8$ is considered to be a strong correlation.

Step 3. Perform statistical comparisons. Determine the criterion for statistical difference (e.g., $P < 0.01$ or < 0.05). Then, choose from among various software programs for statistically comparing two test groups (e.g., paired t-test) or two or more test groups influenced by multiple factors (e.g., analysis of variance, ANOVA).

Step 4. Compute statistical power. Power analysis can be done after the study to ensure there were enough specimens per test group to detect all statistical differences that were actually present (i.e., was type II statistical error avoided?). Statistical power >80% is usually considered to indicate there were enough specimens per test group. Note that if good predictions of averages and standard deviations are available from prior studies, then the number of specimens and/or tests can be chosen before the study begins to ensure a power >80%.

ENGINEER'S TOOLBOX

Classic beam theory can be used to estimate stress and strain at the cross-section of cortical bone specimens used for, respectively, tension and compression tests. Assume a longitudinal force is applied to cortical bone specimens, which have the following characteristics: isotropic (i.e., material properties are the same in all directions), homogeneous (i.e., material properties are the same at all locations), nonhollow solid, and uniform geometry. For tension tests, average engineering stress $\sigma = F/A = F/(wt)$, while the resulting average engineering strain $\varepsilon = \Delta L/L$. For compression tests, average engineering stress $\sigma = F/A = F/(\pi d^2/4)$, while the resulting average engineering strain $\varepsilon = \Delta L/L$. Note that F is applied force, A is cross-sectional area, w is width, t is thickness, d is diameter, ΔL is change in length, and L is original length.

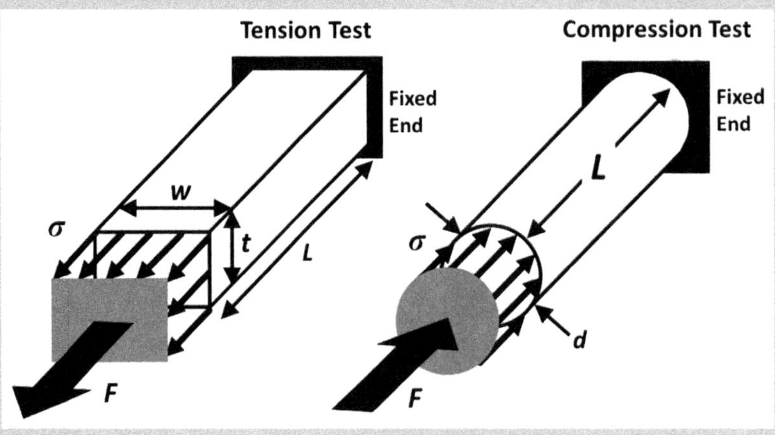

4. RESULTS

Once all tension and compression data collection and analysis have been performed, it is then important to communicate and present the primary results in an understandable and concise manner to the reader of a journal article, conference paper, technical report, or book chapter.

Step 1. Show a stress vs. strain graph. Present a typical graph illustrating stress vs. strain behavior during tension and compression (Fig. 11.7). Several key features can be noted.

First, an initial prestress is applied before the tension or compression test begins to remove any mechanical hysteresis in the test setup or the specimen itself.

Second, the stress grows linearly at first during the elastic region, since the specimen has not yet experienced any permanent damage. The slope of this linear region represents Young's elastic modulus E and is a measure of the intrinsic stiffness of the material.

Third, after reaching the elastic yield stress σ_Y, some permanent deformation begins. The yield point is the transition between elastic (i.e., reversible) and plastic (i.e., irreversible) behavior. It is typically defined by engineers as the intersection between the stress vs. strain curve and a line drawn parallel to the linear elastic portion with a 0.2% strain offset. The yield strain ε_Y is the corresponding strain value for the yield stress point.

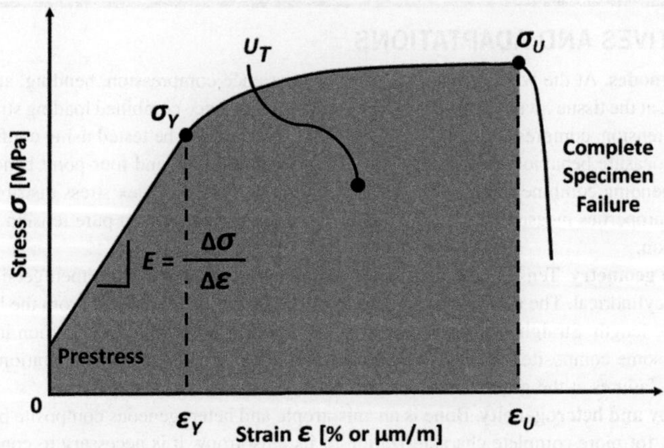

FIGURE 11.7

Typical stress vs. strain curve for tension or compression tests for cortical bone specimens. E, elastic modulus; σ_Y, yield stress; σ_U, ultimate stress; ε_Y, yield strain; ε_U, ultimate strain; U_T, modulus of toughness.

Fourth, the stress eventually reaches the ultimate stress σ_U and drops relatively quickly as the specimen completely fractures and returns to a zero stress state. The ultimate strain ε_U is the corresponding value for the ultimate stress point.

Fifth, the area under the curve up to and including the failure event is known as the modulus of toughness U_T. It represents the cortical bone's ability to absorb energy without failure.

Finally, the main advantage of a stress vs. strain graph vs. a raw force vs. displacement graph, is that it accounts for specimen geometry. Thus, it is more representative of tissue-level properties.

Step 2. Show main findings. The main properties (e.g., E, σ_Y, σ_U, etc.) can be tabulated for tension and compression tests if few test groups are used (e.g., tensile vs. compressive, normal vs. osteoporotic human bone, human vs. bovine bone, etc.) (Table 11.2). However, graphs are more useful if the study investigated properties vs. a large set of quantifiable variables (e.g., BMD, T-score, loading rate, irradiation levels, ribose concentrations, etc.) (Fig. 11.8). Make sure to present all statistics: P values for pairwise comparisons of each property between variables, the equation for each line of best fit (e.g., straight line, a second-order polynomial curve, or an exponential trend), the correlation coefficient R for each line of best fit, the statistical power(s) of the study, etc.

Step 3. Show failed specimen photos. Some photographs should be provided to illustrate the way cortical bone failed, such as fracture line length, location, orientation, etc. (Fig. 11.9). This may have implications about the brittleness or ductility of the specimens.

ALTERNATIVES AND ADAPTATIONS

- ✓ **Loading modes**. At the whole bone level, bones experience compression, bending, and torsion. Therefore, at the tissue level, cortical bone is known to experience combined loading stress states, involving tension, compression, and shear. Cortical bone can also be tested using configurations suited to measure behavior under pure shear, three-point bending, and four-point bending. Because bending combines tension and compression in a more complex stress distribution, materials properties measured in bending differ from those measured in pure tension or pure compression.
- ✓ **Specimen geometry**. Tensile tests can be conducted with "dog bone" specimen geometries that are flat or cylindrical. The radii reduce stress concentrations at the transitions from the larger tabs to the gage length. Straight specimens without the thinner gage length are common in tensile testing of some composites; however, concerns abound regarding stress concentrations and premature failures at the grips.
- ✓ **Anisotropy and heterogeneity**. Bone is an anisotropic and heterogeneous composite biomaterial. Therefore, for more complete characterization of its anisotropy, it is necessary to conduct tests with specimens oriented in multiple directions and under multiple stress states (i.e., tension, compression, shear). Behavior beyond the yield point is very complex and can be challenging to measure and interpret. Fracture patterns are often complex because they are highly dependent upon anisotropy and heterogeneity in addition to the applied stress state. Heterogeneity also adds a source of variance.
- ✓ **Poisson's ratio**. Complete characterization of cortical bone anisotropy also requires measurement of Poisson's ratio in various directions and under different stress states.

Table 11.2 Mechanical properties in tension or compression of cortical bone from stress vs. strain curves. Each numerical value should be given as average ± 1 standard deviation.

Test Group	Elastic Modulus E [GPa]	Yield Stress σ_Y [MPa]	Yield Strain ε_Y [%]	Ultimate Stress σ_U [MPa]	Ultimate Strain ε_U [%]	Toughness U_T [mJ/mm^3]	Fracture Mode Remarks
1							
2							
3							
Etc.							

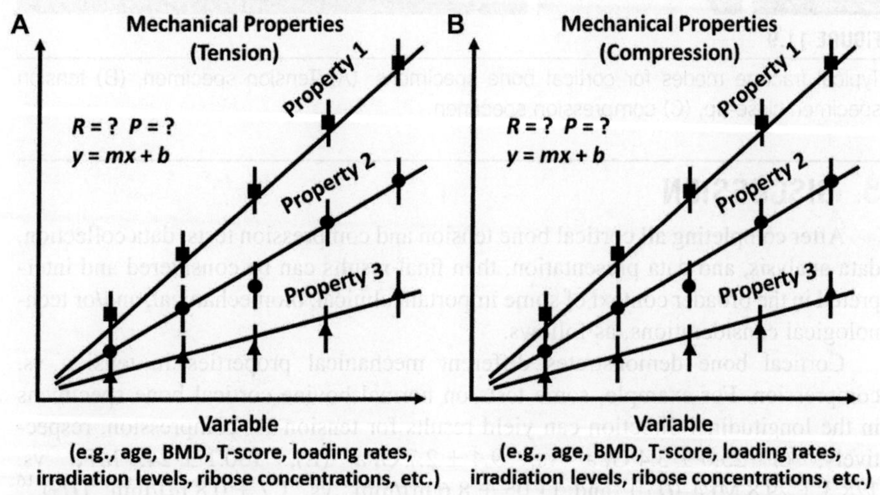

FIGURE 11.8

Cortical bone mechanical properties plotted against variables. (A) Tension tests, (B) compression tests. Each data point is an average ± 1 standard deviation. P values are statistical difference probabilities for each pairwise comparison of properties for different variables. Each line of best fit has an equation $y = mx + b$, a correlation coefficient R, and a statistical P value that proves its slope (i.e., slope $= m$) is different than a horizontal line (i.e., slope $= 0$). Some properties may yield second-order polynomial or exponential lines of best fit.

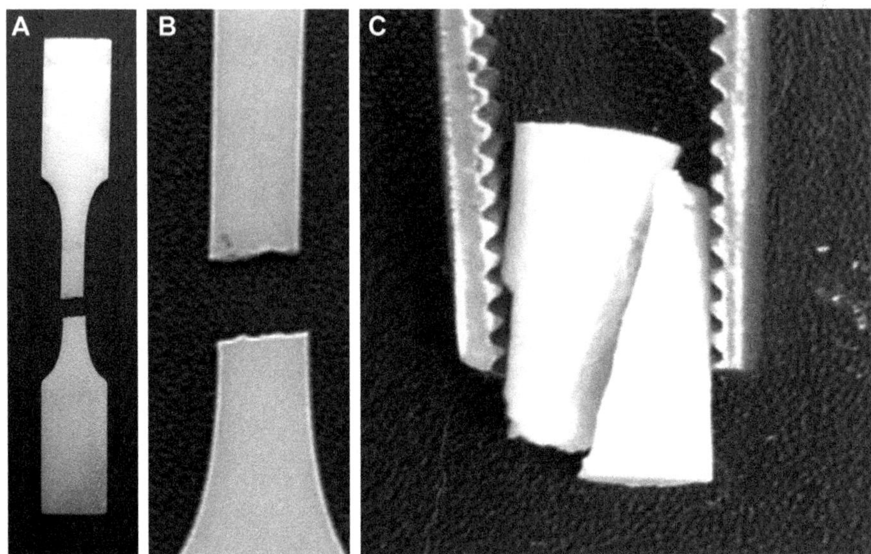

FIGURE 11.9

Typical fracture modes for cortical bone specimens. (A) Tension specimen, (B) tension specimen close-up, (C) compression specimen.

5. DISCUSSION

After completing all cortical bone tension and compression tests, data collection, data analysis, and data presentation, then final results can be considered and interpreted in the broader context of some important clinical, biomechanical, and/or technological considerations, as follows.

Cortical bone demonstrates different mechanical properties in tension vs. compression. For example, some tests on normal bovine cortical bone specimens in the longitudinal direction can yield results for tension vs. compression, respectively, of 25.0 ± 3.4 GPa vs. 9.4 ± 2.7 GPa (E), 130.1 ± 24.9 MPa vs. 128.3 ± 29.8 MPa (σ_U), and 11.05 ± 8.6 mJ/mm^3 vs. 1.7 ± 0.8 mJ/mm^3 (U_T).[16] Conversely, some studies show that E in the longitudinal direction can be around 20 GPa in both tension and compression, while σ_Y and σ_U can be at least two times higher under compression (180 and 210 MPa, respectively) than under tension (75 and 97 MPa, respectively).[17] There are numerous studies on the mechanical properties of cortical bone from human and animal sources. However, a large range in values is reported, which may be caused by interspecimen variability, differences in bone microstructure at different locations in the same bone, diversity in experimental conditions, etc.

Simple uniaxial longitudinally-oriented tension and compression tests are useful for studying the effects of various modifications to the tissue, such as irradiation sterilization.[18] However, they are insufficient for characterizing the anisotropy and heterogeneity of cortical bone mechanical properties. For example, in order to characterize cortical bone's fourth-order stiffness tensor, it is necessary to conduct tests with specimens oriented in multiple directions and under multiple stress states (i.e., tension–compression, shear, or combined). Furthermore, due to its heterogeneous composite microstructure, strain fields can vary within a gage length, especially once critical stress levels are achieved and microcracking commences. Therefore, non-contact full-field strain measurement techniques, such as digital image correlation, are useful.

Biomechanical testing of cortical bone at a tissue level requires standardization since no internationally acknowledged protocols exist. The ASTM provides guidelines for testing plastics in its documents ASTM D638 and D695, which are often used by biomechanical researchers.[12,13] However, these polymers do not necessarily represent cortical bone's complex internal structure, anisotropy, heterogeneity, and viscoelasticity. Although the American National Standards Institute (ANSI) provides guidelines for testing bone in its document ANSI/ASAE S459, this only applies to whole bone specimens in shear and three-point bending.[19] Therefore, this appears to be an important avenue of inquiry for investigators.

6. SUMMARY

- Cortical bone is an anisotropic and heterogeneous material with a hierarchical structure.
- Cortical bone properties affect overall bone and joint health and orthopaedic implant stability.
- Cortical bone can be tested in tension and compression to obtain some mechanical properties
- Extensometers and LVDTs are applied externally for precise displacement measurements.
- Biomechanical properties can be affected by bone type, age, sex, disease, irradiation, etc.
- Standards need to be developed for tension and compression tests on cortical bone.

7. QUIZ QUESTIONS

1. Why is it important to understand the mechanical properties of cortical bone?
2. What is meant by the terms anisotropic and heterogeneous with regard to cortical bone?
3. What is the difference between force vs. displacement and stress vs. strain curves?
4. What is meant by toughness, and how is this different from ultimate stress?

5. Will a cortical bone specimen with a circular cross-section undergo fracture if subjected to a quasi-static compressive force of 2 kN? Assume the following are known: elastic modulus = 15 GPa, ultimate compressive stress = 100 MPa, cross-sectional radius = 5 mm, and length = 20 mm (answer: no fracture, since average stress = 25.5 MPa < ultimate compressive stress).

REFERENCES

1. Nair AK, Gautieri A, Chang SW, Buehler MJ. Molecular mechanics of mineralized collagen fibrils in bone. *Nature Communications* 2013;**4**:1724.
2. Reilly DT, Burstein AH. The elastic and ultimate properties of compact bone tissue. *Journal of Biomechanics* 1975;**8**(6):393–405.
3. Katsamanis F, Raftopoulos DD. Determination of mechanical properties of human femoral cortical bone by the Hopkinson bar stress technique. *Journal of Biomechanics* 1990;**23**(11):1173–84.
4. Nyman JS, Leng H, Dong XN, Wang X. Differences in the mechanical behavior of cortical bone between compression and tension when subjected to progressive loading. *Journal of the Mechanical Behavior of Biomedical Materials* 2009;**2**(6):613–9.
5. Dong XN, Guo XE. The dependence of transversely isotropic elasticity of human femoral cortical bone on porosity. *Journal of Biomechanics* 2004;**37**(8):1281–7.
6. Winwood K, Zioupos P, Currey JD, Cotton JR, Taylor M. Strain patterns during tensile, compressive, and shear fatigue of human cortical bone and implications for bone biomechanics. *Journal of Biomedical Materials Research Part A* 2006;**79**(2):289–97.
7. Turner CH, Wang T, Burr DB. Shear strength and fatigue properties of human cortical bone determined from pure shear tests. *Calcified Tissue International* 2001;**69**(6):373–8.
8. Saha S. Longitudinal shear properties of human compact bone and its constituents, and associated failure mechanisms. *Journal of Materials Science* 1977;**12**(9):1798–806.
9. Dong XLN, Luo Q, Wang XD. Progressive post-yield behavior of human cortical bone in shear. *Bone* 2013;**53**(1):1–5.
10. Park HC, Lakes RS. Cosserat micromechanics of human bone: strain redistribution by a hydration sensitive constituent. *Journal of Biomechanics* 1986;**19**(5):385–97.
11. Zioupos P, Gresle M, Winwood K. Fatigue strength of human cortical bone: age, physical, and material heterogeneity effects. *Journal of Biomedical Materials Research Part A* 2008;**86**(3):627–36.
12. ASTM D638. Standard test method for tensile properties of plastics. West Conshohocken (PA, USA): American Society for Testing and Materials (ASTM); www.astm.org.
13. ASTM D695. Standard test method for compressive properties of rigid plastics. West Conshohocken (PA, USA): American Society for Testing and Materials (ASTM); www.astm.org.
14. Lewiecki EM, Borges JLC. Bone density testing in clinical practice. *Arquivos Brasileiros de Endocrinologia and Metabologia* 2006;**50**(4):586–95.
15. Pettersson U, Nordström P, Lorentzon R. A comparison of bone mineral density and muscle strength in young male adults with different exercise level. *Calcified Tissue International* 1999;**64**(6):490–8.
16. Attia T, Willet TL. Unpublished pilot study data. Toronto (Canada): Musculoskeletal Research Lab, Lunenfeld Tanenbaum Research Institute, Mt. Sinai Hospital.

17. Li S, Demirci E, Silberschmidt VV. Variability and anisotropy of mechanical behavior of cortial bone in tension and compression. *Journal of the Mechanical Behavior of Biomedical Materials* 2013;**21**:109–20.
18. Akkus O, Belaney RM, Das P. Free radical scavenging alleviates the biomechanical impairment of gamma radiation sterilized bone tissue. *Journal of Orthopaedic Research* 2005;**23**(4):838–45.
19. ANSI/ASAE S459. Shear and three-point bending test of animal bone. Washington (DC, USA): American National Standards Institute (ANSI); www.ansi.org.

CHAPTER

Fracture Toughness Testing of Cortical Bone

12

Mitchell Woodside[1], **Thomas L. Willett**[2]

University of Toronto, Toronto, ON, Canada[1]; *University of Waterloo, Waterloo, ON, Canada*[2]

1. BACKGROUND

The strength of materials containing flaws and cracks, such as cortical bone (Fig. 12.1), cannot be completely described by strength-based criteria. A material's mechanical load-bearing capacity is a function of loading conditions, flaw size, and the material's fracture toughness.[1] For a through-thickness crack in a large plate of isotropic, linear elastic material, the crack growth driving force is $G = \sigma^2 \pi a/E$, where G is strain energy release rate per increment of crack growth (dU/da), σ is applied stress, a is crack length, and E is Young's modulus. A crack will grow when G exceeds the material's fracture toughness G_c.[1] Because cortical bone demonstrates nonnegligible irreversible deformation before fracture, it violates linear elastic fracture mechanics (LEFM) assumptions. Consequently, fracture toughness

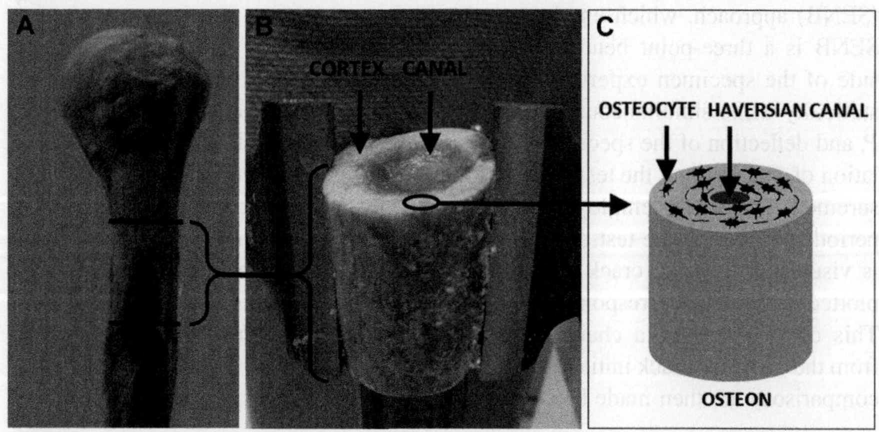

FIGURE 12.1

Cortical bone. (A) Typical human long bone, (B) long bone diaphysis showing the cortex surrounding the marrow-filled intramedullary canal, (C) tube-like osteons are cortical bone's basic structural units, which contain embedded bone cells termed osteocytes (i.e., bone cells).

is better described by the J-integral,[2] which quantifies a strain energy release rate, like G, but it accounts for the energy absorbed by irreversible deformation at and around the crack tip. Therefore, this chapter explains how to perform fracture toughness tests on cortical bone, as well as how to analyze, present, and interpret results.

2. RESEARCH QUESTIONS

Typical research questions might include one or more of the following:

- What elements and features of cortical bone's structure affect its ability to resist crack growth?
- Can the fracture toughness of cortical bone be altered by manipulating its organic phase?
- Are there differences in fracture toughness between the sexes or between species of animals?
- Does the fracture toughness of cortical bone vary with direction (i.e., is it anisotropic)?
- What affects do aging or disease have on cortical bone fracture toughness?
- etc.

3. METHODOLOGY

3.1 GENERAL STRATEGY

Testing is done on cortical bone specimens using a single-edge notched bending (SENB) approach, which employs a classic fracture testing specimen geometry.[2] SENB is a three-point bend test in which an initial crack is machined into the side of the specimen experiencing tensile stress. The specimen is loaded quasi-statically while simultaneously measuring the size of the crack a, the applied load P, and deflection of the specimen δ. These three measurements allow for the calculation of J throughout the test. This test method utilizes an optical crack length measurement technique, employing a high-resolution digital camera taking photos periodically during the test. With appropriate resolution and contrast, the crack tip is visualized, and the crack length is measured with simple image analysis. J is plotted against the corresponding crack growth measurements to get a J-Δa curve. This curve serves as a characterization of the material's fracture behavior, and from the curve the crack initiation fracture toughness J_{Ic} can be determined. Statistical comparisons are then made between different cortical bone specimen test groups.

GLOSSARY

- ✓ **Anisotropic.** Mechanical properties depend on direction, unlike isotropic materials.
- ✓ **Cortical bone.** Hard, dense tissue found as the shell and shafts of bones.
- ✓ **Crack growth initiation fracture toughness J_{Ic}.** The value of J when the crack starts to grow.
- ✓ **Fracture toughness.** A material's intrinsic resistance to crack growth initiation and propagation.
- ✓ **SENB (single-edge notched bending).** A fracture mechanics testing configuration.

SAFETY FIRST

✓ Use appropriate personal protection equipment when handling human and animal tissues.
✓ Use an N95 mask when cutting bone to avoid inhaling biohazardous dusts and aerosols.
✓ Be careful when using a mechanical tester by taking measures to avoid crush injuries.
✓ Only allow one person at a time to load specimens and operate the tester.
✓ Protect against sudden and high energy fractures, which may result in flying debris.

3.2 MATERIALS AND TOOLS LIST

- diamond wire saw
- diffuse white lighting
- digital camera or video camera
- load cell
- mechanical tester (with high frame stiffness)
- metallurgical saw
- phosphate buffered saline
- polishing paper and diamond suspensions
- three-point bending jig
- ultra-sharp razor blades

3.3 SPECIMEN PREPARATION

Step 1. Prepare cortical bone shafts. Beginning with an entire long bone (e.g., human femur), remove the metaphyses using a band saw. Make sure that the rough cuts leave extra diaphysis material while ensuring that the cortices are still thick enough (>7 mm) to obtain final beams of the desired thickness. Cut the diaphysis into lengths greater than that of the desired specimen (Fig. 12.2A) with a little extra material (~20 mm extra). Optionally, this is now a good time to cut the diaphysis sections in half along the long axis to create more manageable pieces and to ease the removal of the bone marrow (Fig. 12.2B). The halving can also be done with a band saw. Tracking of anatomy and orientation is critical here.

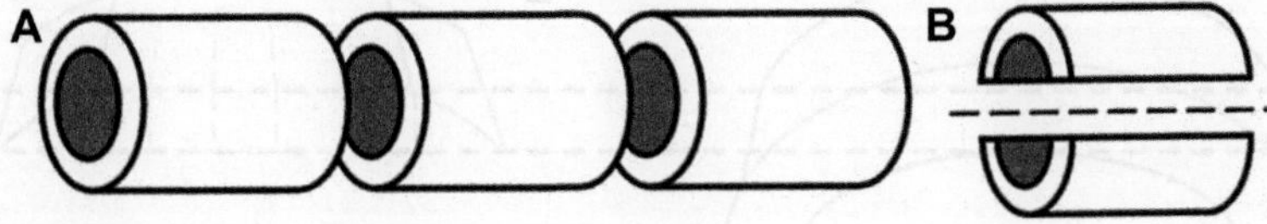

FIGURE 12.2

Cutting cortical bone shafts. (A) A long bone shaft is cut into segments, (B) a long bone segment is cut into halves.

Step 2. Cut the slabs. Use an irrigated abrasion cutting method to cut "slabs" from the straightest and flattest diaphysis sections that are available (Fig. 12.3A). Use a work-holding device to allow for controllable, repeatable, and accurate cuts. Using an abrasion cutting method is desirable because harsher cutting methods create regions of damaged bone around the cuts that will extend into the thickness of the final specimens. Irrigation of the cutting process ensures that the bone does not dry out through the procedure. A water lubricated and cooled metallurgical saw is ideal for this task.

Step 3. Cut the beams. The slabs can now be cut lengthwise to the desired width in successive parallel lengthwise cuts to obtain 4 mm × 4 mm cross-sectional beams (Fig. 12.3B). Because of the composite nature of bone, cut the beams with a span S to thickness W ratio of around 10 to 1. This nonstandard span-to-thickness minimizes contributions due to shear.[3] A minimum width B of around 3 mm (maybe more depending on the specimens) is necessary to obtain the plain strain requirements needed for a valid crack growth initiation fracture toughness J_{Ic} measurement. This requirement is dependent on the material's yield strength and fracture toughness, and can be estimated using preexisting data acquired using similar methods. Machining and propagating the crack in the transverse direction (i.e., the circumferential–radial plane with the longitudinal plane aligned with the long axis of the bone) is recommended because it is more clinically applicable.[4] Specifically, the circumferentially transverse direction is recommended to avoid large crack deflections that accompany radially transverse fractures due to the layered microstructure of cortical bone.

Step 4. Cut the notch and starter crack. According to the American Society for Testing and Materials (ASTM) standards, a proper SENB specimen requires a nominal initial notch length of 0.45–0.7 times the thickness of the specimen,[5] although 0.5 is common in these tests (Fig. 12.4). The tip of the initial notch or crack must also be as sharp as possible (Fig. 12.5). In metals, this is achieved by fatiguing the specimen under low loads so that a crack begins to nucleate. Bone requires a different approach using a razor blade to achieve a microscale crack tip radius (~5 μm) because fatigue loading tends to grow cracks perpendicular to the desired transverse plane–circumferentially directed path.

The notching can be done in two steps. The first is to cut the notch to just under nominal length with a diamond wire histology saw, called macronotching, and the

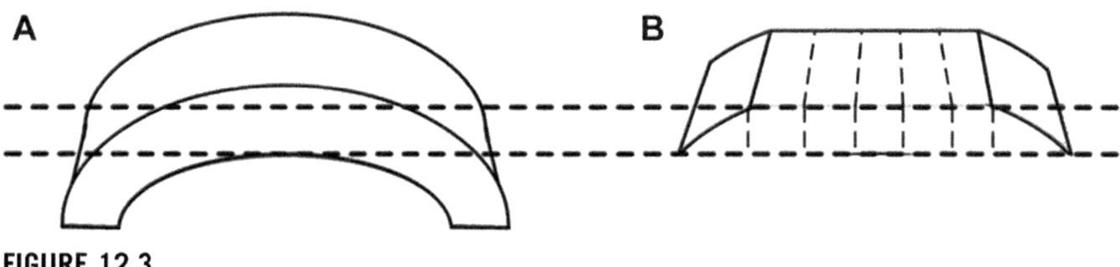

FIGURE 12.3

Cutting cortical bone slabs and beams. (A) A diaphyseal segment half is cut along dotted lines to produce a "slab," (B) "slabs" are cut lengthwise to produce groups of beams.

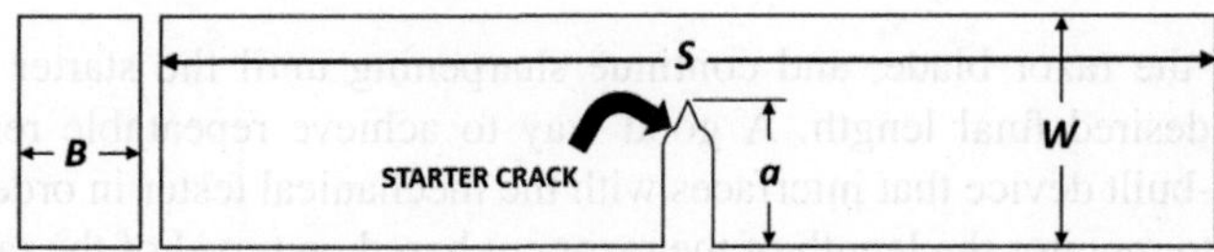

FIGURE 12.4

SENB fracture test specimen geometry. It is simply a rectangular beam with a starter crack machined into the bottom face.

FIGURE 12.5

High-resolution digital photograph of a starter notch.

second is to sharpen the notch the rest of the way with an ultra-sharp razor blade, called micronotching. Cut a macronotch at the midspan of the specimen to a length 0.1–0.2 mm shorter than the desired final length of the starter crack of 2 mm. To micronotch the specimen, insert an ultra-sharp razor into the macronotch along with a couple of drops of 1-μm diamond suspension slurry. Next, apply a small load and slide the specimen back and forth to sharpen the tip of the notch. Monitor the travel of the razor blade, and continue sharpening until the starter crack has reached the desired final length. A good way to achieve repeatable results is to use a custom-built device that interfaces with the mechanical tester in order to apply the load and to monitor the length of the razor cut based on travel of the razor blade.

Step 5. Polish the beam. Finish specimen preparation by polishing the lateral faces with successively finer grit metallurgical polishing paper, starting with 400 grit and finishing with a polishing step in a 1-μm diamond suspension slurry. This serves to eliminate as many stress concentrations on the specimen faces as possible, other than the machined notch. Polishing the top and bottom faces is unnecessary because the top face of the specimen will be under compressive loads, while along the bottom face, the machined notch will be by far the greatest stress concentration, making the effect of all others negligible. The specimen is now prepared for testing.

TIPS AND TRICKS

- ✓ Create a sharpened crack tip using a razor blade; otherwise, the crack will defect early on.
- ✓ Try coating the bone with black ink to increase contrast and better visualize the crack tip.
- ✓ Use cutting jigs and clamps while cutting bone to improve dimensional accuracy and precision.
- ✓ Keep track of specimen orientation to ensure that the crack grows in the correct direction.
- ✓ Keep specimens hydrated by immersion in phosphate buffered solution before testing.

THE "GOLD STANDARD"

No standard methods exist for the fracture testing of cortical bone. The American Society for Testing and Materials (ASTM) documents E1820 (Standard test method for measurement of fracture toughness) and D6068 (Standard test method for determining J-R curves of plastic materials) should be followed as closely as possible. However, these standards were developed for measuring elastic—plastic fracture behavior of metals and polymers, not cortical bone.

3.4 SPECIMEN TESTING

Step 1. Take measurements. Before testing, measure the width B and thickness W of each specimen with a caliper or a micrometer (0.001 mm resolution and $\pm$ 0.004 mm uncertainty are adequate) at the location of the notch.

Step 2. Align the specimen. Place the prepared specimen in the three-point bending jig with the notch opening downward, away from the crosshead of the mechanical tester (Fig. 12.6). Make sure that the notch is aligned directly underneath where the loading nose contacts the top of the specimen.

Step 3. Choose a camera setup. The camera position should allow clear visualization of the propagating crack. For example, a 32$\times$ magnification macroscope lens coupled with a 24-megapixel digital single lens reflex camera oriented perpendicular

FIGURE 12.6

Fracture toughness test setup. The macroscope lens requires bright lighting, so the light bulbs are moved close to the test specimen. The chip that controls the camera shutter can be seen in the bottom right of the photo.

to one side of the SENB specimen works well. It achieves high spatial resolution (~0.75 μm/pixel). The camera and lens are mounted on a stage, and a lead screw allows for fine focus adjustments.

Step 4. Aim the field of view. The camera should visualize the tip of the crack. Make sure that at the beginning of the test, both the top and bottom edges of the specimen are visible in the photo frame, but it is acceptable if the bottom edge disappears from the frame as the beam deflects later in the test. These edges are used later to obtain an accurate crack measurement scale for each test.

Step 5. Run the test. Bring the loading nose barely into contact with the specimen (e.g., load <1 N) and proceed with the test at sufficiently slow, quasi-static speeds (e.g., 0.2 mm/min deflection rate). It is standard practice to conduct tests under displacement control. Ensure time, applied load, and load-line deflection are recorded as the test proceeds.

Step 6. Capture digital photos. Record digital photos at regular time intervals during the test. Different photo recording frequencies will be needed based on the resolution of the camera, the deflection rate to the test, and the measurement resolution requirements of the test. It is essential that the photos are linked to the recorded applied load and load-line deflection data in the time domain. A relatively straightforward and practical way to do this is to send an identical signal to the camera shutter and to the mechanical testing controller and data logger. In this way, the raw data will include a signal pulse that corresponds to each photo in chronological order. Since digital photos will be used to determine the crack length, it is imperative that the exact moment in the test when the photo was taken is known.

Step 7. Complete the test. Test the specimen until fast fracture or a predetermined threshold is achieved. Ensure that the testing procedure is consistent for all specimens.

3.5 RAW DATA COLLECTION

Step 1. Record required data. Tabulate all necessary data (Table 12.1). During the test, the time, load, and deflection data should be recorded by the mechanical tester for later retrieval. Necessary data also include initial dimensions of thickness *W*, width *B*, three-point bending span *S*, and initial crack length a_o, as well as applied load *P*, load-line deflection δ, and crack length a_i vs. time. Additionally, an accurate measure of the Young's modulus *E* is required for *J* calculations.

Step 2. Calculate and record work done. Numerical integration of the *P* vs. δ readout will be required to calculate *J* (Table 12.1). Load-line deflection is not explicitly used in calculations, but is required to perform the numerical integrations to determine work done, *A*, at each point in the test (Table 12.1).

Step 3. Measure crack lengths. At each point in the test, crack lengths are measured in the individual digital photos captured during the test using digital image analysis. Using the known thickness of each specimen or a gage of some type in each photo, a scale can be set. In all individual photos of the test, a line can now be marked from, and perpendicular to, the top surface of the specimen to the observed crack tip. The difference between the length of this line in the first photo of the test and each subsequent photo is the crack growth Δa.

Table 12.1 Fracture toughness data collection table.

Specimen ID	Span *S* [mm]	Width *B* [mm]	Thickness *W* [mm]	Initial Crack Length a_o [mm]	Young's Modulus *E* [GPa]
Time [s]	**Deflection δ [mm]**	**Load *P* [N]**	**Work Done *A* [mJ]**	**Crack Length a_i [mm]**	**Fracture Toughness *J* [mJ/mm²]**
Time 1					
Time 2					
Time 3					
etc.					

3.6 RAW DATA ANALYSIS

Step 1. Compute J. Calculate J for each crack length measurement in the raw data. J is broken down into elastic and plastic components for calculation, and it is an iterative formula building upon the previously calculated point to determine the next. The formulas are as follows[5]:

$$J_{(i)} = J_{\mathrm{el}(i)} + J_{\mathrm{pl}(i)} \tag{12.1}$$

$$J_{\mathrm{el}(i)} = \frac{\left(K_{(i)}\right)^2 (1-\nu^2)}{E} \tag{12.2}$$

$$K_{(i)} = \left[\frac{P_i S}{BW^{\frac{3}{2}}}\right] f \tag{12.3}$$

$$J_{\mathrm{pl}(i)} = \left[J_{\mathrm{pl}(i-1)} + \left(\frac{1.9}{b_{(i-1)}}\right)\left(\frac{A_{\mathrm{pl}(i)} - A_{\mathrm{pl}(i-1)}}{B}\right)\right] \times \left[1 - 0.9\left(\frac{a_{(i)} - a_{(i-1)}}{b_{(i-1)}}\right)\right] \tag{12.4}$$

$$C_{(i)} = \frac{1}{EB}\left(\frac{S}{W-a_i}\right)^2 \left[1.193 - 1.98\left(\frac{a_i}{W}\right) + 4.478\left(\frac{a_i}{W}\right)^2 - 4.443\left(\frac{a_i}{W}\right)^3 + 1.739\left(\frac{a_i}{W}\right)^4\right] \tag{12.5}$$

where J_{el} is elastic component of J [mJ/mm^2], J_{pl} is irreversible "plastic" component of J [mJ/mm^2], K is stress intensity factor [MPa·m]$^{1/2}$, ν is Poisson's ratio (assumed = 0.3), E is Young's modulus [GPa] (assumed ~18 GPa for longitudinal direction, but best if measured), P is applied load [N], S is three-point bending span [mm], B is specimen width [mm], W is specimen thickness [mm], a is crack length [mm], b is unbroken ligament length ($W - a$) [mm], f is the shape factor function which is a function of (a/W) [—], A_{pl} is "plastic" area under the load vs. deflection curve [mJ], $C_{(i)}$ is normalized load-line compliance used to determine A_{pl}, and i is iteration number in the numerical procedure for calculating J_{pl}.

Step 2. Compute A_{pl}. To determine the plastic area under the curve, first plot the load vs. deflection curve (Fig. 12.7A). For each point in the test where a crack length measurement is made, draw a line that intersects the load value at that point and determine the slope of that line by Eq. (12.5) (i.e., Eq. (A1.10) from ASTM E1820) or another accurate load-line compliance equation. The slope of this line should be less than the initial slope of the curve if crack growth has occurred. It is necessary to verify Eq. (12.5) against the initial load-line compliance data to ensure accuracy. The plastic area under the curve A_{pl} is the area under the load vs. deflection curve and above this line. This calculation is repeated for each point at which a crack length measurement was made.

Step 3. Perform curve fitting. Once a J value has been evaluated for each point in the test, another column for J can be appended to the data table (Table 12.1).

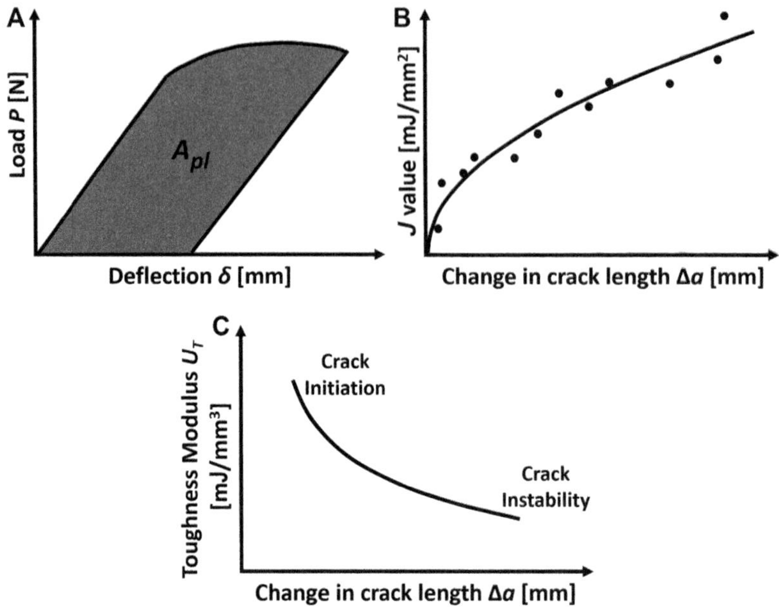

FIGURE 12.7

Fracture toughness parameters. (A) Schematic illustrating how to obtain A_{pl}, (B) fracture resistance curve with the power law fit to individual data points, (C) modulus of toughness graph. Only the regime between crack initiation and instability is depicted.

The J-Δa curves can now be plotted and a model curve fitted as required. On a simple Cartesian grid, plot the value for the J-integral against the corresponding crack growth values. Use a nonlinear least squares approach to fit a power law of the form $J = C_1(\Delta a)^{C_2}$ to the experimental data (Fig. 12.7B). The value of C_2 should be between 0–1.[5]

Step 4. Determine J_{Ic}. Once the curve has been fitted, draw a construction line with a slope of twice the material's apparent yield strength σ_Y passing through 0.2 mm of crack growth. The value of J at the intersection of this line with the power law fit is the crack initiation fracture toughness J_{Ic}.[5] This value is determined by finding the intersection of $\Delta a = J/(2\sigma_Y) + 0.2$ mm and $J = C_1(\Delta a)^{C_2}$. ASTM E1820 defines crack initiation as the moment when the crack has grown by a length of 0.2 mm (offset). This standard has been developed for metals, and therefore, the information may not be completely or directly applicable to cortical bone. It is important to choose an appropriate offset value, as crack initiation can be difficult to pinpoint otherwise (e.g., optically), which is the reason for the existence of the construction line in the standard method. Furthermore, a standard approach to determining an appropriate value for σ_Y is undetermined for cortical bone because, under pure longitudinal tension, bone forms microcracks as part of the yielding process, and it does not plastically deform the same way ductile metals and polymers do. It is also possible to use different fits for the J-Δa curve other than a power law.

Straight lines have also been used.[6] A simpler means of determining J_{Ic} is to use the value of J at $\Delta a = 0.2$ mm or some other meaningful amount of crack growth.

Step 5. Determine the modulus of toughness U_T. Each J-Δa curve is defined by its curve fit and its point J_{Ic} value. Another characterization can be done. By plotting the slope of the J-Δa curve along the same axis as the original curve, the material's rate of fracture resistance change can be explored. This rate of change of crack growth resistance is known as the modulus of toughness $U_T = \mathrm{d}J/\mathrm{d}a$ [mJ/mm^3], which is influenced by toughening mechanisms that engage once crack growth starts. It is acquired by calculating the derivative of the previously fitted power law and plotting it between the crack initiation point and the instability point as a function of crack growth. It is important to note that the modulus of toughness is meaningless before crack initiation or after crack instability (Fig. 12.7C).

Step 6. Perform statistical analysis. Summarize each numerical result as an average $\pm$ 1 standard deviation, and choose the criterion for detecting statistical differences (e.g., $P < 0.01$ or < 0.05). Then compare fracture toughness data for different patient or disease groups (e.g., men ≤ 60 years old vs. men > 60 years old, "normal" women vs. "osteoporotic" women, etc.), bone types (e.g., femur vs. tibia, human vs. bovine, etc.), etc. There are various software programs for comparing two test groups (e.g., paired t-test) or two or more test groups influenced by multiple factors (e.g., analysis of variance, ANOVA). Moreover, power analysis can be done after (but preferably before) the study to ensure power was $> 80\%$ (i.e., type II statistical error was avoided), indicating there were enough specimens per group to detect all statistical differences actually present.

ENGINEER'S TOOLBOX

The stress intensity factor K is a function of the shape factor function f. This function accounts for how the particular test specimen configuration differs from the ideal infinite plate with central crack. For an SENB specimen according to ASTM E1820, the formula is

$$f = \frac{3\left(\frac{a_i}{W}\right)^{\frac{1}{2}}\left[1.99 - \left(\frac{a_i}{W}\right)\left(1 - \frac{a_i}{W}\right)\left(2.15 - 3.93\left(\frac{a_i}{W}\right) + 2.7\left(\frac{a_i}{W}\right)^2\right)\right]}{2\left(1 + 2\frac{a_i}{W}\right)\left(1 - \frac{a_i}{W}\right)^{\frac{3}{2}}}$$

where a is crack length and W is specimen thickness. This standard, however, assumes a span to thickness ratio (S:W) of 4:1. Herein, an S:W of 10:1 is used to promote pure bending and avoid contributions from shear. Therefore, another shape function should be used to reduce error. Furthermore, an alternate load-line compliance equation should be used and be confirmed against experimental data.

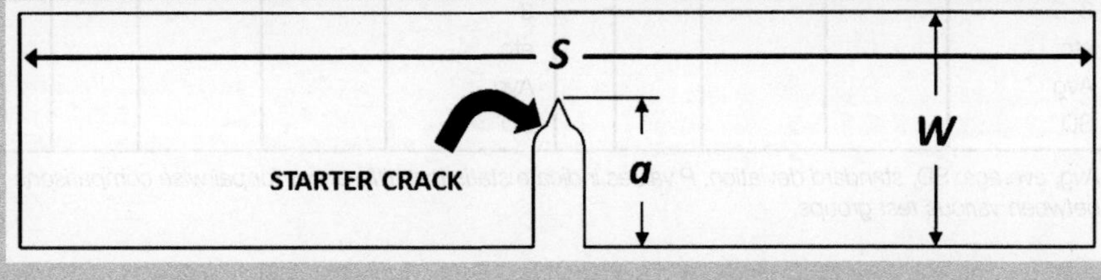

4. RESULTS

Once all fracture toughness data collection and analysis have been performed, it is then important to communicate and present the primary results in an understandable and concise manner to the reader of a journal article, conference paper, technical report, or book chapter.

Step 1. Generate representative curves. Generate and present representative fracture test curves, such as applied load vs. load-line deflection curves (Fig. 12.8A) and J-Δa curves (Fig. 12.8B).

Step 2. Present data tables. Present the main numerical findings of the study in a table or column chart (Table 12.2). Tables are preferred because they report actual numbers, including the average and standard deviation for J_{Ic} and U_T for each experimental group.

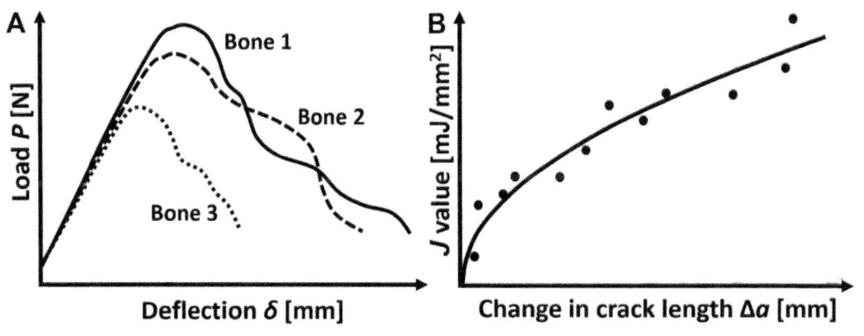

FIGURE 12.8

Typical experimental curves for testing cortical bone. (A) Applied load vs. load-line deflection curves, (B) J-Δa curve.

Table 12.2 Fracture toughness results for cortical bone.

	J_{Ic} [mJ/mm^2]					U_T [mJ/mm^3]			
Specimen ID	Group 1	Group 2	etc.	*P*	Specimen ID	Group 1	Group 2	etc.	*P*
1					1				
2					2				
3					3				
etc.					etc.				
Avg					Avg				
SD					SD				

Avg, average; SD, standard deviation. P values indicate statistical differences for pairwise comparisons between various test groups.

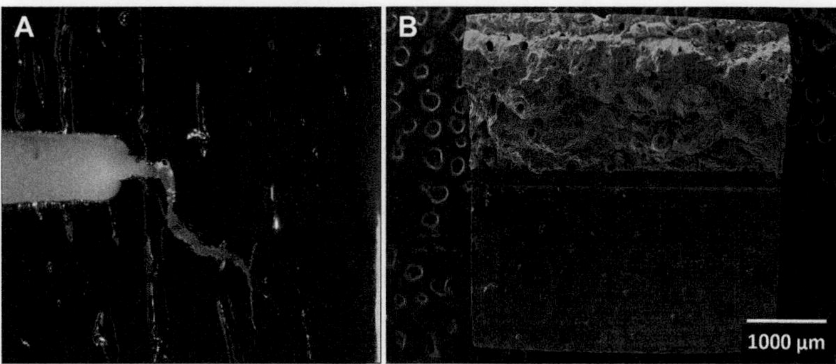

FIGURE 12.9

Human cortical bone images. (A) Digital photograph taken during SENB fracture testing in which the crack defects significantly near the crack tip, forming two tips heading in opposite directions, (B) a scanning electron micrograph of a fracture surface (top) and transverse view (bottom) of the notches that were cut to prepare the specimen for testing.

Step 3. Present visual evidence. Pictures of tested specimens should be presented to communicate their characteristics, the ways in which they fractured, and to document visible mechanisms, such as defection (Fig. 12.9).

ALTERNATIVES AND ADAPTATIONS

- ✓ **Alternate test geometries.** Other test geometries can be used when testing cortical bone. This includes the compact test specimen and chevron notched beam specimen.
- ✓ **Simple calculation of *J*.** A much simpler, though not standard, method to calculate J from experimental data is to use $J_i = 2A_i/[W(B - a_i)]$, where i is some point in the test, A_i is total work done to the specimen up to the point i, W is specimen thickness, B is specimen width, and a_i is the crack length at point i in the test.
- ✓ **Effects of anisotropy.** Bone is an anisotropic and heterogeneous composite biomaterial. Crack growth varies depending on the orientation of the initial crack and the applied loading (e.g., when the crack is oriented in a different plane).

5. DISCUSSION

After completing all fracture toughness tests, data collection, data analysis, and data presentation, then final results can be considered and interpreted in the broader context of some important clinical, biomechanical, and/or technological considerations, as follows.

Cortical bone fracture toughness has been examined numerous times in the literature. Both linear elastic and nonlinear elastic−plastic approaches have been used,[4] with the former being more widespread historically, despite the fact that cortical bone exhibits a substantial amount of postyield behavior before fracture.[7] This violates the underlying assumptions of LEFM. For this reason, elastic–plastic fracture mechanics is a more appropriate method. Results citing *J*-integral values will be more representative of the true fracture toughness of bone.

A fairly large range of critical J_{Ic} values has been reported. The fracture resistance of bone will vary depending on crack orientation, direction, loading mode, and other testing parameters. Values in the range of 2.3–10.0 mJ/mm^2 have been reported for bovine bone[3,8] and 1.34–6.58 mJ/mm^2 for human bone.[9,10]

Cortical bone's fracture toughness has direct implications for skeletal structural integrity. Bone material contains internal microcracks and other types of defects that serve as, or are themselves, crack nucleation sites. As the flaw size and loading condition combination increases, so does the risk of failure. The material's ability to engage fracture toughness mechanisms is essential to mitigate this risk. Bone's fracture toughness mechanisms include microcracking, crack deflection, and crack bridging,[7] and any increase or decrease in fracture toughness signifies an increase or decrease in the effectiveness of these mechanisms. As bone's ability to arrest crack growth diminishes, the tolerable flaw size under given loading conditions (or the other way around) diminishes as well, increasing risk of fracture and failure.

Crack growth resistance is also important in a clinical context. Aging and disease can have deleterious effects on the fracture toughness of bone. A deeper understanding of bone toughness may help develop therapies, treatments, or preventive steps for an array of harmful conditions. The gamma irradiation-based sterilization process for structural cortical bone allograft material also greatly reduces the fracture resistance of the cortical bone. In situations where the graft is structural, fracture rates are unacceptably high.[11–13] Various studies report the effects of gamma irradiation-based sterilization on cortical bone fracture toughness, while the methods to protect against these negative effects have been tested.[8,10,14–17]

6. SUMMARY

- An elastic–plastic approach is more suitable to study cortical bone than a linear elastic approach.
- Fracture toughness quantifies the strain energy release required per crack growth increment.
- Small flaws and cracks mean that fracture toughness is essential for structural integrity.
- Cortical bone displays a rising J-Δa curve because mechanisms engage once crack growth starts.
- Understanding bone fracture toughness and mechanisms is needed to understand fragility.
- No standard methods exist for the fracture toughness testing of cortical bone.

7. QUIZ QUESTIONS

1. When is a straight line fit to the J-Δa data better for determining J_{Ic} than a power curve?
2. What is the importance of an accurate value for Young's modulus in determining J?
3. What is the benefit of a rising J-Δa curve?
4. Why should nonlinear elastic–plastic fracture mechanics methods be used for cortical bone?
5. If at some point in a fracture test of an SENB specimen with $W = 4$ mm and $B = 4$ mm, the total work done is 18 mJ and the crack length is 2.05 mm, what is J? (answer: 4.6 mJ/mm^2).

REFERENCES

1. Hertzberg RW. *Deformation and fracture mechanics of engineering materials.* 4th ed. Hoboken (NJ, USA): John Wiley & Sons, Inc.; 1995.
2. Anderson T. *Fracture mechanics fundamentals and applications.* Boca Raton (FL, USA): Taylor & Francis Group; 2005.
3. Yan J, Mecholsky Jr JJ, Clifton KB. How tough is bone? Application of elastic-plastic fracture mechanics. *Bone* 2006;**40**(2):479–84.
4. Koester K, Ager III J, Ritchie R. The true toughness of human cortical bone measured with realistically short cracks. *Nature Materials* 2008;**7**(8):672–7.
5. ASTM E1820. Standard test method for measurement of fracture toughness. West Conshohocken (PA, USA): American Society for Testing and Materials (ASTM); www.astm.org.
6. Ritchie RO, Nalla RK, Kruzic JJ, Ager III JW, Balooch G, Kinney JH. Fracture and ageing in bone: toughness and structural characterization. *Strain* 2006;**42**(4):225–32.
7. Launey M, Buehler M, Ritchie R. On the mechanistic origins of toughness in bone. *Annual Review of Materials Research* 2010;**40**:25–53.
8. Willett T, Burton B, Woodside M, Wang Z, Gaspar A, Attia T. y-irradiation sterilized bone strengthened and toughened by ribose pre-treatment. *Journal of the Mechanical Behavior of Biomedical Materials* 2015;**44**:147–55.
9. Zioupos P, Currey J, Hamer A. The role of collagen in the declining mechanical properties of aging human cortical bone. *Journal of Biomedical Materials Research* 1999;**45**(2): 108–16.
10. Willett TL, Woodside M. Ribose pre-treatment protects fracture toughness of γ-irradiated sterilized bone allograft. In: *32nd Annual Meeting of the Canadian Biomaterials Society,* Toronto, Canada; May 27–30, 2015. http://biomaterials.ca/static/abstracts/view/113457.
11. Mankin HJ, Friedlaender GE, Tomford WW. Massive allograft transportation following tumor resection. In: *Bone grafts and bone substitutes.* Rosemont (IL, USA): American Academy of Orthopaedic Surgeons; 2006. p. 39–47.
12. Pietrzak WS, editor. *Musculoskeletal tissue regeneration: biological materials and methods.* New York (NY, USA): Humana Press; 2008.

13. Thompson Jr R, Garg A, Clohisy D, Cheng E. Fracture in large-segment allografts. *Clinical Orthopaedics and Related Research* 2000;**370**:227–35.
14. Barth H, Launey M, MacDowell A, Ager J, Ritchie R. On the effect of X-ray irradiation on the deformation and fracture behavior of human cortical bone. *Bone* 2010;**46**(6): 1475–85.
15. Barth H, Zimmermann E, Schaible E, Tang S, Alliston T, Ritchie R. Characterization of the effects of x-ray irradiation on the hierarchical structure and mechanical properties of human cortical bone. *Biomaterials* 2011;**32**(34):8892–904.
16. Akkus O, Belaney R, Das P. Free radical scavenging alleviates the biomechanical impairment of gamma radiation setrilized bone tissue. *Journal of Orthopaedic Research* 2005; **23**(4):838–45.
17. Burton B, Gaspar A, Josey D, Tupy J, Grynpas M, Willett T. Bone embrittlement and collagen modifications due to high-dose gamma-irradiation sterilization. *Bone* 2014; **61**:71–81.

CHAPTER

Multiscale Biomechanical Characterization of Bioceramic Bone Scaffolds

13

Juan F. Vivanco[1], Joshua Slane[2], Ameet Aiyangar[3]

University Adolfo Ibáñez, Viña del Mar, Chile[1]; Institute for Orthopaedics Research and Training at KU Leuven and University Hospitals Leuven, Pellenberg, Belgium[2]; EMPA, Swiss Federal Laboratories for Materials Science and Technology, Duebendorf, Switzerland[3]

1. BACKGROUND

An important idea in bone tissue engineering is the creation of porous three-dimensional (3D) bioceramic bone scaffolds,[1] which are implanted into the defect site to provide mechanical and biological functions facilitating integration with

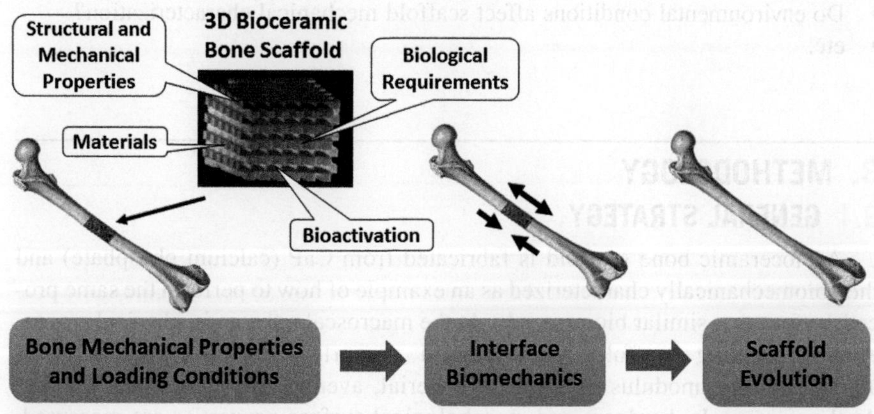

FIGURE 13.1

Biomechanical paradigm in bioceramic bone scaffold engineering. Upon implantation, it is critical that the biomechanical properties of the scaffold should closely match surrounding host bone and loading conditions to reduce the stress shielding effect. Interface biomechanics should allow the scaffold–bone interface to permit enhanced osteointegration of the scaffold. As the scaffold degrades, ingrowing bone tissue will begin to support the mechanical load of the bone scaffold.

Experimental Methods in Orthopaedic Biomechanics.

surrounding bone. A scaffold structure should be able to support and stimulate new bone ingrowth and become completely integrated into the native tissue (Fig. 13.1). Porous scaffolds can either induce formation of bone from the surrounding tissue or act as a carrier or template for implanted bone cells. In addition, scaffolds are required to meet several criteria, such as architectural features, structural mechanics, mass transport, surface properties, degradation products, and cell–material interaction properties. Micro- and nanostructural properties of scaffolds dictate the capacity to induce bone formation; thus, biomechanical properties should be evaluated at multiple scales before implantation.[2] Therefore, this chapter explains how to perform multiscale (i.e., macro, micro, and nano) biomechanical characterization of bioceramic bone scaffolds, as well as how to analyze, present, and interpret results.

2. RESEARCH QUESTIONS

Typical research questions could include one or more of the following:

- How can scaffold geometrical and mechanical properties be controlled during fabrication?
- How is multiscale biomechanical characterization to be properly conducted?
- What is the effect of interscale property relationships on scaffold performance?
- Which scaffold parameters are needed to optimize bone regeneration?
- Do environmental conditions affect scaffold mechanical characterization?
- etc.

3. METHODOLOGY

3.1 GENERAL STRATEGY

A bioceramic bone scaffold is fabricated from CaP (calcium phosphate) and then biomechanically characterized as an example of how to perform the same procedure for other similar biomaterials. At the macroscale, the main physical parameters controlling scaffold biomechanical properties are measured, namely, apparent elastic modulus of the base material, average pore size, and porosity. At the microscale, hardness and morphological surface properties are measured. At the nanoscale, reduced (indentation) modulus and indentation hardness are measured. Such multiscale characterization provides a comprehensive understanding of the required mechanical properties at different scales relevant to biological interactions with the native tissue before implantation.[2] Finally, statistical comparisons are made of the geometric volumes and the macro-, micro-, and nanobiomechanical properties of the scaffolds fabricated across the various environmental conditions.

GLOSSARY

✓ **Bioceramics.** A type of biomaterial commonly used as bone replacement grafts made from materials, such as calcium phosphate.
✓ **Bone scaffold.** A bioporous structure implanted into the defect site to provide temporary mechanical and biological properties.
✓ **Scanning electron microscope (SEM).** Advanced imaging equipment that produces high-resolution images to characterize surface morphology on bone scaffolds.
✓ **Sintering.** A manufacturing method to fabricate bioceramics from powder, which compacts and adheres solid particles to each other at high temperatures around 1000°C.
✓ **X-ray diffraction (XRD).** A laboratory-based technique commonly used for identification of crystalline materials, such as bioceramics.

SAFETY FIRST

✓ The sintering fabrication process must be conducted by an experienced operator.
✓ It is safe to fabricate bioceramic bone scaffolds by traditional 3D printing techniques.
✓ Obtain training for using X-ray diffractometers and SEM microscopes.

3.2 MATERIALS AND TOOLS LIST

- bioceramic bone scaffolds
- digital balance
- grinding and polishing system
- mechanical tester
- microhardness tester
- nanoindentation tester
- scanning electron microscope (SEM)
- Vernier calipers
- X-ray diffractometer (XRD)

3.3 SPECIMEN PREPARATION

Step 1. Design scaffold architecture. Common bioceramic bone scaffold designs consist of a microarchitecture similar to cancellous bone, which presents porosity and pore size in the range, respectively, of 50–90% and 200–500 μm (Fig. 13.2).[3] Architectural matching to real bone permits better scaffold integration with surrounding host bone when the scaffold is implanted into live patients.

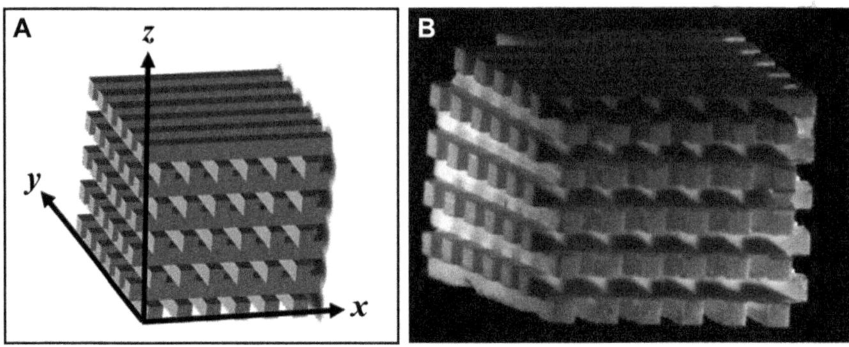

FIGURE 13.2

Bioceramic bone scaffold specimen fabricated from CaP. (A) Theoretical design, (B) photograph of an actual specimen (~5 mm^3).

Step 2. Design scaffold size. Use commercially available design software to prepare the scaffold design with a geometry of six beams with a cross-sectional area of about 500 μm^2. Stack the beams upon one another in orthogonal directions to form a cubic structure with a nominal or bulk volume of about 5 mm^3 (Fig. 13.2). The reason for this size is that it represents the dimensions of a scaffold that can be implanted.

Step 3. Choose scaffold material. Specimens are commonly made from CaP (calcium phosphate) because of clinical applications due to their inherent biocompatibility, osteoconductivity, and osteointegrity (Fig. 13.2). However, other potential materials may be polymers and composites.

Step 4. Choose sintering temperatures. Scaffolds made of bioceramics are manufactured by sintering, a well-known method to fabricate ceramics from powder. This method compacts and adheres solid particles to each other at high temperatures around 1000°C. Thus, scaffolds should be sintered at target temperatures of $T_1 = 950°C$, $T_2 = 1050°C$, and $T_3 = 1150°C$ in air using a common bioceramic heating scheme. The sintering temperature during the fabrication process can be an important determinant of microstructural material properties.

Step 5. Fabricate the scaffold. Different techniques are available to fabricate scaffolds: (1) traditional techniques, such as solvent casting, particulate leaching, gas foaming, and melt molding, which are limited in controlling internal architecture[4]; (2) rapid prototyping (RP) techniques, which combine computer-aided design (CAD) with RP for increased precision[4]; and (3) injection molding, which is a cost-effective alternative to RP for large-scale productions.[5] Fabricate at least five specimens to do a proper study on biomechanical characterization.

TIPS AND TRICKS

- ✓ Use the fluid displacement method to measure total volume of porous bone scaffold.
- ✓ Use ethanol as a displacement liquid, as it penetrates the scaffold to fill small pores.
- ✓ Remove trapped bubbles before weighing the specimen to avoid measurement errors.
- ✓ Nanoindentation is an advanced technique requiring training and/or expert assistance.
- ✓ Evaluate machine compliance using a material with known properties (e.g., fused silica).

THE "GOLD STANDARD"

The American Society for Testing and Materials (ASTM) has testing standards for some aspects of the biomechanical characterization of bioceramics, such as ASTM C1239 (Standard test method for reporting uniaxial strength data and estimating Weibull distribution parameters for advanced ceramics) and ASTM C1326 (Standard test method for Knoop indentation hardness of advanced ceramics). However, researchers should also consult previously published peer-reviewed journal articles to provide guidelines specifically for bioceramic bone scaffold fabrication and testing.

3.4 SPECIMEN TESTING

Measure Architectural and Geometrical Properties

Step 1. Analyze the crystal structure. Determine the crystallographic phase and crystallite size for each sintering temperature T_1, T_2, and T_3 using XRD. Use a 2D area detector with settings at 40 kV and 40 mA. Record the XRD patterns with an angle of diffraction θ in the range of $15° < 2\theta < 65°$ with a step size of 0.02° and step duration of 0.5 s. The experimental XRD spectra can be identified by computer matching with the ones based on the structural data of similar apatite bioceramics available in the International Centre for Diffraction Data database.[6,7]

Step 2. Analyze the morphology. Investigate the morphology of the scaffold surfaces by SEM. Mount specimens on aluminum stubs with double-sided carbon tape. Sputter coat the surface with gold for 30 s using a current of 45 mA. Perform the morphology characterization using field emission SEM (or equivalent) that is operating at 5 kV.

Step 3. Measure scaffold dimensions. Measure the dry bulk outer dimensions of length, width, and height of the scaffold using a digital Vernier caliper (Fig. 13.3A).

Step 4. Measure scaffold weight. Measure the dry weight of the specimen with a digital balance that has a precision of 0.1 g. Then, in order to evaluate water absorption, submerge specimens in distilled water for 24 h, and then centrifuge them at 3000 rpm for 3 min to eliminate excess water on their surface. Finally, measure the weight of the wet specimen $W_{scaffold}$.

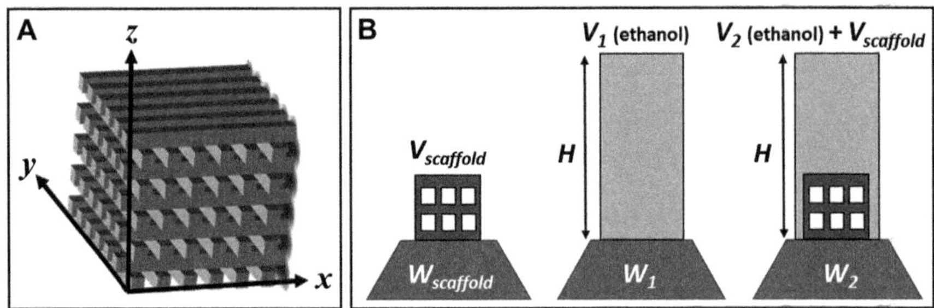

FIGURE 13.3

Geometrical aspects of the bioceramic bone scaffold. (A) Scaffold computer-aided design model (~5 mm^3), (B) measurement of scaffold densities and volumes by fluid displacement.

Step 5. Measure scaffold volume. Apply Archimedes' principle to measure volume using the fluid displacement method with a fluid of known density (Fig. 13.3B).[8,9] To do so, fill a container of volume V_1 completely with a 70% ethanol solution mixed with water, and record the height of the liquid level H. Measure the weight W_1 of the liquid, and subtract the weight of the container. Immerse the water-saturated scaffolds in the container with the ethanol solution, and adjust the volume of alcohol so that the container with the scaffold is completely filled again to the same liquid level H. For each specimen, agitate the ethanol until no air bubbles emerge from the scaffold and all pores are filled. This new volume is composed of the scaffold $V_{scaffold}$ and the remaining ethanol V_2, so that $V_{scaffold} = V_1 - V_2$.

Measure Macrobiomechanical Properties

Step 1. Choose a mechanical tester. Any standard mechanical tester can be used to perform a quasi-static mechanical test, including screw-driven, servohydraulic, or electromechanical testers.[10] Alternatively, a custom-made loading system that applies controlled uniaxial forces and calculates deformations can be designed and calibrated to measure stiffness in the range of trabecular bone.[11]

Step 2. Choose anisotropic and environmental conditions. To assess the effects of anisotropy, evaluate the mechanical behavior of specimens in uniaxial compression along the x and z direction for each sintering temperature T_1, T_2, and T_3 (Fig. 13.3A). To assess the effects of environmental conditions, also evaluate the mechanical behavior using dry specimens at a room temperature of $T_{dry\ room} = 23°C$ and wet specimens in a physiological saline solution at a body temperature of $T_{wet\ body} = 37°C$. Thus, a total of four test conditions is possible.

Step 3. Precondition the specimens. Subject scaffolds to quasi-static uniaxial compression loads for the purpose of preconditioning. The aim is to improve reproducibility by applying two load–unload cycles and allow full contact of specimen surfaces. For example, by applying two preconditioning trials with a preload of 10 N followed by a compressive displacement of 5 μm, these values correspond to a specific stress and strain for the dimensions of the chosen specimens.

Step 4. Perform macroquasistatic tests. Perform a quasi-static uniaxial compression test on each scaffold until the specimen fails. To do so, apply displacement (~5 mm/min) gradually until a maximum force F_{max} is reached.

Step 5. Identify the failure event. Failure, or partial cracking, of the specimen can be identified in the force vs. displacement curve as the peak force followed by a sudden decrease in force.

Measure Microbiomechanical Properties

Step 1. Mechanically grind the specimens. Grind the scaffold's outer surfaces using a series of silicon carbide paper of 600 to 800 to 1200 grit in ascending order by using a belt sander. Then, polish surfaces with a 0.25-μm diamond suspension on a wet grinding wheel.

Step 2. Choose test conditions. Conduct microbiomechanical testing on specimens fabricated at T_1, T_2, and T_3. For practical reasons, conduct the experiments on the *xy* plane. In this particular case, testing should be performed in the dry condition.

Step 3. Conduct a Knoop hardness test. Measure the Knoop hardness of each scaffold specimen following the ASTM C1326 standards for advanced ceramics.[12] To do so, conduct microhardness tests on the cross-section of the scaffolds with a digitally controlled microindentation hardness tester. At room temperature, apply a test load of 5 N and a hold time of 15 s.

Step 4. Calculate Knoop hardness. Knoop hardness is a function of the test force divided by the projected area of the indentation. Measure the longest diagonal *d* of the indentation using a digital caliper usually placed in the instrument. Calculate Knoop hardness using the formula $H_K = 14{,}229F/d^2$, where H_K is Knoop hardness [kg/mm^2], F is indentation force [g], and d is longest diagonal of the indentation [μm]. Convert H_K values to GPa units by multiplying by 9.806×10^{-3}.

Step 5. Perform multiple tests. Perform at least 10 measurements, but maintain a minimum spacing of 2.5*d* between each successive microindentation test to prevent a strain field around the last indentation, and so two neighboring measurements can be considered independent.

Measure Nanobiomechanical Properties

Step 1. Choose the specimens. The same specimens used for microtesting can also be used for nanomechanical characterization if previous indentations do not affect the new ones and are located farther away. However, if using new specimens, grind the scaffold's outer surfaces using a series of silicon carbide paper of 600 to 800 to 1200 grit in ascending order by using a belt sander. Then, polish surfaces with a 0.25-μm diamond suspension on a wet grinding wheel.

Step 2. Measure tester compliance. Quantify the nanoindentation tester compliance as a calibration standard before starting any measurement. Tester compliance is obtained by performing a series of 100 indents ranging from

0.2–10 mN on a fused silica calibration standard. The majority of commercially available nanoindentation testers have specific preestablished protocols for performing compliance measurements.

Step 3. Choose test conditions. For nanobiomechanical testing, use specimens fabricated at sintering temperatures T_1, T_2, and T_3. For practical reasons, conduct the experiments on the *xy* plane. In this particular case, testing should be performed in the dry condition.

Step 4. Perform nanoindentation tests. Conduct static indentation tests to evaluate linear elastic properties. This can be done using a trapezoidal loading function with a defined peak load (e.g., 200 μN), a low load–unload rate (e.g., 40 μN/s to minimize viscoelastic effects), and a low dwell time (e.g., 5 s).

Step 5. Calculate nanoproperties. Compute the nanohardness H_n and elastic modulus E_n at each load from the resulting force vs. displacement curves. Then, average the values together using the equations that are based on traditional methods.[13] Most commercial equipment has software that calculates H_n and E_n for each individual measurement.

Step 6. Compensate for specimen structural compliance (flexing). If specimen-scale flexing occurs during nanoindentation of porous bone scaffolds, an advanced procedure can be performed to remove this potential artifact, known as structural compliance.[14] Thus, load control multiload indents can be conducted, which consist of loading segments, holding at the partial load, and unloading segments with load increasing for each cycle.[15] For example, each scaffold can be subjected to a series of 14 partial unload tests, which consist of 12 steps ranging from 1–12 mN with loading–unloading segments of 3 s and hold segments of 1 s.

3.5 RAW DATA COLLECTION

Step 1. Record scaffold characteristics. For each sintered scaffold group, record the geometrical dimensions and weights (Table 13.1).

Table 13.1 Raw data collected for physical properties for bioceramic bone scaffolds.

Fabrication Temp. [°C]	Specimen ID	Dimensions [mm]			Scaffold Weight $W_{scaffold}$ [g]		Liquid Weight, W_1 [g]	Weight of Remaining Liquid Plus Scaffold, W_2 [g]	Liquid Density [g/mm^3]
		x	*y*	*z*	Dry	Wet			
T_1 (T_2, T_3)	1								
	2								
	etc.								
Avg	—								
SD	—								

T_1, T_2, and T_3, sintering temperatures used for scaffold fabrication. Avg, average; SD, standard deviation.

Table 13.2 Raw data collected for macromechanical behavior of bioceramic bone scaffolds at different sintering temperatures, environmental conditions, and compressive loading directions.

Fabrication Temp. [°C]	Specimen ID	Load Direction [x, y, or z]	Environmental Test Conditions [Dry or Wet, and T_{room} or T_{body}]	Stiffness K [N/μm]	Failure Force F_{max} [N]
T_1 (T_2, T_3)	1				
	2				
	etc.				
Avg	—	—			
SD	—				

T_1, T_2, *and* T_3, *sintering temperatures used for scaffold fabrication.* Avg, *average;* SD, *standard deviation.*

Step 2. Record scaffold macromechanical properties. For each scaffold orientation and environmental condition, store raw compressive force vs. displacement data from which related mechanical properties, such as stiffness and maximum force, will be obtained (Table 13.2).

Step 3. Record scaffold micro- and nanomechanical properties. Micromechanical measurements of hardness are obtained by the applied load, and the diagonals of the indented area measured directly by the instrument (Table 13.3). Nanomechanical properties, such as nanohardness and elastic modulus, are obtained directly from the software during testing when standard tests with a single maximum load are performed (Table 13.3). Finally, identify the location where testing was conducted on the specimen surface (Table 13.3).

Table 13.3 Raw data collected for micro- and nanomechanical properties of bioceramic bone scaffolds fabricated at different sintering temperatures.

			Microlevel	Nanolevel	
Fabrication temp. [°C]	Specimen ID	Indent Location [Corner, Edge, or Center]	Knoop Hardness H_K [GPa]	Hardness H_n [GPa]	Elastic Modulus E_n [GPa]
T_1 (T_2, T_3)	1				
	2				
	etc.				
Avg	—	—			
SD					

T_1, T_2, *and* T_3, *sintering temperatures used for scaffold fabrication.* Avg, *average,* SD, *standard deviation.*

3.6 RAW DATA ANALYSIS

Step 1. Calculate scaffold volume. Scaffold volume is $V_{scaffold} = V_1 - V_2 = [W_1 - (W_2 - W_{scaffold})]/\rho_{ethanol}$, where W_2 is weight of remaining fluid plus the scaffold, W_1 is weight of the liquid, $W_{scaffold}$ is weight of the scaffold in the dry condition, and $\rho_{ethanol}$ is density of ethanol.

Step 2. Determine macro structural stiffness. Use the slope of the linear region of the force vs. displacement curve to calculate stiffness $K = \Delta F/\Delta\delta$, where F is applied force and δ is compressive displacement.

Step 3. Determine macro apparent elastic modulus. Use the bulk dimensional measurements to convert stiffness to apparent elastic modulus as $E_{app} = KL/A$, where K is stiffness, L is bulk length of the specimen in the direction of load, and A is bulk area of the top surface.

Step 4. Determine macro compressive strength. Compute the compressive strength at specimen failure as $\sigma_c = F_{max}/A$, where F_{max} is maximum applied force and A is measured initial bulk cross-sectional area of the specimen.

Step 5. Compute macro compressive strength distribution. Compressive strength of bioceramic materials can be analyzed by the two-parameter Weibull distribution.[16] Use this statistical distribution to determine the variation in compressive strength of bioceramic scaffolds by calculating failure probability $P_f = 1 - \exp[(-\sigma_c/\sigma_o)^m]$, where σ_c is compressive strength, σ_o is Weibull scale parameter, and m is Weibull modulus.

Step 6. Compute micro- and nanomechanical properties. For specimens fabricated at each sintering temperature, calculate microproperties, such as Knoop hardness H_K, and obtain nanoproperties like hardness H_n and elastic modulus E_n from the nanoindenter equipment.

Step 7. Perform statistical analysis. To investigate the effect of sintering temperature T_1, T_2, and T_3 on physical properties, one-way analysis of variance (ANOVA) with a Tukey's honestly significant difference posthoc test can be performed. To evaluate if there are significant differences in apparent elastic modulus E_{app} as an effect of either sintering temperature (i.e., T_1, T_2, and T_3) or testing environment (i.e., $T_{dry\ room}$ and $T_{wet\ body}$), a two-way ANOVA can be performed. When only two groups are compared, a two-sample student's t-test can be performed. In all cases, the significance level of $P < 0.05$ is commonly accepted, whereas assumptions of normality and equality of variance can be checked by the Anderson–Darling normality test and the F-test, respectively.

Step 8. Compute statistical power. Power analysis can be done after the study to ensure there were enough specimens per test group to detect all statistical differences that were actually present (i.e., was type II statistical error avoided?). Statistical power >80% is usually considered to indicate there were enough specimens per test group. Note that if good predictions of averages and standard deviations are available from prior studies, then the number of specimens and/or tests can be chosen before the study begins to ensure a power >80%.

ENGINEER'S TOOLBOX

A nanoindentation tester equipped with a Berkovich diamond probe can be used to evaluate several important nanomechanical properties. First, the reduced modulus for the specimen and indenter materials is measured as $E_r = (K_u/2\beta)(\pi/A_c)^{1/2}$, where K_u is unloading stiffness measured by taking the derivative of a nonlinear least-squares power law fit to the unloading curve at maximum force P_{max}, β is nondimensional parameter = 1.034 for a Berkovich indenter, and A_c is projected contact area of the indentation. Second, the contact area is calculated from the ideal indenter geometry as a function of contact depth $h_c = h_{max} - \mathcal{E}P_{max}/K_u$, where h_{max} is indenter depth at P_{max} and $\mathcal{E}$ is geometric constant = 0.75 for most indenters. Third, indentation modulus E_s of the specimen is determined from the reduced modulus using $1/E_r = (1 - \nu_s^2)/E_s + (1 - \nu_i^2)/E_i$, where ν_s is Poisson's ratio of the specimen material, ν_i is Poisson's ratio of the indenter material = 0.07 for diamond, and E_i is elastic modulus of the indenter material = 1141 GPa for diamond. Finally, nanoindentation hardness of the specimen is computed as $H_n = P_{max}/A_c$.

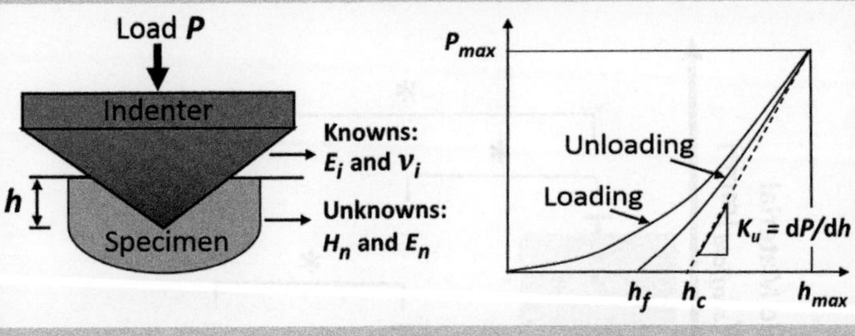

4. RESULTS

Once all bone scaffold data collection and analysis have been performed, it is then important to communicate and present the primary results in an understandable and concise manner to the reader of a journal article, conference paper, technical report, or book chapter.

Step 1. Show images of micro- or nanoindentation. Present images of micro- and nanoindentation to verify that the experiment was conducted according to standards and that the measurement was correctly read. A typical microhardness indentation has a size scale or bar, which is mandatory in any microscope image (Fig. 13.4A). In addition, it is useful to show a representative SEM micrograph to illustrate scaffold surface morphology characteristics, such as grain size and grain boundary (Fig. 13.4B).

Step 2. Show scaffold geometrical properties. Report scaffold geometrical features either in a table or using box plots, such as volume as a function of the sintering fabrication process (Fig. 13.5).

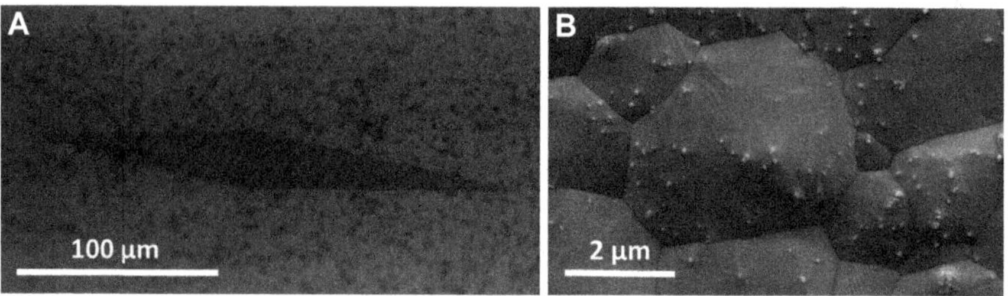

FIGURE 13.4

Micro- or nanoindentation images. (A) Representative Knoop hardness microindentation on a bioceramic bone scaffold, (B) representative SEM micrograph of the morphology of the cross-section of a bioceramic bone scaffold.

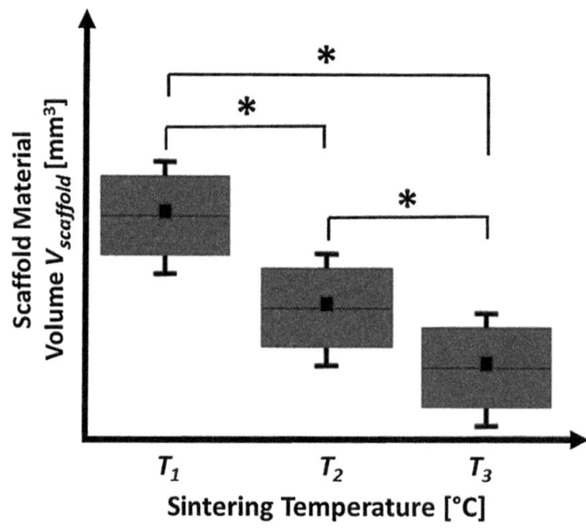

FIGURE 13.5

Box plot of volumes $V_{scaffold}$ of bioceramic bone scaffolds sintered at different temperatures T_1, T_2, and T_3. Volumes are evaluated by the fluid displacement method. Error bars represent ±1 standard deviation, vertical line limits are maximum and minimum values, and shaded squares are average values. Statistical differences (i.e., P <0.01 or <0.05) between test groups are indicated by symbols, such as asterisks (*).

Step 3. Show raw macromechanical test results. Provide a representative force vs. displacement curve obtained during uniaxial compression test of a porous bioceramic scaffold (Fig. 13.6). Bioceramic porous scaffolds under compression tests commonly have force vs. displacement behavior with a linear elastic region followed by a collapse plateau, which is characteristic of brittle fracture of elastic cellular solid structures.

Step 4. Show multiscale mechanical properties. Sintering temperatures (i.e., T_1, T_2, and T_3) and environmental conditions (i.e., dry and wet) might have a

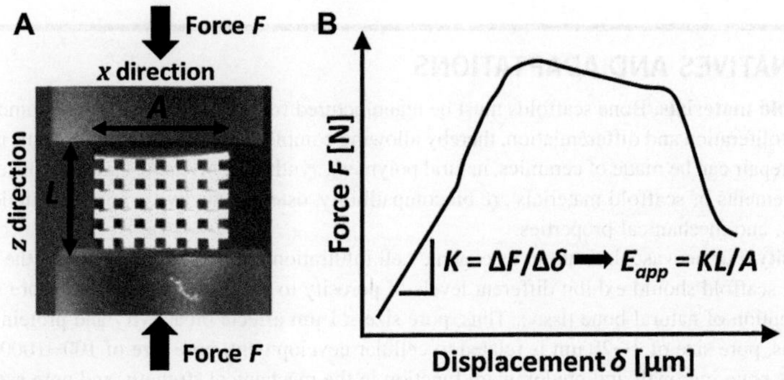

FIGURE 13.6

Compressive macromechanical testing of a bioceramic bone scaffold. (A) Test configuration, (B) representative force vs. displacement curve showing the linear region used to calculate specimen stiffness K and apparent elastic modulus E_{app}. For this test configuration, cross-sectional area A is measured in the xy plane, whereas the length L of the specimen used to calculate compressive displacement is measured along the z direction.

statistically significant effect on mechanical properties at multiple scales. The interaction of both effects also might have a significant influence. Hence, it is common to show several bar plots, which, in turn, statistically compare a particular mechanical property at the macro- (i.e., E_{app}, K, F_{max}), micro- (i.e., H_K), or nano- (i.e., H_n, E_n) scale across sintering temperatures and environmental conditions (Fig. 13.7).

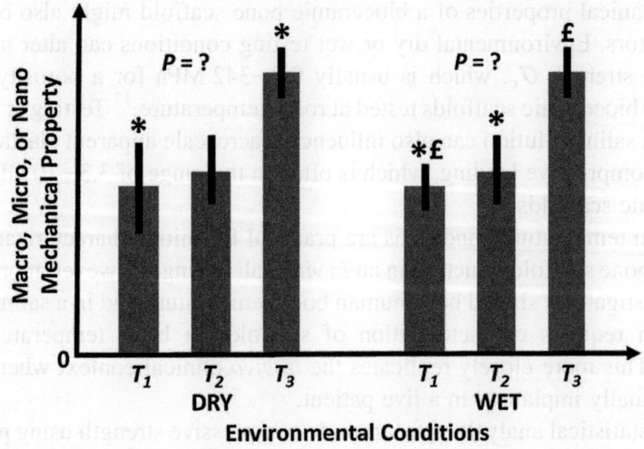

FIGURE 13.7

Representative bar plot for a given macro-, micro-, or nanomechanical property. The bioceramic scaffolds are tested at different sintering temperatures T_1, T_2, and T_3 and environmental conditions. Each result is an average ±1 standard deviation. Statistical differences (i.e., $P < 0.01$ or < 0.05) between test groups are indicated by symbols, such as asterisks (*), pounds (£), etc.

ALTERNATIVES AND ADAPTATIONS

- ✓ **Scaffold materials.** Bone scaffolds must be manufactured from base materials that promote bone cell proliferation and differentiation, thereby allowing complete integration. Biomaterials used in bone repair can be made of ceramics, natural polymers, synthetic polymers, and composites. The requirements of scaffold materials are biocompatibility, osteoconductivity, porosity, biodegradability, and mechanical properties.
- ✓ **Porosity.** For neovascularization, osteogenic cell infiltration, and bone ingrowth into the defect site, a scaffold should exhibit different levels of porosity to mimic the hierarchical pore size distribution of natural bone tissue. Thus, pore size <1 μm affects bioactivity and protein interactions, pore size of 1–20 μm is related to cellular development, pore size of 100–1000 μm affects bone ingrowth and predominant function in the mechanical strength, and pore size >1000 μm is related to implant functionality and implant esthetics.

5. DISCUSSION

After completing all bioceramic bone scaffold tests, data collection, data analysis, and data presentation, then final results can be considered and interpreted in the broader context of some important clinical, biomechanical, and/or technological considerations, as follows.

Increasing the sintering temperatures T_1, T_2, and T_3 used to fabricate bioceramic bone scaffolds might affect scaffold geometrical and physical characteristics, such as decreasing pore size, decreasing porosity, and increasing biomechanical properties.

Biomechanical properties of a bioceramic bone scaffold might also be affected by other factors. Environmental dry or wet testing conditions can alter macroscale compressive strength σ_c, which is usually 0.8–342 MPa for a porosity range of 23–80% for bioceramic scaffolds tested at room temperature.[17] Testing at body temperature in a saline solution can also influence macroscale apparent elastic modulus E_{app} under compressive loading, which is often in the range of 3.3–10 GPa for useful bioceramic scaffolds.[9]

Dry room temperature conditions are practical for initial characterization of the bioceramic bone scaffold structure in an *in vitro* lab setting. However, more comprehensive investigations should be at human body temperature and in a saline environment, which requires characterization of scaffolds at body temperature in wet conditions. This more closely replicates the *in vivo* clinical context when the scaffold is eventually implanted in a live patient.

Weibull statistical analysis of macroscale compressive strength using probability P_f can be performed for several loading configurations (e.g., x and z directions) and environmental conditions (i.e., $T_{dry\ room}$ and $T_{wet\ body}$). Weibull analysis characterized by the Weibull modulus m indicates the reliability of sintering fabrication on macroscale compressive strength. Consequently, successful bioceramic bone scaffolds have Weibull probabilities P_f of 2–17 and moduli m of 0.85–5.55.[9]

The macroscale apparent elastic modulus E_{app} of bioceramic bone scaffolds with controlled macroporosity can be comparable to human bone; therefore, these systematically manufactured scaffolds may be used for bone tissue regeneration while providing structural integrity. E_{app} for human bones and good quality scaffolds are typically in the range of 0.1–1.0 GPa.[9]

Clinically successful bioceramic bone scaffolds may stimulate bone regeneration and scaffold integration while providing structural integrity. Many biological studies are required since physical and biological functionalities of scaffolds can specifically influence the cell–material interactions at the nanoscale. Furthermore, scaffold bioactivity is important to consider since the nanoscale mechanical properties influence the scaffold's bioactivity, which is related to the protein adhesion and subsequently cellular ingrowth on the scaffold surface.

6. SUMMARY

- Bone scaffolds are porous structures that provide temporary support at a defect site.
- Bone scaffold engineering combines cell biology, biomaterials, and biomechanics.
- Multiscale testing offers a better understanding of scaffold behavior before cell seeding.
- Mechanical testing should consider more realistic environmental conditions.
- Sintering temperature of bioceramics affects structural properties at all scales.
- Bioceramic compressive strength is better represented by a Weibull distribution.
- Bioscaffolds can be customized by controlling fabrication and architectural parameters.

7. QUIZ QUESTIONS

1. What is the purpose of a bone scaffold?
2. How is elastic modulus determined for a porous biocompatible structure?
3. Why is it important to conduct a multiscale characterization of bioceramic scaffolds?
4. What are the common fabrication techniques that are used to produce scaffolds?
5. Estimate the length of the longest diagonal in μm of a microindentation test conducted on a bioceramic scaffold surface when a load of 5 N was applied for 15 s and the H_K value resulted in 2.53 GPa (answer: 167.69 μm).

REFERENCES

1. Hollister S. Scaffold engineering: a bridge to where? *Biofabrication* 2009;**1**:012001.
2. Bohner M, Loosli Y, Baroud G, Lacroix D. Commentary: deciphering the link between architecture and biological response of a bone graft substitute. *Acta Biomaterialia* 2011;**7**(2):478–84.

3. Karageorgiou V, Kaplan D. Porosity of 3D biomaterial scaffolds and osteogenesis. *Biomaterials* 2005;**26**(27):5474–91.
4. Bartolo PJS, Almeida H, Laoui T. Rapid prototyping and manufacturing for tissue engineering scaffolds. *International Journal of Computer Applications and Technology* 2009; **36**(1):1–9.
5. Kramschuster A, Turng L-S. An injection molding process for manufacturing highly porous and interconnected biodegradable polymer matrices for use as tissue engineering scaffolds. *Journal of Biomedical Materials Research (Part B): Applied Biomaterials* 2009;**92**(2):366–76.
6. *International Centre for Diffraction Data (ICDD)*, Newtown Square (PA, USA). www.icdd.com.
7. Vivanco J, Slane J, Nay R, Simpson A, Ploeg HL. The effect of sintering temperature on the microstructure and mechanical properties of a bioceramic bone scaffold. *Journal of the Mechanical Behavior of Biomedical Materials* 2011;**4**(8):2150–60.
8. Vivanco J, Garcia-Rodriguez S, Smith E, Ploeg H. Material and mechanical properties of Tricalcium phosphate-based (TCP) scaffolds. *American Society of mechanical engineering (ASME) Summer Bioengineering conference*. June 2009. Lake Tahoe, CA, USA, paper no. SBC2009–204934.
9. Vivanco J, Aiyangar A, Araneda A, Ploeg HL. Mechanical characterization of injection-molded macro porous bioceramic bone scaffolds. *Journal of the Mechanical Behavior of Biomedical Materials* 2012;**9**:137–52.
10. Roeder R. Mechanical characterization of biomaterials. In: Bandyopadhyay A, Bose S, editors. *Characterization of biomaterials*. Amsterdam, Netherlands: Elsevier, Inc.; 2013 [Chapter 3].
11. Garcia-Rodriguez S, Smith EL, Ploeg HL. A calibration procedure for a bone loading system. *Journal of Medical Devices* 2008;**2**(1):011006-6.
12. ASTM C1326. Standard test method for Knoop indentation hardness of advanced ceramics. West Conshohocken (PA, USA): American Society for Testing and Materials (ASTM). www.astm.org.
13. Oliver WC, Pharr GM. An improved technique for determining hardness and elastic modulus using load and displacement sensing indentation experiments. *Journal of Materials Research* 1992;**27**:1564–83.
14. Vivanco J, Jakes J, Slane J, Ploeg HL. Accounting for structural compliance in nanoindentation measurements of bioceramic bone scaffolds. *Ceramics International* 2014; **40**(8):12485–92.
15. Slane J, Vivanco J, Ebenstein D, Squire M, Ploeg HL. Multiscale characterization of acrylic bone cement modified with functionalized mesoporous silica nanoparticles. *Journal of the Mechanical Behavior of Biomedical Materials* 2014;**37**:141–52.
16. ASTM C1239. Standard test method for reporting uniaxial strength data and estimating Weibull distribution parameters for advanced ceramics. West Conshohocken (PA, USA): American Society for Testing and Materials (ASTM). www.astm.org.
17. Wagoner Johnson AJ, Herscherl BA. A review of the mechanical behavior of CaP and CaP/polymer composites for applications in bone replacement and repair. *Acta Biomaterialia* 2011;**7**(1):16–30.

CHAPTER 14

Measuring Bone Cell Response to Fluid Shear Stress and Hydrostatic/Dynamic Pressure

Kevin A.J. Middleton, Yu-Heng (Vivian) Ma, Lidan You
University of Toronto, Toronto, ON, Canada

1. BACKGROUND

Mechanical stimulation of bone regulates bone remodeling via Wolff's Law, which can be applied to bone disorders (e.g., osteoporosis) to improve bone strength.[1–4] Tissue level loading of bone transduces to cellular forces in a complex manner (Fig. 14.1). Bone-forming osteoblasts follow bone-resorbing osteoclasts in a bone remodeling unit and are exposed to fluid shear stresses.[5,6] Mechanosensing osteocytes reside within a lacunar–canalicular space and are exposed to fluid flow,[7,8] hydrostatic pressure,[9,10] and strain.[11] Bone marrow stromal cells (BMSCs) reside within the marrow and undergo intramedullary pressures along with intermedullary and interstitial fluid flow.[12] Understanding bone cell responses to mechanical

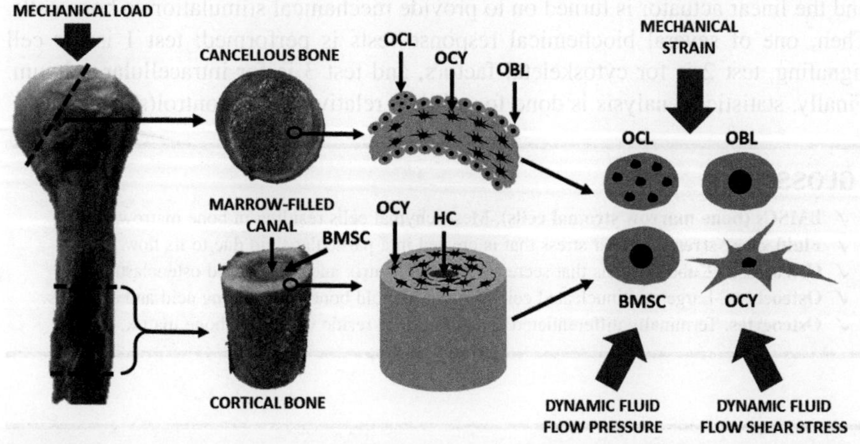

FIGURE 14.1

Transduction of mechanical load to bone cells. *BMSC*, Bone marrow stromal cell; *HC*, haversian canal; *OBL*, osteoblast; *OCL*, osteoclast; *OCY*, osteocyte.

Experimental Methods in Orthopaedic Biomechanics.

stimulation helps identify signaling pathways and cellular interactions for bone homeostasis, thus leading to new therapies for bone disorders. Therefore, this chapter explains how to perform *in vitro* mechanical stimulation tests on bone cells, as well as how to analyze, present, and interpret results.

2. RESEARCH QUESTIONS

Typical research questions might include one or more of the following:

- How do bone cells respond to mechanical loading using biochemical signals?
- How do changes in shear, frequency, and pressure affect cellular response?
- How do effector cells respond to the signals produced by stimulated bone cells?
- What are the major signaling pathways involved in cell mechanotransduction?
- etc.

3. METHODOLOGY

3.1 GENERAL STRATEGY

Bone cells are mechanically stimulated in order to understand their biochemical response. To do so, bone cells are seeded on glass slides. Then, glass slides are placed into flow chambers, experimental parameters are selected for the type of mechanical stimulation (i.e., fluid flow shear stress or hydrostatic/dynamic pressure), and the required linear actuator displacement is calculated and set. Next, flow chambers are connected to the linear actuator via tubing and syringes, and the linear actuator is turned on to provide mechanical stimulation to bone cells. Then, one of several biochemical response tests is performed: test 1 is for cell signaling, test 2 is for cytoskeletal factors, and test 3 is for intracellular calcium. Finally, statistical analysis is done for all data relative to the control(s).

GLOSSARY

✓ **BMSCs (bone marrow stromal cells)**. Mesenchymal cells residing in bone marrow.
✓ **Fluid shear stress**. A shear stress that is created in a particular fluid due to its flow.
✓ **Osteoblasts**. Cuboidal cells that secrete new bone matrix and trail behind osteoclasts.
✓ **Osteoclasts**. Large multinucleated cells that remove old bone by secreting acid and enzymes.
✓ **Osteocytes**. Terminally differentiated osteoblasts that reside within the bone matrix.

SAFETY FIRST

✓ Make sure to wear a lab coat and gloves while working in the biosafety cabinet (BSC) and with biological samples.
✓ Disinfect all tools with 70% ethanol before placing them into the BSC.
✓ Perform all tasks in a fume hood when working with formaldehyde.

3.2 MATERIALS AND TOOLS LIST

- accessories (i.e., connectors, dye, gaskets, grease, screws, syringes, tubing)
- bone cells (human or animal osteoblasts, osteoclasts, osteocytes, or BMSCs)
- bone cell–specific media
- BSC
- flow chambers
- fluorescence microscope
- glass slides
- incubator
- linear actuator
- ruler or Vernier calipers

3.3 SPECIMEN PREPARATION

Bone Cell Preparation for Test 1 (Cell Signaling) and Test 2 (Cytoskeleton)

Step 1. Prepare the glass slides. Some cell lines may require collagen coating of glass slides. To do this, soak the glass slides in 70% ethanol for 15 min, allow them to dry, place them in sterile petri dishes in the BSC, cover them with 1 mL of collagen solution (i.e., 5% rat tail collagen (weight/volume) in 0.02 N acetic acid solution), let them sit for 1 h, and remove excess collagen solution. If glass slides will not be used immediately, store them at 4°C in sterile dishes up to 30 days.

Step 2. Prepare the cells. Cells should be harvested and suspended at about 150,000 cells/mL in media, which varies depending on cell type and incubation time. Apply 1 mL of cell suspension to glass slides while ensuring full coverage but without spillage over the edges. Place dishes with slides in the incubator for at least 1 h to let cells attach, while not too long to avoid media drying. Fill dishes with 10 mL of fresh media in the BSC, and put them back into the incubator. Incubate cells until they are confluent (i.e., 70–80% coverage of the glass slide).

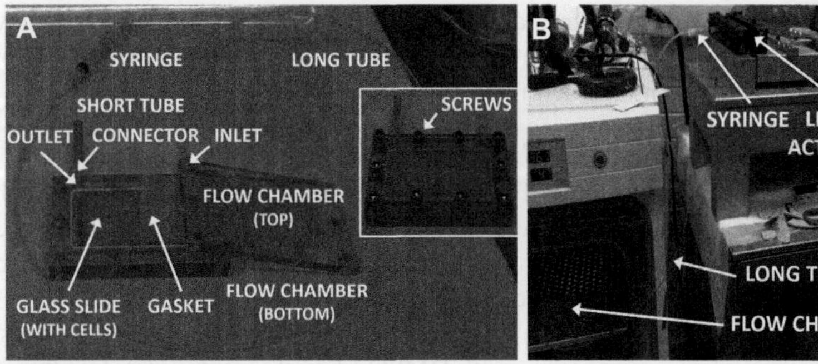

FIGURE 14.2

Experimental apparatus for test 1 (cell signaling) and test 2 (cytoskeleton). (A) Unassembled flow chamber (inset photo: assembled flow chamber), (B) assembled flow chamber placed in the incubator and connected to the linear actuator.

Step 3. Assemble the flow chambers. Autoclave the chambers, including bottoms/tops, connectors, gaskets, screws, and tubes (Fig. 14.2A). Assemble each flow chamber as follows. Attach connectors to inlets and outlets of the chamber bottom. Position the chamber bottom flat with the slot side facing up. Fill the inlet and outlet of the chamber bottom with media using syringes and short tubing. Use a tweezer to slowly place a glass slide (with cell-seeded side facing down) onto the chamber bottom slot while adding media into the slot. (Note: Avoid air bubbles between the slide and the chamber by tilting the slide with one edge touching an edge of a chamber bottom slot, then slowly lowering the glass slide while adding growth media through the inlet near the touching edge.) Place a gasket on top of the slide, and press down on the gasket to clear air bubbles and create a tight seal. Place the chamber top onto the chamber bottom, and tighten the screws.

Step 4. Connect flow chamber to syringe. Connect a syringe to the long tubing. Fill them both with growth media. Replace the short tubing on the outlet with the long tubing connected to the syringe. Leave short tubing on the outlet (Fig. 14.2A). Note that for static controls, remove short tubing on both the inlet and outlet, but cap an inlet to maintain a constant pressure and avoid leakage.

Step 5. Insert chambers into the incubator. Flow chambers are placed in the incubator as soon as assembled (Fig. 14.2B). To do this, flip the flow chamber over so that the cells are facing up. Then, place the assembled chamber in the incubator with the syringe on a linear actuator.

Step 6. Prepare the linear actuator. For custom-made linear actuators, calculate the correct linear actuator stroke length, and adjust it accordingly (Fig. 14.2B). For example, a typical shear stress of 1 Pa and a 3 mL syringe with inner diameter of 8.66 mm requires a stroke length of 3.4 mm. Then, place the syringe into the slots on the linear actuator. If the syringe stroke length needs to be adjusted to fit into the slots, apply pressure on the syringes slowly to minimize shear stress.

Step 7. Apply fluid flow. Start the linear actuator for the necessary duration for the experiment (Fig. 14.2B). For commercial linear actuators, this is done by computer control, whereas for custom-made linear actuators, the instrument would have to be turned on and off at the beginning and the end of the flow duration. A flow duration of 1–2 h is recommended for osteocytes.

Bone Cell Preparation for Test 3 (Calcium)

Step 1. Prepare the glass slides. Follow exactly the same procedure to prepare the glass slides as described earlier for test 1 (cell signaling) and test 2 (cytoskeleton).

Step 2. Prepare the cells. Follow exactly the same procedure to prepare the bone cells as described earlier for test 1 (cell signaling) and test 2 (cytoskeleton).

Step 3. Stain the cells. Prepare a Fura 2-AM calcium dye in complete darkness. To do this, rinse a glass syringe twice with DMSO (dimethyl sulfoxide). Use the syringe to add 50 μL of DMSO to 50 μg of Fura 2-AM dye, and mix thoroughly. Add this solution to 5 mL of cell-specific media without phenol red (since it interferes with fluorescence), and mix thoroughly. Aspirate all media from the cells in the BSC, and rinse twice with DPBS (Dulbecco's phosphate buffered saline). Coat

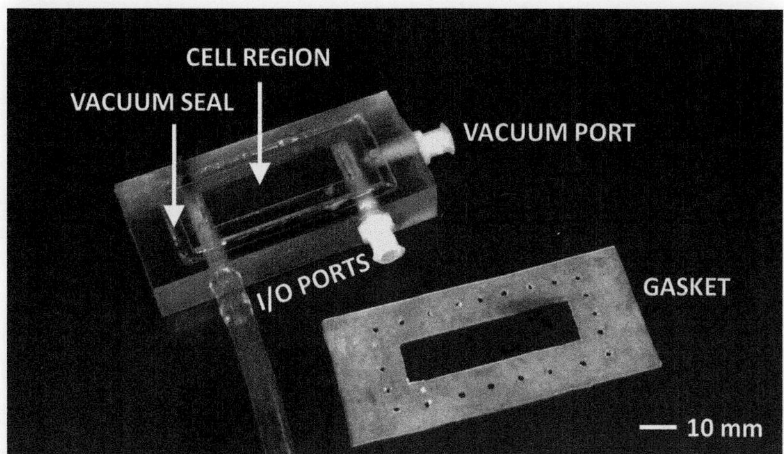

FIGURE 14.3

Unassembled small flow chamber used for test 3 (calcium).

each glass slide with 1 mL of dye, cover the dish with tin foil, and incubate for 30–60 min at room temperature. Remove the dye, rinse cells with DPBS, and fill the dish with working media.

Step 4. Assemble the flow chambers. Chamber setup should be done at the same time that incubation occurs, as described above in step 3 (Fig. 14.3). Uniformly coat the gasket in sealing grease. Uniformly coat the chamber in sealing grease, but ensure the grease does not touch the cell chamber region. Firmly press the gasket onto the chamber, ensuring no air gaps.

Step 5. Insert the tubing. Use a syringe to fill the tubing at both the inlet and outlet of the flow chamber with working media. Calculate and adjust the stroke length of the linear actuator based on the desired shear stress for the test. Attach the syringe to the syringe pump.

Step 6. Position the glass slide. Dry the outer region of the glass slide, that is, the region that will not sit in the cell culture region. Press the glass slide into the gasket, ensuring no leaks. Connect the chamber to a vacuum pump to generate a good seal. Place the chamber on the fluorescence microscope on a heated stage.

Step 7. View the cells. View cells using "brightfield" microscopy to set the correct focal length, then turn off the brightfield setting. Switch the view from the eyepiece to the camera, turn on fluorescence through a ratiometric imaging program, and set the brightness so cells are visible (Fig. 14.4). If the image is grainy, increase the image exposure time. Then, turn off exposure, and leave cells in a static condition for at least 30 min so their responsivity returns.

Step 8. Apply fluid flow. Turn on the exposure, identify the cells of interest, and start recording in real time. Then, for the first 3 min, image the cells in a static condition to get a baseline calcium response. Next, turn on the linear actuator to apply

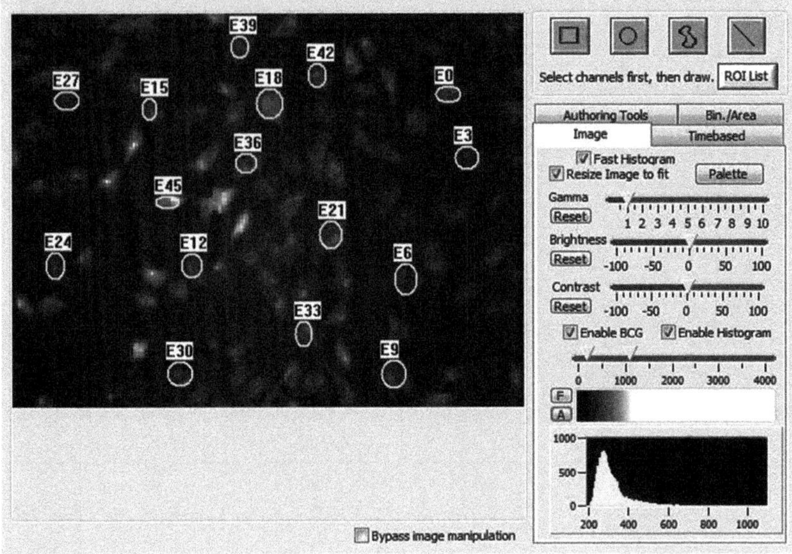

FIGURE 14.4

Viewing cells using fluorescence microscopy and software.

Software window courtesy of PTI Easy Ratio Pro, www.pti-nj.com.

fluid flow for 3 min, and continue imaging the cells in a static condition for a final 3 min.

TIPS AND TRICKS

- ✓ Prelabel tape and then simply put tape on the appropriate assembled flow chambers.
- ✓ Flow chamber tops with screws inserted may be stacked vertically prior to experiments.
- ✓ Only use tweezers to touch glass slides, gaskets, and flow chambers to avoid contamination.
- ✓ Place a cell-seeded glass slide in a dish (instead of a flow chamber) as a control specimen.
- ✓ For statistical analysis, pair one flow sample with one no flow control in a randomized order.

THE "GOLD STANDARD"

No international standards exist for the duration, frequency, and magnitude of fluid shear stress for fluid flow tests on bone cells. Thus, researchers are encouraged to consult previously published peer-reviewed journal articles and textbooks to serve as precedents.

3.4 SPECIMEN TESTING

Bone Cell Test 1 (Cell Signaling)

Step 1. Disassemble the flow chamber. To maintain pressure, clamp the short tubing on the outlets of flow chambers that underwent flow. Disconnect long tubing from flow chambers. Transfer the main chamber to the BSC. Remove flow chamber tops. Lift off gaskets.

Step 2. Incubate the cells. Place the glass slide in a petri dish (with cell-seeded side facing up) of 10 mL of fresh growth media. Place the petri dish in the incubator for an optimal time, such as 24 h for quantification of RANKL (receptor activator of nuclear factor kappa-B ligand) release from MLO-Y4 osteocytes. Media obtained from flow chambers may be saved to examine the expression of factors released during flow.

Step 3. Clean the setup. Rinse all parts with tap water and then deionized water. Make sure to rinse the inside of the tubing, as well as the inlets and outlets.

Step 4. Collect the media. Collect media after incubating cell-seeded slides for an optimal time. Collected media may be used fresh or stored at −20°C for later use. Collected media can be used in conditioned media studies (i.e., apply to other cells to assess downstream signaling pathways) or analyzed for secreted factors with ELISA (enzyme-linked immunosorbent assay) or other assays.

Step 5. Count the cells. Count the number of cells on the slide by detaching cells from the glass slide using trypsin and staining them with trypan blue. Counting can be done manually with a hemocytometer or automatically with a commercially available machine.

Step 6. Lyse the cells. Lyse the cells in order to assay gene or intracellular protein expression by applying cell-lysing reagents to incubated cells on glass slides. Lysed cells can be scraped off the glass slides, analyzed immediately, or stored at −80°C for later use.

Step 7. Normalize to the number of cells. Normalize assay results with respect to the cell number on each slide (i.e., divide the assay results by the number of cells for each slide). This will account for the effect of cell numbers on total protein expression.

Step 8. Normalize to control. Normalize the number of cells on the collected media with respect to a control slide that has not been placed in chambers in order to combine results from different cell test groups. This accounts for uncontrollable variables, such as cell passage number.

Bone Cell Test 2 (Cytoskeleton)

Step 1. Disassemble the flow chamber. Take apart the flow chamber in exactly the same manner as described above for test 1 (cell signaling).

Step 2. Clean the setup. Clean all of the components of the test setup in exactly the same manner as described above for test 1 (cell signaling).

Step 3. Fix the cells. Fix the cells on the glass slide with 4% formaldehyde using 10% volume/volume formalin in DPBS. Let the cells fix for 10 min at room temperature and then rinse the cells three times with DPBS, letting the cells sit for 2 min after each rinse.

Step 4. Stain the cells. Perform the specific cytoskeleton stain as prescribed. For actin staining, first permealize the cells with a 0.1% volume/volume Triton-X solution in DPBS for 10 min, followed by rinsing cells with DPBS three times. Stain actin filaments with a phalloidin solution (320 μL DPBS + 8 μL Alexa Fluor 488) for 20 min in darkness, and again rinse the cells three times with DPBS. Finally, DAPI (i.e., 4′,6-diamidino-2-phenylindole) stain (1 mL distilled water + 1 μL

DAPI) the nucleus for 10 min in darkness, then rinse with DPBS followed by distilled water. Use Fluoromount to glue a cover slip to the glass slide.

Step 5. Image the cells. View the slide of stained cells using a fluorescent microscope in a darkroom (Fig. 14.5A). Find the correct focal length to view the cells using the brightfield setting. For each stain applied to the cells (i.e., DAPI and phalloidin), set the excitation wavelength of the monochromator and microscope emission filter for the specific stain being imaged, and record the image. Make sure to image at the same location on the slide for each stain in order to allow for an image overlay. Move the stage to image a new location on the glass slide and repeat.

Bone Cell Test 3 (Calcium)

Step 1. Clean the test setup. Remove the chamber from the microscope. Bleach and then dispose of the cell coated slide. Wash the greasy gasket and chamber with dish detergent.

Step 2. View the cells. For each slide, go through the recording on the computer and select all cells in the viewing region to be viewed. Play through the entire recording to collect data for all cells, and save it as a text file. During playback, pay attention to cell debris or bubbles that might flow over the cells and disturb data so as to exclude them from data analysis.

Step 3. Select time points. Go through the data in the computer text file to determine the point that coincides with flow "on" (at t = 180 s) and flow "off" (at t = 360 s) to determine a baseline.

Step 4. Measure the response. Run each test through a custom-made computer software algorithm. If cells are responding, then use the microscope to measure the magnitude of the fluorescent intensity peaks. The methodology of this script will be to first determine maximum peak in the baseline region and then the magnitude of the largest peak in the flow region. Compare the magnitude to determine if the flow peak is significant (i.e., greater than the baseline peak by a threshold multiple of about four times).

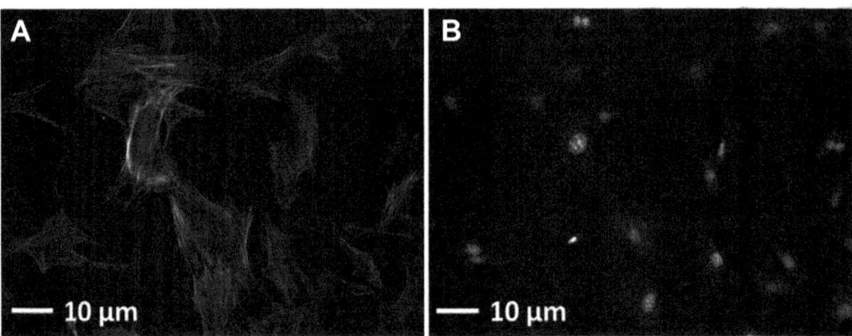

FIGURE 14.5

Typical fluorescent image of stained cells. (A) Actin fibers for test 2 (excitation wavelength: 495 nm, emission wavelength: 518 nm), (B) cell nuclei for test 3 (excitation wavelength: 340 nm, emission wavelength: 488 nm).

Table 14.1 Raw data table.

Test 1 (Cell Signaling)	Absolute Protein Concentration [pg/mL]	Normalized Protein Concentration [pg/mL/cell]
Slide 1		
Slide 2		
Slide 3		
etc.		
Test 2 (Cytoskeleton)	**Relative Cytoskeletal Angle (FF) [°]**	**Relative Cytoskeletal Angle (HS or HD) [°]**
Slide 1		
Slide 2		
Slide 3		
etc.		
Test 3 (Calcium)	**Percentage of Responding Cells [%]**	**Magnitude of Responding Cells [—]**
Slide 1		
Slide 2		
Slide 3		
etc.		

FF, *fluid flow shear stress;* HD, *hydrodynamic pressure;* HS, *hydrostatic pressure.*

Step 5. Validate the response. If the response is a real peak, there should be a distinctive increase in the 340-nm channel and a simultaneous decrease in the 380-nm channel (Fig. 14.5B).

3.5 RAW DATA COLLECTION

Step 1. Record data for test 1 (cell signaling). Tabulate absolute and normalized values for protein concentrations and mRNA expression from biochemical analysis of bone cells (Table 14.1).

Step 2. Record data for test 2 (cytoskeleton). Tabulate cytoskeletal angles relative to fluid flow direction or hydrostatic/dynamic pressure from biochemical analysis of bone cells (Table 14.1).

Step 3. Record data for test 3 (calcium). Tabulate the percentage of responding cells as well as the magnitude of responding cells (i.e., magnitude = fluorescent intensity peak at 340 nm/fluorescent intensity peak at 380 nm) from biochemical analysis of bone cells (Table 14.1).

3.6 RAW DATA ANALYSIS

Step 1. Perform statistical analysis. Choose the statistical difference criterion (e.g., $P < 0.01$ or < 0.05). Then compare data between different test groups (e.g., fluid shear stress 1 vs. 2 vs. 3, pressure 1 vs. 2 vs. 3, bone cell type 1 vs. 2 vs. 3, etc.). There are various software programs for comparing two test groups (e.g., paired *t*-test) or two or more test groups influenced by multiple factors (e.g., analysis of variance, ANOVA).

Step 2. Compute statistical power. Power analysis can be done after the study to ensure there were enough bone cell slides per test group to detect all statistical differences that were actually present (i.e., was type II statistical error avoided?). Statistical power >80% is usually considered to indicate there were enough specimens per test group. Note that if good predictions of averages and standard deviations are available from prior studies, then the number of bone cell test groups and/or tests can be chosen before the study begins to ensure a power >80%.

ENGINEER'S TOOLBOX

Stroke length of the linear actuator to generate a specific wall shear stress can be calculated as $L = (4\tau h^2 w)/(3\pi\omega\mu d^2)$, where τ is desired wall shear stress [Pa], h is flow chamber height [mm], w is flow chamber width [mm], μ is dynamic viscosity of the flowing fluid [Pa·s], ω is angular frequency of the linear actuator [rad/s], and d is inner diameter of the syringe [mm].

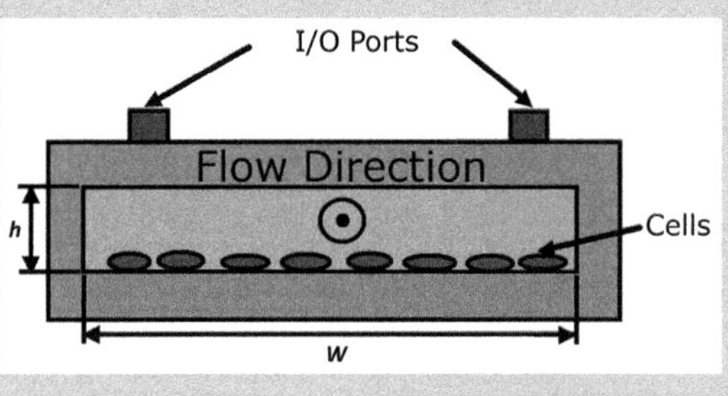

4. RESULTS

Once all bone cell data collection and analysis have been performed, it is then important to communicate and present the primary results in an understandable and concise manner to the reader of a journal article, conference paper, technical report, or book chapter.

Step 1. Show test 1 (cell signaling) results. For assays that only give qualitative results (e.g., western blot or PCR), present pictures of raw data. For assays that give quantitative results (e.g., ELISA or real-time PCR), present protein or gene expression of static and flow groups (Fig. 14.6).

Step 2. Show test 2 (cytoskeleton) results. Present representative colorized images of the static control and flow stain for all conditions examined (Fig. 14.7). Then, if investigating how the cytoskeleton reorientates with respect to the loading axis, prepare a rose plot of the actin filament directions for both the static and flow case for each (Fig. 14.8).

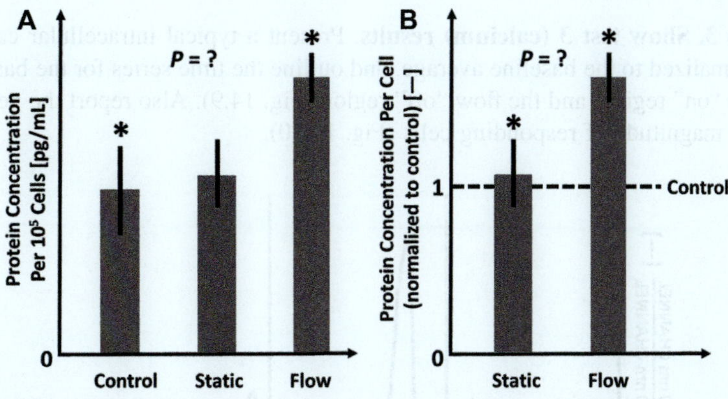

FIGURE 14.6

Typical protein expression results. (A) Absolute protein concentration, (B) protein concentration normalized with respect to control. Each bar indicates an average ± 1 standard deviation. *P* values should be given for all significant differences between groups and identified by symbols, such as asterisks (*).

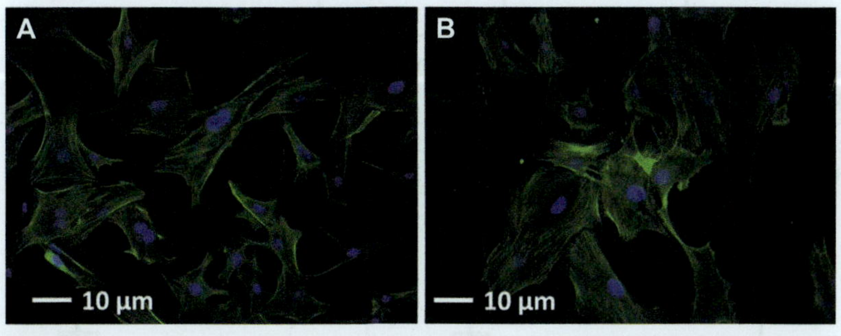

FIGURE 14.7

Typical images of filaments and cells. (A) Hydrostatic pressure control, (B) fluid flow shear stress condition. Actin filaments are green, while cell nuclei are blue for cells.

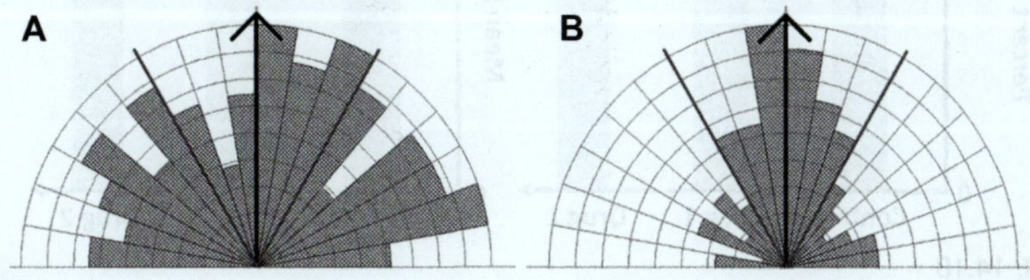

FIGURE 14.8

Typical rose plots of cytoskeleton angles with respect to flow direction. (A) Hydrostatic pressure control, (B) fluid flow shear stress condition. Cytoskeletal alignment of $\pm 30°$ is deemed to be aligned with the flow. If significantly more than one-third of flow elements are aligned with flow direction, then cells are considered to be preferentially aligned.

Step 3. Show test 3 (calcium) results. Present a typical intracellular calcium plot normalized to the baseline average, and outline the time series for the baseline, the flow "on" region, and the flow "off" region (Fig. 14.9). Also report the percentage and magnitude of responding cells (Fig. 14.10).

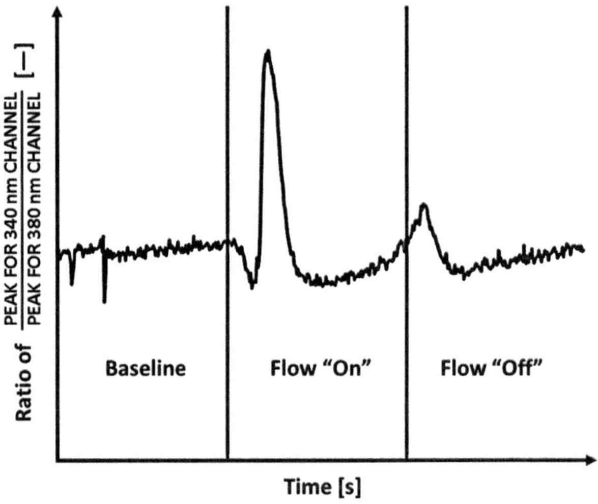

FIGURE 14.9

Typical plot of calcium response. The time points when flow is turned "on" and "off" are clearly displayed.

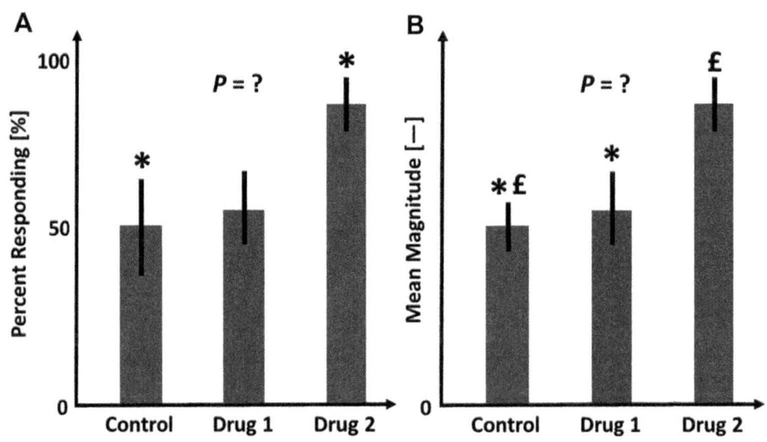

FIGURE 14.10

Typical results for responding cells. (A) Percentage of responding cells, (B) mean magnitude of response, where magnitude = fluorescent intensity peak at 340 nm/fluorescent intensity peak at 380 nm. Each bar indicates an average ± 1 standard deviation. *P* values should be given for all significant differences between treatment groups and identified by symbols, such as asterisks (*) and pounds (£).

ALTERNATIVES AND ADAPTATIONS

✓ **Pulsatile flow.** Wall shear stress can be varied cyclically from zero to peak using a peristaltic pump. This method has reduced physiological relevance compared to oscillatory flow, although mechanical stimulation is variable as would occur physiologically.

✓ **Steady flow.** Wall shear stress can be maintained at a fixed level using a unidirectional syringe pump. This has reduced physiological relevance vs. oscillatory or pulsatile flow since mechanical stimulation and mass transport are unidirectional.

✓ **Shaker plates.** An alternative method for fluid shear stress tests involves seeding cells in well plates on a moving shaker. Fluid flows over the cell only in a unidirectional circular pattern, and no calcium experiments can be done. However, the test setup is very easy.

5. DISCUSSION

After completing all bone cell tests, data collection, data analysis, and data presentation, then final results can be considered and interpreted in the broader context of some important clinical, biomechanical, and/or technological considerations, as follows.

In vitro bone cell mechanotransduction tests can address a variety of research questions, albeit under highly controlled laboratory conditions. Blocking experiments with neutralizing antibodies can be done to identify different pathways and mechanisms through which bone cells respond to mechanical stimulation. Also, the effects of diseases and drugs can be examined on bone cell mechanosensitivity and remodeling. Moreover, tests can evaluate the response of different cells (e.g., osteoclasts, endothelial cells, cancer cells, etc.) to signals produced by mechanically stimulated bone cells by applying the resulting conditioned medium.

In vitro bone cell mechanotransduction tests are dependent on experimental flow conditions and the type of cells used. The effects of flow type and frequency on human fetal osteoblast calcium response have been investigated, and the results show that pulsatile flow is the most responsive, and increasing flow frequency reduces cell responsivity.[13] The difference between osteoblast and osteocyte calcium response has also been assessed, and it has been observed that osteocytes are typically more responsive in terms of percent response (95% vs. 60–95%), number of peaks (1–5 vs. 1–3), and magnitude (2–3× vs. 1.5–3×) depending on the shear stress applied.[14] Moreover, a thorough analysis of the different flow variables on cell mRNA in response to flow has been performed.[15] The findings show that increasing shear stress (from 0.5 to 5 Pa) causes an increase in osteocyte expression of COX-2 (1.2× to 3× that of no flow) and a reduction in the RANKL/OPG ratio (from 0.8× to 0.2× that of no flow at a 1 Hz frequency), where OPG is osteoprotegerin. Furthermore, it has been reported that the frequency of the flow has a limited effect on COX-2 mRNA levels, but it brings about a decreasing RANKL/OPG ratio with increasing flow frequency (from 0.8× to 0.4× that of no flow at a 1 Pa shear stress).[15] Also, the difference between osteoblast and osteocyte flow response for different flow profiles has been measured, and the data show an increase in normalized COX-2 expression for osteoblasts compared to osteocytes under flow (2× vs. 1.2× for steady flow, and 1.8× vs. 0.8× for oscillatory flow).[16] Finally, cytoskeletal alignment observations demonstrate no major cytoskeleton alignment in osteocytes under any type of flow profile, but there is a slight directional alignment of osteoblast F-actin in the flow direction.[16]

In vitro bone cell mechanotransduction tests have limits. Cells are studied in a 2-D culture vs. a 3-D physiological context, which affects the type and degree of cellular response. As well, cells in a physiological context will communicate back and forth with other cell populations, creating highly complex biochemical environments. Furthermore, cells are only stimulated using one type of mechanical stimulation (i.e., fluid shear stress), whereas cells are physiologically exposed to strain, hydrostatic pressure, and vibration. Finally, for conditioned medium tests, important low half-life signals are often lost (e.g., nitric oxide), causing a loss in those interaction effects. However, new technologies are being developed to better mimic the *in vivo* mechanical stress context.[17–20]

6. SUMMARY

- Mechanical stimulation of bone cells maintains overall bone homeostasis and strength.
- Identifying mechanisms and signaling pathways can improve strategies against bone disease.
- *In vitro* mechanical bone cell tests can apply physiologically relevant mechanical loads.
- Bone cell response can be measured using protein expression and calcium signaling.
- New technologies may potentially enhance the relevance of *in vitro* tests on bone cells.

7. QUIZ QUESTIONS

1. What are the basic principles underlying Wolff's Law as it relates to overall bone health?
2. Without extra tools, how can flow rate be maintained at the correct level during fluid flow tests?
3. How could the air bubbles in the tubing effect the fluid shear stress level during flow tests?
4. Why might bone cells appear fuzzy or invisible in the fluorescence images for a calcium test?
5. For a fluid flow test, compute linear actuator stroke length for a wall shear stress = 2 Pa and frequency = 1 Hz. Assume flow chamber width = 3.75 cm, flow chamber height = 200 μm, dynamic fluid viscosity = 0.7×10^{-3} Pa·s, and 1 mL syringe inner diameter = 4.88 mm (answer: 12 mm).

REFERENCES

1. Chen JH, Liu C, You L, Simmons CA. Boning up on Wolff's Law: mechanical regulation of the cells that make and maintain bone. *Journal of Biomechanics* 2010;**43**(1):108–18.
2. Gilsanz V, Wren TA, Sanchez M, Dorey F, Judex S, Rubin C. Low-level, high-frequency mechanical signals enhance musculoskeletal development of young women with low BMD. *Journal of Bone and Mineral Research* 2006;**21**(9):1464–74.
3. Turner C, Robling A. Mechanisms by which exercise improves bone strength. *Journal of Bone and Mineral Metabolism* 2005;**23**(1):16–22.
4. Nikander R, Sievänen H, Heinonen A, Daly RM, Uusi-Rasi K, Kannus P. Targeted exercise against osteoporosis: a systematic review and meta-analysis for optimising bone strength throughout life. *BMC Medicine* 2010;**8**(1):47.
5. Kapur S, Baylink DJ, William Lau KH. Fluid flow shear stress stimulates human osteoblast proliferation and differentiation through multiple interacting and competing signal transduction pathways. *Bone* 2003;**32**(3):241–51.
6. Basso N, Heersche JNM. Characteristics of in vitro osteoblastic cell loading models. *Bone* 2002;**30**(2):347–51.
7. Weinbaum S, Cowin SC, Zeng Y. A model for the excitation of osteocytes by mechanical loading-induced bone fluid shear stresses. *Journal of Biomechanics* 1994;**27**(3): 339–60.
8. Price C, Zhou X, Li W, Wang L. Real-time measurement of solute transport within the lacunar-canalicular system of mechanically loaded bone: direct evidence for load-induced fluid flow. *Journal of Bone and Mineral Research* 2011;**26**(2): 277–85.
9. Zhang D, Weinbaum S, Cowin SC. On the calculation of bone pore water pressure due to mechanical loading. *International Journal of Solids and Structures* 1998;**35**(34–35): 4981–97.
10. Cowin SC, Gailani G, Benalla M. Hierarchical poroelasticity: movement of interstitial fluid between porosity levels in bones. *Philosophical Transactions of the Royal Society A: Mathematical, Physical, and Engineering Sciences* 2009;**367**(1902):3401–44.
11. Fritton SP, McLeod KJ, Rubin CT. Quantifying the strain history of bone: spatial uniformity and self-similarity of low-magnitude strains. *Journal of Biomechanics* 2000;**33**(3): 317–25.
12. Gurkan UA, Akkus O. The mechanical environment of bone marrow: a review. *Annals of Biomedical Engineering* 2008;**36**(12):1978–91.
13. Jacobs CR, Yellowley CE, Davis BR, Zhou Z, Cimbala JM, Donahue HJ. Differential effect of steady versus oscillating flow on bone cells. *Journal of Biomechanics* 1998; **31**(11):969–76.
14. Lu XL, Huo B, Chiang V, Guo XE. Osteocytic network is more responsive in calcium signaling than osteoblastic network under fluid flow. *Journal of Bone and Mineral Research* 2012;**27**(3):563–74.
15. Li J, Rose E, Frances D, Sun Y, You L. Effect of oscillating fluid flow stimulation on osteocyte mRNA expression. *Journal of Biomechanics* 2012;**45**(2):247–51.
16. Ponik SM, Triplett JW, Pavalko FM. Osteoblasts and osteocytes respond differently to oscillatory and unidirectional fluid flow profiles. *Journal of Cellular Biochemistry* 2007;**100**(3):794–807.

17. Guo XE, Takai E, Jiang X, Xu Q, Whitesides GM, Yardley JT, et al. Intracellular calcium waves in bone cell networks under single cell nanoindentation. *Molecular and Cellular Biomechanics* 2006;**3**(3):95–107.
18. Wei C, Fan B, Chen D, Liu C, Wei Y, Huo B, et al. Osteocyte culture in microfluidic devices. *Biomicrofluidics* 2015;**9**(1):014109.
19. Middleton K, Mei X, You L. A microfluidic system to study cross-talk between osteocytes and osteoclasts. Las Vegas (NV, USA): Orthopaedic Research Society; March 28–31, 2015 [poster #1436].
20. Liu C, Sun F, You L. Mechanical loading system to apply concurrent physiological pressure and shear to osteocytes. Las Vegas (NV, USA): Orthopaedic Research Society; March 28–31, 2015 [poster #1433].

Joints: Human and Artificial

2

CHAPTER

Electromagnetic Tracking of the Kinematics of Articulating Joints

15

G. Daniel G. Langohr, Jacob Reeves, James A. Johnson
Western University, London, ON, Canada

1. BACKGROUND

The tracking of the kinematics of articulating joints (Fig. 15.1) during *in vitro* testing is of interest in determining the motion of healthy intact joints, as well as the changes after joint realignment, joint reconstruction, and/or soft tissue degeneration.[1] Joint reconstruction can alter the kinematics of the joint, often resulting in a reduced range of motion and/or a significant change to soft tissue and interarticular loading. *In vitro* cadaveric testing, coupled with motion tracking of the bones of the joint, can provide insight into the changes when the native kinematics are compared to the reconstructed joint,[2] permit the digitization of biologic structures using tracing techniques,[3] and allow for the measurement of the relative motion of one structure (e.g., implanted or biological) with respect to another.[4] The ability to track the motion of the bones associated with the articulating joint is of paramount importance in the study of joint motion. Numerous techniques have been employed including goniometers, optical tracking systems, and electromagnetic tracking (ET) systems. However, ET offers the advantage of rapid real-time data acquisition in six degrees of freedom and eliminates line-of-sight issues common to optical techniques. Therefore, this chapter explains how to perform ET testing on articulating joints, as well as how to analyze, present, and interpret results.

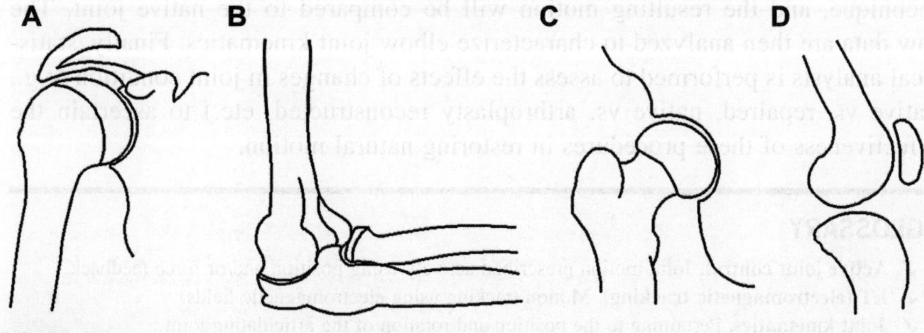

FIGURE 15.1

Human articulating joints. (A) shoulder, (B) elbow, (C) hip, (D) knee.

Experimental Methods in Orthopaedic Biomechanics.

2. RESEARCH QUESTIONS

Typical research questions could include one or more of the following:

- What is the motion profile of the healthy native intact joint?
- How does soft tissue damage or degeneration affect joint motion?
- What soft tissue repair technique most closely replicates the motion of the healthy joint?
- What effect does total joint replacement geometry have on postoperative joint kinematics?
- What is the stability of an implant or fracture fixation device relative to bone?
- etc.

3. METHODOLOGY

3.1 GENERAL STRATEGY

ET is described for the *in vitro* biomechanical testing of human elbow joints to illustrate the application of the technique to any articulating joint. Fresh or fresh–frozen cadaveric human or animal elbow joints are obtained. Care is taken to include sufficient bone stock on either side of the elbow joint to allow for tracker fixation and ensure that soft tissue surrounding the joint is minimally disturbed. Bony structures both proximal and distal to the elbow are sufficiently exposed to allow rigid fixation of the trackers. The elbow is clamped to a fixture, which stabilizes one side of the joint and is subjected to either passive (i.e., manually actuated) or active (i.e., actively controlled based on position feedback) motion, while the position and rotation of the bony structures of the joint are tracked.[5,6] A tracker (or receiver) is affixed to the moving limb, and the transmitter is attached to the clamped side. If some relative motion between the clamped bone and the clamp is suspected, then a tracker is affixed to that bone. Depending on study design, in certain cases, soft tissues will be systematically removed and the joint kinematics reassessed, or the elbow joint or soft tissues will undergo a reconstructive technique, and the resulting motion will be compared to the native joint. The raw data are then analyzed to characterize elbow joint kinematics. Finally, statistical analysis is performed to assess the effects of changes in joint condition (e.g., native vs. repaired, native vs. arthroplasty reconstructed, etc.) to ascertain the effectiveness of these procedures in restoring natural motion.

GLOSSARY

✓ **Active joint control.** Joint motion prescribed actively using position and/or force feedback.
✓ **ET (electromagnetic tracking).** Motion tracking using electromagnetic fields.
✓ **Joint kinematics.** Pertaining to the position and rotation of the articulating joint.
✓ **Musculotendinous junction.** The transition between the muscle and the tendon.
✓ **Passive joint control.** Joint motion prescribed manually.

SAFETY FIRST

✓ Remember to always wear gloves, a lab coat, and safety glasses for protection.
✓ Be careful not to cut fingers during articulating joint specimen preparation.
✓ Clean the work area and all tools with bleach or disinfectant after testing.

3.2 MATERIALS AND TOOLS LIST

- clamping fixture for passive testing
- computer and dedicated software
- drill bits and handheld drill
- electromagnetic receivers and transmitter
- human or animal articulating joints
- joint simulator for active testing
- manual surgical screwdriver
- screwdriver bits that match screw heads
- screws for affixing trackers

3.3 SPECIMEN PREPARATION

Step 1. Store the specimen. Fresh or fresh−frozen elbow joint specimens need to be sealed in plastic bags for proper freezer storage before testing. Storage temperature should be −20°C or lower, and it should remain at that temperature until testing.

Step 2. Thaw the specimen. Approximately 12 h prior to the desired start of testing, specimens should be removed from the freezer and placed in a sink at room temperature for thawing. Care should be taken to ensure desiccation of the tissues does not occur by leaving the specimen in its plastic bag or wrapped in coverings soaked in saline solution to prevent dehydration.

Step 3. Perform surgery. Specimen surgery should be performed, depending on study design. If the study includes reconstruction of the joint or surrounding soft tissues, then this should be implemented. Typically for both active and passive motion simulation, tone loads are applied to certain tendons to simulate the musculature acting across the joint. To allow for actuation of these structures, the musculotendinous junctions of each muscle should be identified, and sutures should be attached to these structures to allow forces to be applied to them during testing (Fig. 15.2A). A running-locking stitch is suitable for this purpose.

Step 4. Identify tracker fixation sites. To allow for tracking of the elbow joint's bony structures, identify suitable surfaces on which the trackers can be mounted. These locations should be far enough away from the joint to prevent negative affects on the motion of the soft tissue surrounding the joint while permitting the rigid fixation of the tracker. Any relative motion between the tracker and the bone will introduce error into the results, and hence, care should be taken to ensure adequate fixation. The tracker should be affixed bicortically using cortical

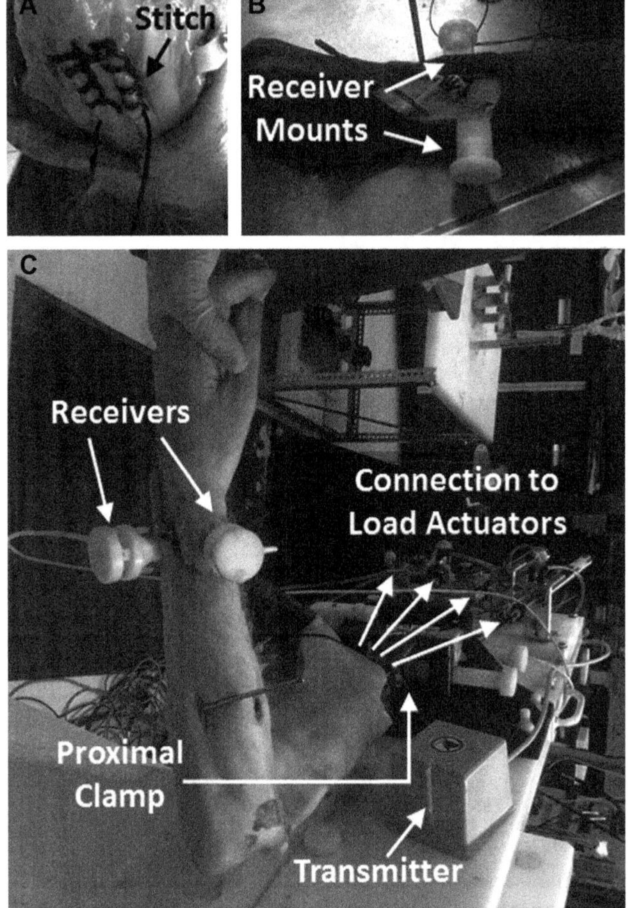

FIGURE 15.2

Elbow joint preparation prior to *in vitro* cadaveric elbow joint kinematic testing. (A) A running-locking stitch is attached to the musculotendinous junction, (B) ET receiver mounts are rigidly affixed to distal bones that will be tracked with minimally invasive exposure, (C) a cadaveric elbow is mounted to an active joint simulator using a proximal clamp, which holds a metal rod inserted into the intramedullary canal of the proximal bone. Connections between the musculotendinous junctions and the load actuators permit active motion simulation during testing.

bone screws into the diaphyseal region (Fig. 15.2B, C and 15.3A). If the trackers must be affixed to the metaphyseal or epiphyseal regions, cancellous bone fixation will be required, and hence, cancellous screws are recommended. If sufficient screw purchase is a concern, bone cement can be introduced into the pilot holes before inserting the screws to aid in screw rigidity.

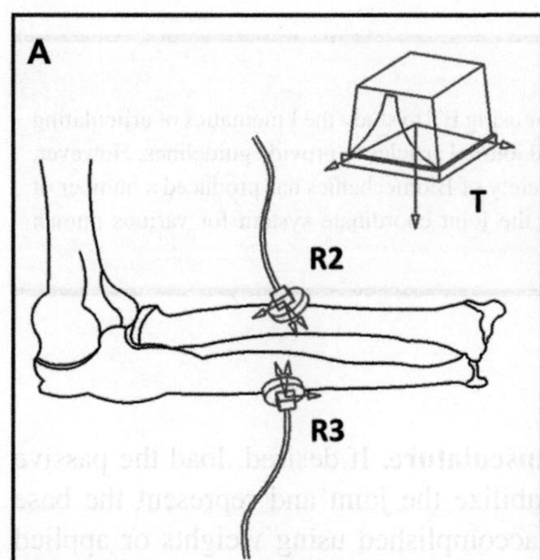

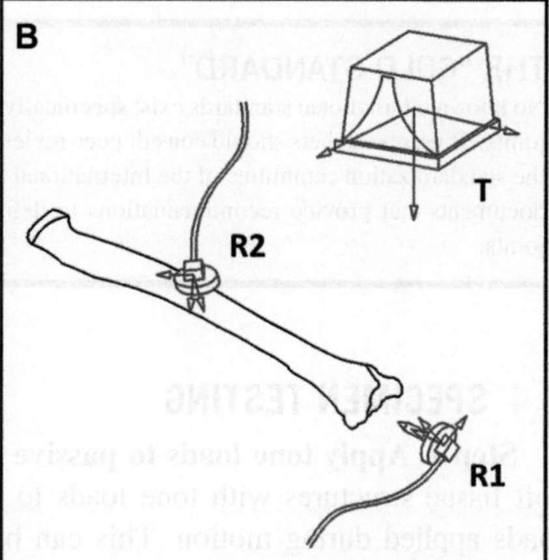

FIGURE 15.3

ET articulating joint test setup. (A) ET system, whereby two receivers (R2 and R3) track the two distal bones, while the transmitter (T) is affixed to the base of the joint simulator to which the proximal bone is also attached, (B) digitization process, whereby receivers for the stylus (R1) and bone (R2) help yield the relative position of each receiver with respect to the transmitter (T). The position of the stylus (R1) can then be transformed with respect to the bone receiver (R2) so that the relative position of important landmarks can be used in postprocessing of the joint kinematic data. Left arm specimens are shown.

Step 5. Digitize pertinent landmarks. If the axes of motion can be defined by bony landmarks, then make recordings of and/or digitize these locations so they are available during posttesting data analysis. A separate receiver attached to a plastic stylus will be required to perform these digitizations (Fig. 15.3B). It is critical to ensure that the tracker is rigidly affixed to the bone undergoing digitization prior to data acquisition with the stylus.

Step 6. Expose bony region to be affixed. The bony structures of one side of the joint, typically the proximal side, should now be affixed to the testing apparatus. This can be achieved via clamping a long bone to the apparatus (Fig. 15.2C) or potting the bone in a cement mantle, which is then affixed to the apparatus.

Step 7. Attach actuators to musculature. To allow for force application during testing, the sutures that were employed for the pertinent musculotendinous junctions need to be attached to load actuators. After being wrapped around pulleys, weights can also be used if static loads are desired (Fig. 15.2C).

TIPS AND TRICKS

- ✓ Predrilling pilot holes before affixing trackers can make screw insertion easier.
- ✓ A benchtop vice can be used to hold the specimen during preparation.
- ✓ Repeated testing of the same joint is a powerful tool in cadaveric testing.
- ✓ Bone cement can contribute to screw fixation in the case of insufficient bone stock.

THE "GOLD STANDARD"

No known international standards exist specifically for using ET to study the kinematics of articulating joints. Thus, researchers should consult peer-reviewed journal articles to provide guidelines. However, the standardization committee of the International Society of Biomechanics has produced a number of documents that provide recommendations to define the joint coordinate system for various human joints.

3.4 SPECIMEN TESTING

Step 1. Apply tone loads to passive musculature. If desired, load the passive soft tissue structures with tone loads to stabilize the joint and represent the base loads applied during motion. This can be accomplished using weights or applied via computer-controlled actuators (Fig. 15.2C). Tone load level should be selected to best suit the joint being tested (e.g., 13–72 N for a normal elbow).

Step 2. Initiate motion recording. Before applying motion to the joint, start the tracking process so that the position and rotation of each tracker will be measured and recorded. This is most commonly performed using computer software, which may be integrated into a larger program that also controls the joint simulator (Fig. 15.2C).

Step 3. Apply motion to the joint. If active testing is being performed, then the loads applied to simulate the musculature across the joint will be increased and proportioned to drive the desired motion. The motion can be controlled to have constant angular velocity, if desired, by using a feedback control loop incorporating real-time joint angles. If passive testing is being done, then the joint should be manually moved through the desired motion using operator intervention against the tone loads applied to the joint musculature. In either case, the full range of clinically relevant motion will depend on the particular joint (e.g., 0°–170° of flexion–extension for a normal elbow).

Step 4. Perform repeated trials. If repeated trials are desired, repeat the motion for three to five cycles consecutively to obtain an average value of the motion runs.

Step 5. Alter specimen configuration. If the study design involves investigating the effects of joint reconstruction or soft tissue repair, perform this and then repeat steps 1–4 for each study parameter (e.g., intact, soft tissue deficient, etc.).

Step 6. End motion recording. After the desired joint motion has been achieved, stop the tracking process within the computer program, and name the file appropriately. The resulting raw output file will include a position vector in displacement (x, y, z) and the rotation in quaternion format (q_w, q_x, q_y, q_z).

3.5 RAW DATA COLLECTION

Step 1. Record demographics. For each individual specimen, enter basic demographic information, such as the species (e.g., human, bovine, etc.), joint (e.g., elbow, hip, etc.), condition (e.g., normal, injured, etc.), age, sex, and tone load used to compress or distract the joint prior to applying motion (Table 15.1).

Table 15.1 Raw data recorded for joint kinematic tests for each specimen. Position and rotation vector data are collected for each receiver R1, R2, and R3.

Species	Joint	Condition		Age [Years]	Sex [M, F]	Tone Load [N]	
	Position Vector			Rotation vector			
Time [s]	x [mm]	y [mm]	z [mm]	q_w [°]	q_x [°]	q_y [°]	q_z [°]
t_0							
t_1							
t_2							
etc.							

Step 2. Record position. During testing, the computer software program will automatically store each receiver's position vector coordinates (x, y, z) vs. time (Table 15.1).

Step 3. Record rotation. During testing, the computer software program will automatically store each receiver's rotation vector coordinates (q_w, q_x, q_y, q_z) vs. time (Table 15.1).

3.6 RAW DATA ANALYSIS

Step 1. Understand matrix math. Transformation matrix math is composed of three primary matrices, that is, the position matrix P, the rotation matrix R, and the transformation matrix T. When expressing any one of these matrices, it is important to understand which object (e.g., bone) is being referenced, as well as its relationship to other objects. For example, the position of object A with respect to object B is expressed as ${}^{B}P_{A}$, the rotation of object A with respect to object B is expressed as ${}^{B}_{A}R$, and the transformation of object A with respect to object B is expressed as ${}^{B}_{A}T$.

Step 2. Convert data to matrix format. The raw output file for each tracker can be expressed in multiple formats, such as the position vector in displacement (x, y, z) and the rotation in quaternion format (q_w, q_x, q_y, q_z). Convert the rotation data from quaternions to rotation matrix format R using the appropriate equation (Eq. (15.1)).[7]

$$R = \begin{bmatrix} 1 - 2q_y^2 - 2q_z^2 & 2q_xq_y + 2q_wq_z & 2q_xq_z - 2q_wq_y \\ 2q_xq_y - 2q_wq_z & 1 - 2q_x^2 - 2q_z^2 & 2q_yq_z + 2q_wq_x \\ 2q_xq_z + 2q_wq_y & 2q_yq_z - 2q_wq_x & 1 - 2q_x^2 - 2q_y^2 \end{bmatrix} \tag{15.1}$$

Step 3. Determine the reference bone. In analyzing the movements of bones relative to one another, a common and necessary assumption is that each bone acts as a rigid body (i.e., it does not deform throughout motion). Initially, choose one bone, which will act as the primary reference. The coordinate system of the primary reference bone will typically be the one about which the joint angles are calculated.

Step 4. Construct the reference bone's coordinate system. First, select three easily identifiable landmarks whose positions will be digitized relative to the transmitter using a stylus probe (Fig. 15.3B). While digitizing the points, ensure that the bone and affixed receiver do not move relative to the transmitter, and that the digitized points and receiver data are collected. The landmarks must be geometrically meaningful and should correspond to established standards.[7]

Second, once the landmarks are chosen, use these points to construct a coordinate system that is equivalent to established standards[6] by constructing three orthogonal and normal vectors ($\widehat{X}$, $\widehat{Y}$, $\widehat{Z}$), which form the rotation matrix ${}^{Transmitter}_{Bone}R$ of the bone with respect to the transmitter. The selection of which axis will be $\widehat{X}$, $\widehat{Y}$, and $\widehat{Z}$ is important for joint angle calculations.

Third, to finalize the bone's coordinate system, choose a single point of significance ${}^{Transmitter}P_{Bone}$ to be the origin for the bone's coordinate system. This point and the rotation matrix are then combined into a transformation matrix ${}^{Transmitter}_{Bone}T$. In essence, the position matrix P is a 3×1 matrix formed by the x, y, and z positions of point A relative to point B (Eq. (15.2)), the rotation matrix R is a 3×3 matrix whose columns are formed by three orthogonal and normal vectors (Eq. (15.3)), and the transformation matrix T is a 4×4 matrix, which is a combination of the position and rotation matrices (Eq. (15.4)).

$$
{}^{B}P_{A} = \begin{bmatrix} x \\ y \\ z \end{bmatrix} \tag{15.2}
$$

$$
{}^{B}_{A}R = \begin{bmatrix} r_{11} & r_{12} & r_{13} \\ r_{21} & r_{22} & r_{23} \\ r_{31} & r_{32} & r_{33} \end{bmatrix} = \begin{bmatrix} \widehat{X} & \widehat{Y} & \widehat{Z} \end{bmatrix} \tag{15.3}
$$

$$
{}^{B}_{A}T = \begin{bmatrix} r_{11} & r_{12} & r_{13} & x \\ r_{21} & r_{22} & r_{23} & y \\ r_{31} & r_{32} & r_{33} & z \\ 0 & 0 & 0 & 1 \end{bmatrix} = \begin{bmatrix} {}^{B}_{A}R & & & {}^{B}P_{A} \\ 0 & 0 & 0 & 1 \end{bmatrix} \tag{15.4}
$$

For example, to determine the transformation matrix that defines a coordinate system for a radius (i.e., ${}^{RadReceiver}_{Rad}T$), four points on the radius are determined through digitization (Fig. 15.4). Point A is digitized on the radial styloid, B on the volar tip of the distal radioulnar joint (DRUJ), D on the dorsal tip of the DRUJ, and O_{Rad} at the center of the radial head (which is determined through sphere fitting). A final point, C, is defined as the centroid of points A, B, and D. All of these points are with respect to the transmitter, as is the transformation matrix of the receiver mounted on the radius (i.e., ${}^{Transmitter}_{RadReceiver}T$). Then, the internal–external rotation axis is defined as $\overrightarrow{X}$ running from C to O_{Rad}, the flexion–extension axis is defined as

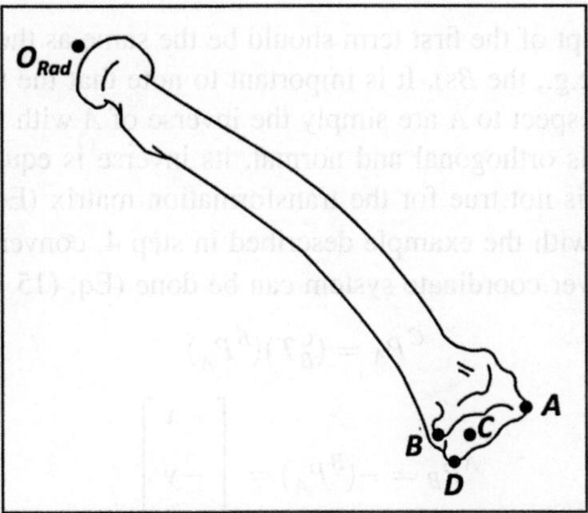

FIGURE 15.4

Digitization of points on the radius, all with respect to the transmitter. A left arm bone specimen is shown.

$\overrightarrow{FE}$ running from the bisector of B and D to point C, and the varus–valgus axis $\overrightarrow{Y}$ is defined as the cross product of $\overrightarrow{X} \times \overrightarrow{FE}$. Finally, a new orthogonal flexion–extension axis $\overrightarrow{Z}_{Rad}$ is calculated as the cross product of $\overrightarrow{X} \times \overrightarrow{Y}$. To complete the rotation matrix ${}^{Transmitter}_{Rad}R$, the terms $\overrightarrow{X}$, $\overrightarrow{Y}$, and $\overrightarrow{Z}$ are normalized to form $\widehat{X}$, $\widehat{Y}$, and $\widehat{Z}$. The corresponding transformation matrix is then determined (Eq. (15.5)).

$$ {}^{Transmitter}_{Rad}T = \begin{bmatrix} {}^{Transmitter}_{Rad}R & {}^{Transmitter}O_{Rad} \\ 0\,0\,0 & 1 \end{bmatrix} \tag{15.5} $$

Step 5. Convert coordinate system relative to bone receiver. Once the bone coordinate system transformation matrix is established with respect to the transmitter, this can be converted to be with respect to the bone receiver.

In general terms, this is done by multiplying by the single instance of ${}^{Transmitter}_{Receiver}T$ that is captured at the time the points are digitized. The output of this step is a transformation matrix that describes the bone's coordinate system with respect to the bone's receiver ${}^{Receiver}_{Bone}T$. Repeat this process for all other bones as needed to quantify rotations, being careful to construct the coordinate system relative to established standards.[7]

In specific terms, use the transformation matrices to transform a point from one coordinate system into another so that it can be expressed with respect to the desired reference. This is done through sequential matrix multiplication with the right-most term being the position, and it is dependent on what information is available. For example, if the position of A with respect to B and the transformation of B with respect to C are known, the position of A can be expressed with respect to C (Eq. (15.6)). Verify if the sequence is correct by checking the subscripts and superscripts such that the subscript of the first term should be the same as the superscript of the second, and so on (e.g., the *B*s). It is important to note that the transformation and rotation of *B* with respect to *A* are simply the inverse of *A* with respect to *B*. Since the rotation matrix is orthogonal and normal, its inverse is equal to its transpose. However, the same is not true for the transformation matrix (Eqs. (15.7)–(15.9)). Finally, continuing with the example described in step 4, conversion of ${}^{Transmitter}_{Rad}T$ into the radial receiver coordinate system can be done (Eq. (15.10)).

$$ {}^{C}P_{A} = ({}^{C}_{B}T)({}^{B}P_{A}) \tag{15.6}$$

$$ {}^{A}P_{B} = -({}^{B}P_{A}) = \begin{bmatrix} -x \\ -y \\ -z \end{bmatrix} \tag{15.7}$$

$$ {}^{A}_{B}R = ({}^{B}_{A}R)^{-1} = ({}^{B}_{A}R)^{\mathrm{T}} \tag{15.8}$$

$$ {}^{A}_{B}T = ({}^{B}_{A}T)^{-1} = \begin{bmatrix} ({}^{B}_{A}R)^{T} & -({}^{B}_{A}R^{T})^{B}P_{A} \\ 0\ \ 0\ \ 0 & 1 \end{bmatrix} \tag{15.9}$$

$$ {}^{RadReceiver}_{Rad}T = ({}^{RadReceiver}_{Transmitter}T)({}^{Transmitter}_{Rad}T) = ({}^{Transmitter}_{RadReceiver}T)^{-1}({}^{Transmitter}_{Rad}T) \tag{15.10}$$

Step 6. Calculate relative rotations. To calculate the rotation of one bone relative to another, the transformation matrix is needed of the moving bone with respect to the reference bone's coordinate system ${}^{ReferenceBone}_{MovingBone}T$. Then, the rotation of the moving bone about an axis of the reference bone can be calculated from the terms of the rotation matrix ${}^{ReferenceBone}_{MovingBone}R$ using ZYX Euler angle analysis (Eqs. (15.11)–(15.13)). (Note: Positive angles are given by the "right hand rule," where the thumb aligns with the positive direction of the axis, and the curl of the remaining fingers indicates positive rotation. In addition ZYX Euler angle analysis is based on a sequence of rotations applied in a set order about the reference axes ($\overrightarrow{Z}$, $\overrightarrow{Y}$, then $\overrightarrow{X}$). The axis order is important for preventing "gimbal lock," which occurs when two axes are driven into a parallel configuration, effectively locking the system into rotation, and should be chosen such that the first rotation is the largest, the second is the smallest, and the remaining is the third.)

$$ \text{Rotation about } \widehat{X}: \alpha = \text{Arctan2}\left(\frac{r_{21}}{r_{11}}\right) \tag{15.11}$$

$$ \text{Rotation about } \widehat{Y}: \beta = \text{Arctan2}\left(\frac{-r_{21}}{\sqrt{r_{11}^{2} + r_{21}^{2}}}\right) \tag{15.12}$$

$$\text{Rotation about } \widehat{Z}: \gamma = \text{Arctan2}\left(\frac{r_{32}}{r_{33}}\right) \tag{15.13}$$

Step 7. Perform statistical analysis. Summarize each numerical result as an average ± 1 standard deviation, and choose the criterion for detecting statistical differences (e.g., $p < 0.01$ or < 0.05). Then, compare ET kinematic data (e.g., position and rotation) for different test groups (e.g., normal, diseased, young, elderly, etc.). There are various software programs for comparing two test groups (e.g., paired *t*-test) or two or more test groups influenced by multiple factors (e.g., analysis of variance, ANOVA). Moreover, power analysis can be done after (but preferably before) the study to ensure power was greater than 80%, indicating there were enough specimens per group to accurately detect statistical differences (i.e., type II statistical error is avoided).

ENGINEER'S TOOLBOX

The loads acting at a moving joint can be estimated using kinematic and other data. Consider that (1) joint motion occurs in a 2D plane, (2) joint friction is negligible, (3) the proximal limb is fixed, while the distal limb moves, (4) the moving limb can be modeled as a rigid, solid, uniform cylinder, (5) the external loads acting on the moving limb are tangential, normal, inertial, and gravitational in nature, and (6) the orthogonal reference axes are normal (n) and tangential (t). Consequently, the loads acting on the joint are normal force $F_n = \pm mL\omega^2 \pm mg(\sin\theta)$, tangential force $F_t = \pm mL\alpha \pm mg(\cos\theta)$, and moment $M = \pm mgL(\cos\theta) \pm mL^2\alpha \pm I_{CG}\alpha$, where m is mass of the moving limb, L is distance from joint center to limb center of gravity CG, g is gravitational acceleration, θ is angular position relative to the horizontal, ω is angular velocity, α is angular acceleration, and $I_{CG} = m(3r^2 + 4L^2)/12$ = moment of inertia of the limb around CG, where r is limb radius. (Note: After drawing a free body diagram of all loads, choose a ± sign convention and use it consistently for linear and angular directions. This ensures that the terms in the formulas are properly added or subtracted.)

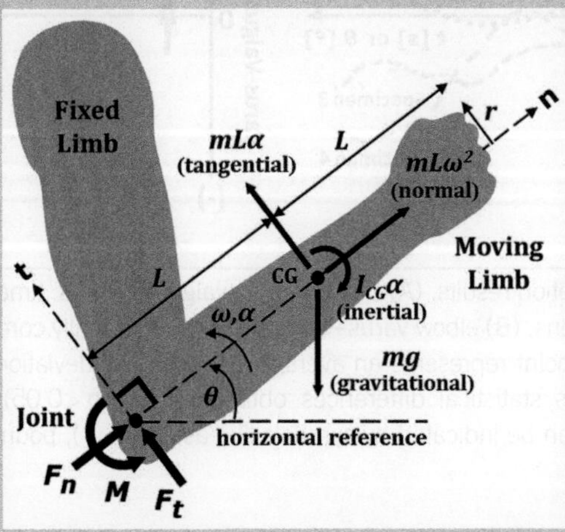

4. RESULTS

Once all ET data collection and analysis have been performed, it is important to communicate and present the primary results in an understandable and concise manner to the reader of a journal article, conference paper, technical report, or book chapter.

Step 1. Plot the joint angles. Depending on the study design, the independent variables determined during data analysis (e.g., joint angle) can be plotted against the dependent variable (e.g., time, another joint angle, etc.) for each specimen, which may have different demographics (e.g., age, sex, species, etc.) or conditions (e.g., normal, injured, diseased, surgically repaired, etc.) (Fig. 15.5A). If multiple specimens are tested, then the curves can be shown as an average curve with a $\pm$ 1 standard deviation envelope surrounding the curve.

Step 2. Show statistical comparisons. If the motion of the elbow joint is of interest, a curve fit can be performed on the resulting motion profile. If multiple joint demographics (e.g., age, sex, species, etc.) or joint conditions (e.g., normal, injured, diseased, surgically repaired, etc.) are investigated, the statistical differences between all test groups can be shown (Fig. 15.5B).

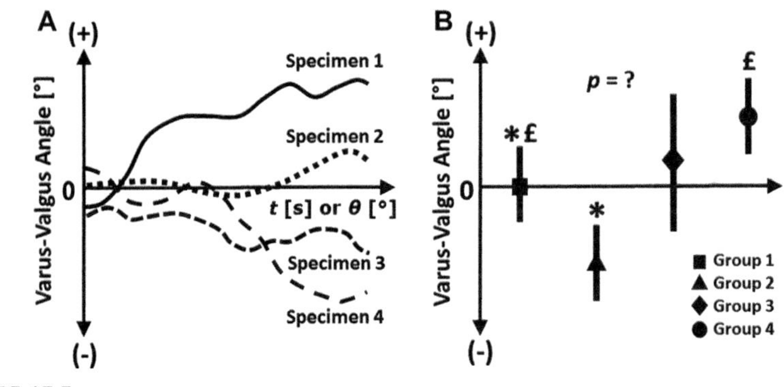

FIGURE 15.5

Typical elbow joint motion results. (A) Elbow varus–valgus angle vs. time t or flexion angle θ for individual specimens, (B) elbow varus–valgus angles statistically compared between test groups, where each point represents an average $\pm$ 1 standard deviation. Multiple p values represent the various statistical differences obtained (e.g., $p < 0.05$) for each pairwise comparison, which can be indicated by symbols like asterisks (*), pounds (£), etc.

ALTERNATIVES AND ADAPTATIONS

✓ **Direct line-of-sight motion tracking.** This method can be used instead of ET if the testing modality permits an uninterrupted view of the motion trackers.

✓ **Unrestricted motion.** While less common, some approaches employ unrestricted motion, such as flexion and extension of the knee while articulating at the hip and ankle. Hence, trackers are required on both sides with a transmitter fixed in a convenient location.

✓ **Other articulating joints.** ET can be applied to other human articulating joints and artificial joints with minor adjustments to the methodology, including the location of clamping, positioning of ET receivers, and coordinate system construction.

5. DISCUSSION

After completing all ET tests, data collection, data analysis, and data presentation, then final results can be considered and interpreted in the broader context of some important clinical, biomechanical, and/or technological considerations, as follows.

ET tests can be used for a variety of different human joint applications, including the shoulder, elbow, hip, knee, etc. The resulting joint kinematic data can be valuable in assessing the effectiveness of joint replacement techniques in restoring natural joint motion, in addition to ascertaining the importance of soft tissue structures to joint motion. It is important to ensure that all coordinate systems are constructed so that their axes are oriented in a clinically meaningful way regardless of the joint under analysis.

ET can be used for various *in vitro* applications. For instance, ET shows that radial head excision in the elbow following radial head fracture can increase varus–valgus laxity from $18 \pm 3.2°$ to $35.6 \pm 10.3°$ and that metallic radial head hemiarthroplasty can provide increased elbow constraint compared to radial head excision.[8] In addition the use of ET to digitize biological structures is an important component of computer-assisted surgery, as it assists in the optimal positioning of joint replacement implants. Moreover, using ET during stylus tracing to determine the center of rotation shows that the average error of the estimated center of rotation is 2.3 ± 1.5 mm for the humeral head and 4.3 ± 3.6 mm for the femoral head when compared to the true center position.[3] This error is hypothesized to be a combination of the nonspherical anatomical structures of both surfaces, some variation in surface tracing, and the error associated with the tracking technique. Finally, the use of ET to test plate fixation of olecranon fractures shows no significant difference between lateral and posterior plate fixation in the relative translation and rotation of the reaffixed bone fragment.

ET is advantageous for joint tracking since it does not require a direct line of sight of the trackers compared to optical tracking methods, which require all markers to be visible during testing. This means that for joints that exhibit large ranges of motion, which would obstruct the direct line of sight, ET may alleviate dropping of tracking data. However, it does require that there are no metallic materials in the test envelope, as this will produce tracking errors.

6. SUMMARY

- Motion trackers rigidly affixed to the bones of the joint allow for the recording of joint kinematics.
- Care must be taken to eliminate all relative motion between the bone and attached trackers.
- In most instances, the kinematics of the intact joint are compared to other conditions.
- Raw motion data are converted to T-matrices and transformed relative to one structure.
- Pertinent joint angles are calculated based on these T-matrices.
- Joint angles and changes in joint angles are useful in graphical format.
- Statistical analysis can ascertain the effects of various joint states on kinematics.

7. QUIZ QUESTIONS

1. What is joint kinematics, and why is it important?
2. Why should motion trackers not move relative to bone?
3. What are the main elements of the T-matrix?
4. Where are position and rotation data stored?
5. Compute the elbow joint loads if the forearm rotates upward at a constant angular velocity, while the upper arm is fixed. The distance from the elbow joint center to the forearm center of gravity is 14 cm, forearm mass is 1 kg, forearm angular velocity is 2 rad/s, and forearm angular position to the horizontal is 20° (answer: $F_n = 2.79$ N, $F_t = 9.22$ N, $M = 1.29$ N·m).

REFERENCES

1. Milne AD, Chess DG, Johnson JA, King G. Accuracy of an electromagnetic tracking device: a study of the optimal operating range and metal interference. *Journal of Biomechanics* 1996;**29**(6):791–3.
2. King GJ, Zarzour ZD, Rath DA, Dunning CE, Patterson SD, Johnson JA. Metallic radial head arthroplasty improves valgus stability of the elbow. *Clinical Orthopaedics and Related Research* 1999;**368**:114–25.

3. Whitney KD, Ferreira LM, King GJW, Johnson JA. The effect of surface area digitizations on the prediction of spherical anatomical geometries for computer-assisted applications. *Journal of Biomechanics* 2009;**42**(8):1158–61.
4. King GJ, Lammens PN, Milne AD, Roth JH, Johnson JA. Plate fixation of comminuted olecranon fractures: an in vitro biomechanical study. *Journal of Shoulder and Elbow Surgery* 1996;**5**(6):437–41.
5. Dunning CE, Gordon KD, King GJW, Johnson JA. Development of a motor-controlled in vitro elbow testing system. *Journal of Orthopaedic Research* 2003;**21**(3):405–11.
6. Wu G, van der Helm FC, Veeger HE, Makhsous M, Van Roy P, Anglin C, et al. ISB recommendation on definitions of joint coordinate systems of various joints for the reporting of human joint motion–Part II: shoulder, elbow, wrist and hand. *Journal of Biomechanics* 2005;**38**(5):981–92.
7. Shoemake K. Animating rotation with quaternion curves. *ACM SIGGRAPH Computer Graphics* 1985;**19**(3):245–54.
8. Sabo MT, Shannon H, Ng J, Ferreira LM, Johnson JA, King GJ. The impact of capitellar arthroplasty on elbow contact mechanics: implications for implant design." *Clinical Biomechanics* 2011;**26**(5):458–63.

CHAPTER

Fujifilm Measurements of Interfacial Contact Area and Stress in Articulating Joints

16

Radovan Zdero[1], Ziauddin Mahboob[2], Habiba Bougherara[2]

Western University, London, ON, Canada[1]; Ryerson University, Toronto, ON, Canada[2]

1. BACKGROUND

Human articulating joints (e.g., shoulder, hip, knee) are junctions of two cartilage-covered bones moving against each other at a common interface. Joints help transfer loads from bone to bone to permit human movement. However, osteoarthritis is a disease that damages cartilage, resulting in bone-on-bone articulation, which is painful and debilitating.[1] Each year, one million surgeries are done globally to insert a total joint replacement (TJR),[2] which replaces human articulating surfaces with artificial components made from ceramic, metal, and/or polymer. TJRs, however, can experience forces and motions that create wear debris, leading to osteolysis, device loosening, and revision surgery as early as 6–8 years.[1] Consequently, interfacial contact area and stress are important factors for joint health and survival. Fujifilm (FF) is a single-use, easy-to-use, inexpensive, nontoxic, nondestructive, pressure-sensitive film with a well-established test protocol (Fig. 16.1).[3–5] It has been widely used to study the *in vitro* contact mechanics of orthopaedics applications, especially human and artificial shoulder joints,[6,7] hip joints,[8–10] and knee joints.[11–16] Therefore, this chapter explains FF testing, as well as how to analyze, present, and interpret results.

2. RESEARCH QUESTIONS

Typical research questions might include one or more of the following:

- Do articulating joint shape, texture, and material properties influence contact area and stress?
- Do applied force, loading speed, loading duration, etc., affect the contact area and stress?

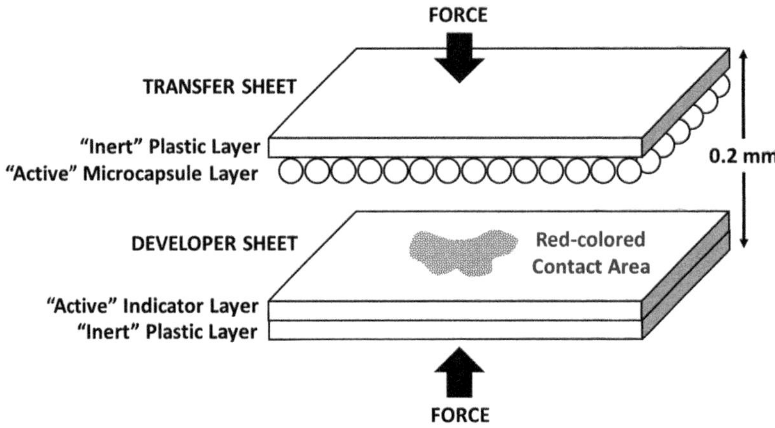

FIGURE 16.1

Schematic of FF structure. When the sheets are pressed together between two articulating objects, the transfer sheet's microcapsules burst, releasing a mild acid, which reacts with the developer sheet's indicator layer. The resulting red-colored contact area has a size, shape, and density that depends on applied force, loading rate, loading duration, interfacial geometry, temperature, humidity, etc. Outer inert layers have a glossy appearance, while inner active layers have a matte appearance.

- Which regions of an articulating joint are at risk of failure due to excessively high stress?
- How do FF test results compare with other computational and experimental techniques?
- etc.

3. METHODOLOGY

3.1 GENERAL STRATEGY

FF is used to assess the contact mechanics of a total knee replacement (TKR) as a typical example of FF testing of human joints and TJRs. A well-established protocol is employed, which has been developed by the FF manufacturer and prior researchers.[3–16] Specifically, artificial sawbone material mimicking human bone is shaped to receive a TKR. Then, FF is cut to shape and placed onto the TKR articulating surfaces. Next, the TKR is mounted in a mechanical tester for load application duplicating a physiological force. FF is removed, scanned, and analyzed by dedicated computer software to obtain total contact area, average contact stress, and peak contact stress. Data are used to identify any regions of the TKR that may be at risk of failure due to high contact stress. Finally, statistical comparisons are made between test groups, and correlation coefficients are computed to determine important influencing factors.

GLOSSARY

✓ **Articulating joint**. Two cartilage-covered bones moving at a common interface.
✓ **Contact area**. Total surface area in direct contact at the interface of an articulating joint.
✓ **Contact stress**. Ratio of applied mechanical force experienced per unit of contact area.
✓ **FF (Fujifilm)**. Pressure-sensitive double-sheet film used to measure contact area and stress.

SAFETY FIRST

✓ Secure the sawbones and/or implants before applying load in the mechanical tester.
✓ Place a protective plastic shield around the mechanical test area to contain flying debris.
✓ Remember to always wear goggles and gloves for protection if testing cadaveric joints.
✓ Clean the work area and all tools with bleach or disinfectant if testing cadaveric joints.

3.2 MATERIALS AND TOOLS LIST

- adhesive tape
- artificial sawbones
- bone cement
- FF
- mechanical tester
- scissors
- software for image analysis
- TKR
- tweezers

3.3 SPECIMEN PREPARATION

Step 1. Clean the articulating surfaces. TKR articulating surfaces to be tested must be clean and smooth to avoid FF artifacts, which may incorrectly appear as points of high or low stress. Loose debris should be removed with a nonabrasive dry cloth to avoid scratching the surface.

Step 2. Shape the sawbones. Obtain artificial sawbone materials, which mimic the biomechanical behavior of human bone, such as a sawbone "femur" and a sawbone "tibia" block (Fig. 16.2A and B).[11] Shape the sawbones by cutting, drilling, and rasping to match the geometry of the backside of the metal femoral and tibial components of the TKR.

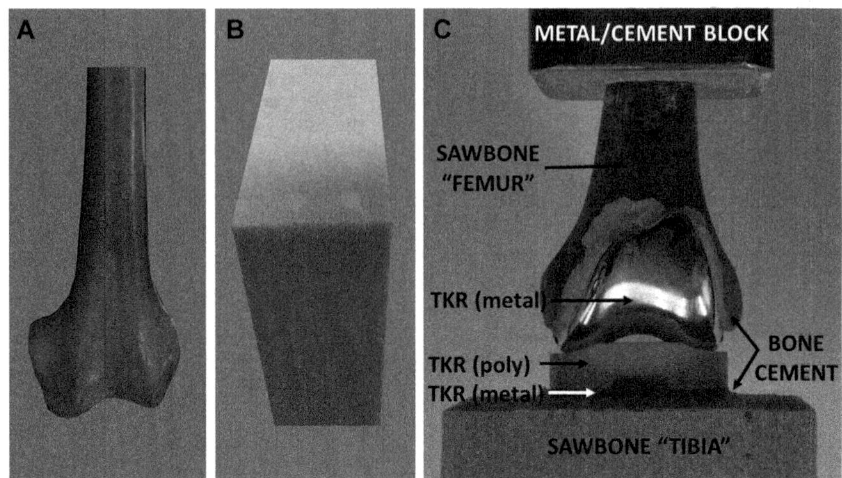

FIGURE 16.2

TKR assembly. (A) Artificial sawbone femur replicating a human femur, which is potted in a hollow metal cube filled with cement, (B) artificial sawbone block replicating a human tibia, (C) TKR components mounted into sawbone material.

Step 3. Apply the bone cement. Place the sawbones into a fume hood in order to absorb the pungent and toxic fumes produced by surgical bone cement (i.e., poly methyl methacrylate, PMMA). Apply a 1–2 mm layer of surgical bone cement to the shaped sawbone surfaces, use a rubber hammer or press firmly by hand to position the metal femoral and tibial components into place, and let the bone cement cure while the fume hood is still operating. This will reasonably simulate the clinical situation of host bone surrounding a TKR (Fig. 16.2C).

Step 4. Pot the sawbones. In some cases, it may be necessary to pot the sawbone "femur" and/or "tibia" into surgical bone cement, industrial anchoring cement, or liquid metal in order to mount it securely during later mechanical loading. To do this, place the sawbones into a fume hood, mount the sawbones in a multiaxial clamp, insert one end of the sawbones into a hollow metal cube, align the sawbones in the appropriate orientation using a leveling gage, pour the potting medium into the hollow metal cube, and allow the potting medium to cure.

Step 5. Choose the correct FF. There are various commercially available FF types, each with a different pressure sensitivity over a wide range from 0.05–300 MPa.[3] If only the TKR's total contact area (and hence, average contact stress) is of interest, then choose the FF type with the lowest pressure threshold for microcapsule bursting since it will yield the largest possible contact area (i.e., "extreme low pressure," 0.05–0.2 MPa, or "ultra super low pressure," 0.2–0.6 MPa). However, if the TKR's contact stress profile (and hence, peak contact stress) is needed, then it is critical to choose the FF type whose sensitivity range covers the expected peak contact stress since contact stress profile is based on color density distribution across the contact area.

Step 6. Cut FF to shape. Obtain new unused FF sheets. Put on a pair of latex or vinyl gloves before handling FF so that any dirt, grease, or moisture will not contaminate the FF. Use scissors to cut FF to match the shape of the articulating surfaces where FF will be placed, but ensure that FF shapes are large enough to fully cover the anticipated contact areas. Specifically, the transfer sheets can be cut into rectangular shapes to match the geometry of the TKR's metal femoral condyles (Fig. 16.3A). The developer sheets can be cut into "half moon" or "kidney bean" shapes to exactly match the geometry of the TKR's polymer tibial boundary, which

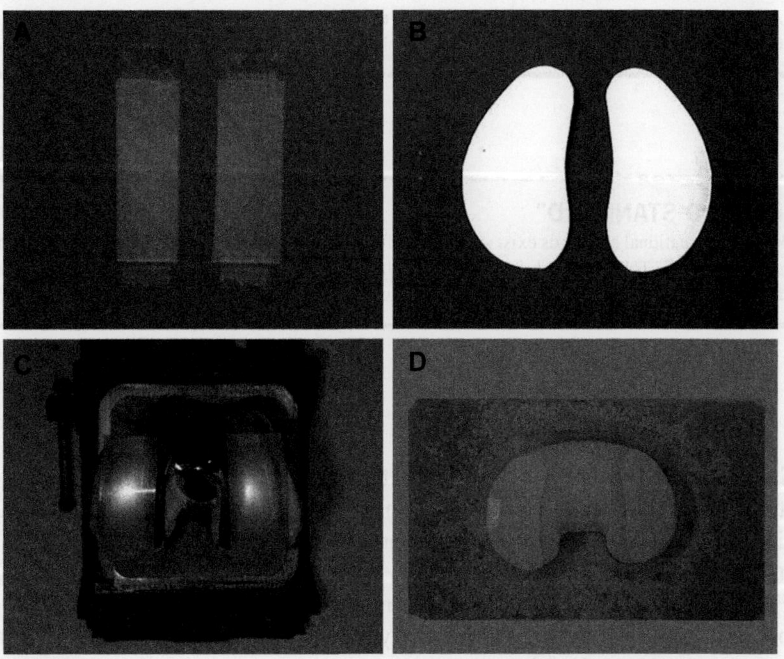

FIGURE 16.3

Placement of FF on the TKR. (A) FF transfer sheets cut to shape, (B) FF developer sheets cut to shape, (C) FF transfer sheets with active matte sides exposed are secured by adhesive tape onto the metal femoral component, (D) FF developer sheets with active matte sides exposed are secured by adhesive tape onto the polymer tibial component.

acts as a fixed reference frame for the relative position of the contact area to be obtained (Fig. 16.3B).

Step 7. Attach FF to surfaces. Place FF shapes on a clean smooth surface with inert glossy sides facing down. Apply small strips of adhesive tape to a few locations around the edges of the FF shapes so some tape is exposed beyond the edges. Use tweezers to position each FF shape (with adhesive tape) onto the appropriate TKR articulating surface so that inert glossy sides are facing the articulating surfaces while the active matte sides remain exposed. By hand, press down onto each strip of exposed adhesive tape so FF is secured into its final location (Fig. 16.3C and D).

TIPS AND TRICKS

- ✓ FF is a single-use technology; thus, new FF must be used for each new test.
- ✓ Use tweezers when handling FF to avoid accidental finger marks or pressurization.
- ✓ Cut FF into a "star" shape to avoid crinkling when testing ball-in-cup geometries.
- ✓ Apply small strips of adhesive tape around the edges of FF to secure it during testing.
- ✓ Cover cadaveric joint surfaces with food wrapping plastic film before inserting FF.

THE "GOLD STANDARD"

No known international standards exist specifically for using FF to study the contact mechanics of human joints or TJRs. Thus, researchers should consult literature from their manufacturer or supplier of FF, as well as peer-reviewed journal articles that have used FF for this application.

3.4 SPECIMEN TESTING

Step 1. Monitor environmental conditions. FF results are sensitive to environmental conditions, so it is important to ensure tests are done within manufacturer specifications. Typical biomechanics laboratory settings, however, fall within the recommended ranges of temperature (20–35°C) and relative humidity (35–80%) for FF.[3] Note that environmental conditions must be maintained within ± 3°C for temperature and ± 3% for relative humidity for a given FF research project to maintain consistency in results.[4,5]

Step 2. Mount the specimen. Secure the TKR's femoral and tibial components (now equipped with FF) separately into a mechanical tester to ensure that opposing FF sheets do not accidentally come into contact. Align the TKR's femoral and tibial components to the desired orientation (e.g., 20° knee flexion, 1 mm medial shift, etc.), and then slowly lower the femoral component to within several millimeters of the tibial component so FF matte sides are facing each other (Fig. 16.4).

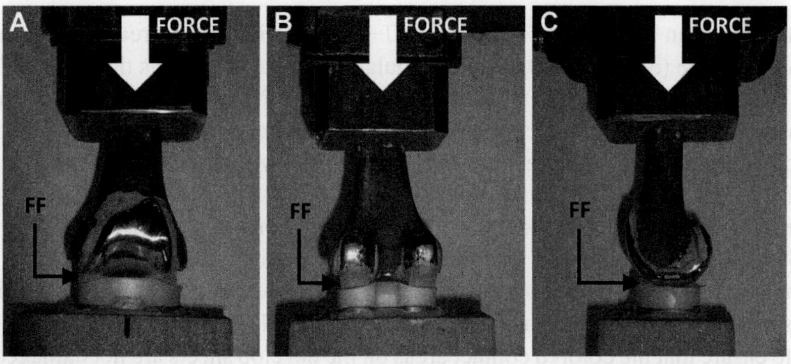

FIGURE 16.4

TKR equipped with FF and mounted in a mechanical tester. (A) Anterior view, (B) posterior view, (C) lateral view.

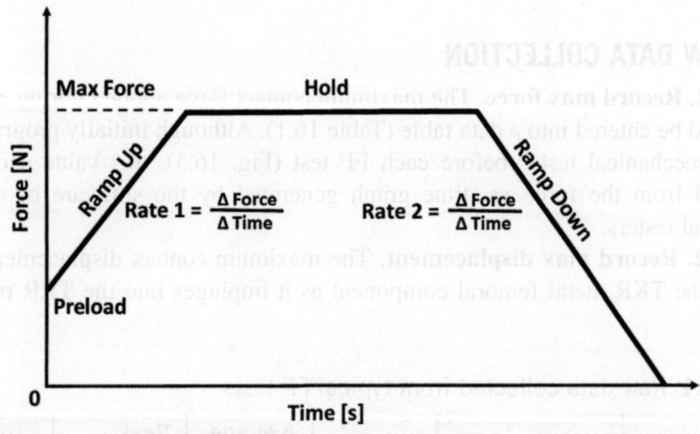

FIGURE 16.5

Loading regime for FF mechanical tests on human articulating joints or TJRs.

Step 3. Position the safety shield. Place a transparent, nonreflective, protective, plastic shield around the TKR test area if there is any potential concern about the specimen fracturing and releasing flying debris during the mechanical loading process.

Step 4. Apply compressive force. Program the mechanical tester using load control with a loading regime composed of several phases (Fig. 16.5). The small initial "preload" (e.g., 5–50 N) eliminates mechanical slack in the test setup (i.e., hysteresis), such as slippage between TKR components.[4,11] Then, the linear force "ramp up" at an appropriate rate (e.g., 10–100 N/s) is used to reach the desired maximum force (e.g., 2–3 kN), which simulates three to four times the body weight of a 75 kg person as occurs during walking.[4,5,10,11,16] Next, the force "hold" phase maintains the maximum force for a suitable duration (e.g., 60–90 s), which allows the polymer component to reach a mechanical steady state.[4,5,10,11,16] Finally, the linear "ramp down" at the same rate (e.g., 10–100 N/s) decreases the compressive force to zero, so TKR components are not in direct contact.[11]

Step 5. Remove the FF. Use the mechanical tester's controls to separate the femoral and tibial components by 20–30 mm to provide adequate room for physical access to the TKR's articulating surfaces. Then, carefully remove the FF using tweezers to ensure that the colored contact areas on the develop sheets are not damaged or contaminated.

Step 6. Rest and repeat. If the same TKR is reused for multiple tests, allow the polymer tibial component a rest of 2 h to recover from elastic deformation[16] (although plastic deformation will be permanent), then attach new unused FF, and repeat steps 3–5 above. However, if a new TKR is used for each subsequent test, then repeat steps 2–5 above.

3.5 RAW DATA COLLECTION

Step 1. Record max force. The maximum contact force applied during each FF test should be entered into a data table (Table 16.1). Although initially programmed into the mechanical tester before each FF test (Fig. 16.5), this value should be confirmed from the force vs. time graph generated by the software of modern mechanical testers.

Step 2. Record max displacement. The maximum contact displacement travelled by the TKR metal femoral component as it impinges into the TKR polymer

Table 16.1 Raw data collected from typical FF tests.

Test	Max Force F [N]	Max Displacement δ [mm]	Total Contact Area A [mm^2]	Average Contact Stress σ_{AVG} [MPa]	Peak Contact Stress σ_{PEAK} [MPa]	Remarks
1						
2						
3						
etc.						
Avg						–
SD						–

Avg, *average;* SD, *standard deviation.*

tibial component should also be entered (Table 16.1). Although this is not initially programmed into the mechanical tester before each FF test, this value can be obtained from the displacement vs. time graph generated by the software of the mechanical tester.

Step 3. Record FF data. Record total contact area, average contact stress, and peak contact stress (Table 16.1). To do so, it is recommended to use the dedicated image analysis software and flatbed scanner provided by the FF manufacturer.[3] Place the FF developer sheet with the red-colored contact area into the flatbed scanner, scan the FF as a color image using a suitable resolution, and import the image into the software. The software will measure contact area in pixels^2 and contact stress using the FF manufacturer's proprietary precalibrated pixel color density scale, after which it converts these, respectively, to area in mm^2 and stress in MPa (Fig. 16.6). There are two options used by the software. The "trace" option allows the user to employ their subjective judgment by using the cursor to manually draw a line around the contact area. The "thresholding" option allows the user to automatically define a minimum or maximum color scale range within which pixels are automatically excluded or included as part of the contact area. (Note: an alternative method is to scan the FF developer sheet using the user's own preferred flatbed scanner, save the image as a color or grayscale image with a

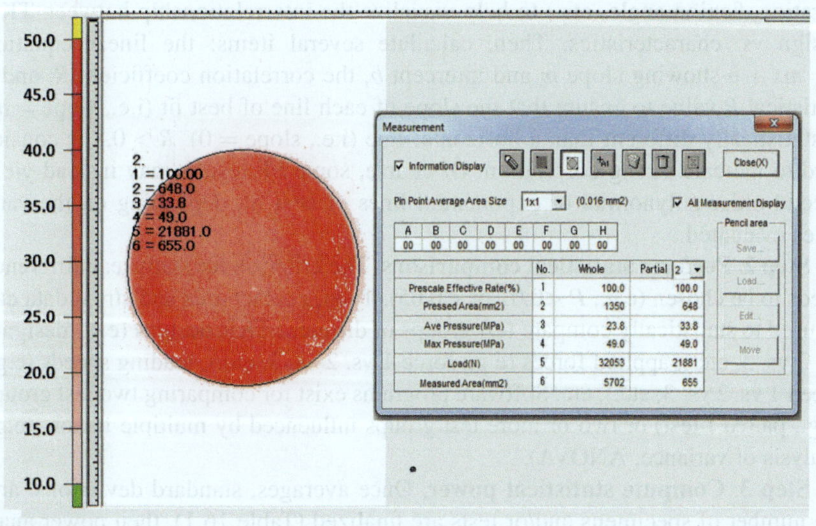

FIGURE 16.6

Typical FF imaging software window. Tabulated parameters include total contact area, average contact stress, maximum contact stress, and applied force. Only a circular FF contact area is shown, but any contact area shape can be analyzed.

Software image courtesy of www.fujifilm.com/products/prescale and www.pressuremetrics.com.

minimum resolution of 1200 ppi or dpi (i.e., pixels or dots per inch), import the image into the user's preferred image analysis software, and perform "trace" or "thresholding" operations to obtain measurements.[4,5,16,17] Some software may not have all desired capabilities; thus, researchers may need to perform the following computations or tests themselves: total contact area [mm^2] = total contact area [$pixels^2$]/image resolution^2 [$pixels^2/in^2$] $\times$ 25.4^2 [mm^2/in^2], average contact stress = max force/total contact area, and prestudy calibration experiments with known loads and geometries to obtain a color or gray density level scale for determining peak contact stress.)

Step 4. Record final remarks. Some final comments should be recorded about any cracking, bending, or unusual movement that may have occurred at the specific location where contact area was measured with FF or at any other location on the TKR (Table 16.1). These remarks can confirm or deny the presence of high-stress regions.

3.6 RAW DATA ANALYSIS

Step 1. Calculate correlation coefficients. Plot the measured total contact area, average contact stress, and peak contact stress for each TKR design versus any characteristics that were varied (e.g., applied load, loading speed, loading duration, flexion angle, etc.) to help visualize the interrelationship between TKR design vs. characteristics. Then, calculate several items: the linear equation $y = mx + b$ showing slope m and intercept b, the correlation coefficient R, and a statistical P value to ensure that the slope of each line of best fit (i.e., slope = m) is statistically different than a horizontal line (i.e., slope = 0). $R > 0.8$ is considered to indicate strong correlation. Of course, some FF results may instead yield second-order polynomial or exponential lines of best fit, depending on the variables evaluated.

Step 2. Perform statistical comparisons. The criterion for statistical difference needs to be chosen (e.g., P <0.01 or <0.05). Then, contact area and stress data can be used to statistically compare test groups of different TKR designs (e.g., design 1 vs. 2 vs. 3, etc.), applied forces (e.g., force 1 vs. 2 vs. 3, etc.), loading speeds (e.g., speed 1 vs. 2 vs. 3, etc.), etc. Software programs exist for comparing two test groups (e.g., paired t-test) or two or more test groups influenced by multiple factors (e.g., analysis of variance, ANOVA).

Step 3. Compute statistical power. Once averages, standard deviations, and the number of specimens and/or tests are finalized (Table 16.1), then power analysis can be done after the study to ensure there were enough specimens per test group to detect all statistical differences that were actually present (i.e., was type II statistical error avoided?). Statistical power >80% is usually considered to indicate there were enough specimens per test group. Note that if good predictions of averages and standard deviations are available from prior studies, then the number of specimens and/or tests can be chosen before the study begins to ensure a power >80%.

ENGINEER'S TOOLBOX

Early pioneering work by Heinrich Hertz resulted in "Hertzian" theoretical formulas for understanding interfacial contact mechanics of two articulating solid bodies while assuming quasi-static loading, frictionless contact, elastic bodies, small deformations, and symmetry. The formulas can be applied to human or artificial joint biomechanics, as shown in the figure, by using approximations for geometry and material properties, where F is applied force, D is diameter, L is length, E is elastic modulus, and ν is Poisson's ratio.

Consequently, for ball-in-cup joints (e.g., shoulder and hip), the formulas are average contact stress $\sigma_{AVG} = F/A$, peak contact stress $\sigma_{PEAK} = (3/2)\sigma_{AVG}$, and

$$\text{total contact area } A = \pi\left[\frac{\left(\frac{3F}{8}\right)\left(\frac{1-\nu_1^2}{E_1}+\frac{1-\nu_2^2}{E_2}\right)}{\left(\frac{1}{D_1}-\frac{1}{D_2}\right)}\right]^{\frac{2}{3}}$$

Similarly, for cylinder-in-groove joints (e.g., knee), the formulas are average contact stress $\sigma_{AVG} = F/A$, peak contact stress $\sigma_{PEAK} = (4/\pi)\sigma_{AVG}$, and

$$\text{total contact area } A = \left[\frac{\left(\frac{8FL}{\pi}\right)\left(\frac{1-\nu_1^2}{E_1}+\frac{1-\nu_2^2}{E_2}\right)}{\left(\frac{1}{D_1}-\frac{1}{D_2}\right)}\right]^{\frac{1}{2}}$$

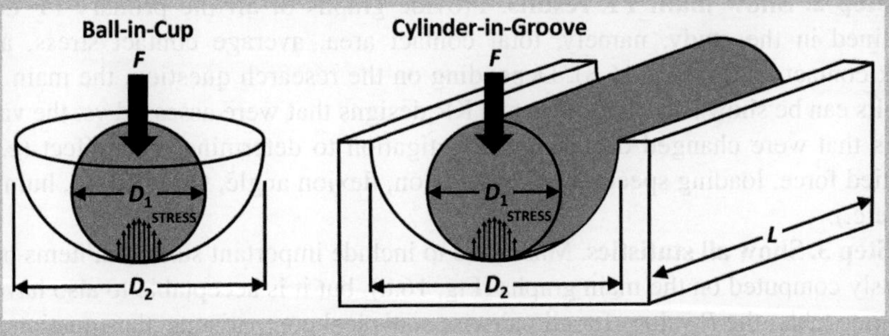

4. RESULTS

Once all FF data collection and analysis have been performed, it is then important to communicate and present the primary results in an understandable and concise manner to the reader of a journal article, conference paper, technical report, or book chapter.

Step 1. Show original FF sheets. Present an original contact area from the FF develop sheets for a typical TKR test (Fig. 16.7A). This helps visually illustrate

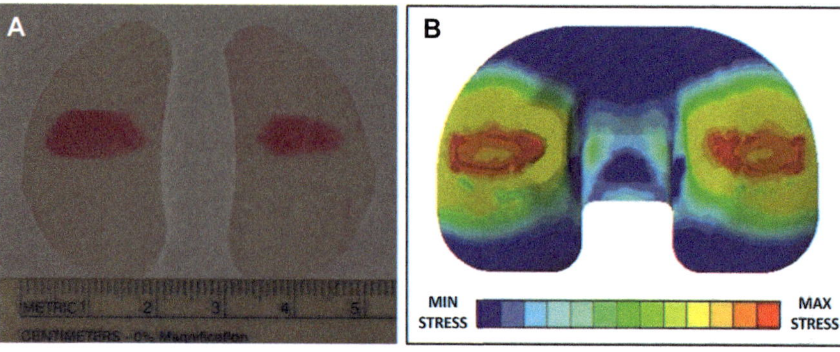

FIGURE 16.7

Typical contact area for the TKR. (A) Contact area obtained from FF tests, (B) contact area (and/or stress) obtained from finite element computer modeling.

the characteristic size and shape of the contact area, as well as any locations of high and low stress. If pertinent, also show the contact area from another computational or experimental method that may have been used for the same TKR for comparison purposes (Fig. 16.7B).

Step 2. Show main FF results. Provide graphs of all the primary FF data obtained in the study, namely, total contact area, average contact stress, and peak contact stress (Fig. 16.8). Depending on the research question, the main FF results can be shown for the different TKR designs that were assessed vs. the variables that were changed during the investigation to determine their effect (e.g., applied force, loading speed, loading duration, flexion angle, temperature, humidity, etc.).

Step 3. Show all statistics. Make sure to include important statistical items previously computed on the main graphs (Fig. 16.8), but it is acceptable to also have a separate table: the *P* values for all pairwise statistical comparisons; the equation for each line of best fit, whether it is a straight line, a second-order polynomial curve, or an exponential trend; the correlation coefficient *R* for each line of best fit; the statistical power(s) of the study; etc.

ALTERNATIVES AND ADAPTATIONS

- ✓ **Ball-in-cup joints.** For human joints or TJRs that are highly curved with a ball-in-cup geometry (e.g., shoulders and hips), FF sheets need to be cut into "star" shapes so they can be properly folded into the joint space without crinkling.
- ✓ **Human articulating joints.** Clean all human joint surfaces of extraneous soft tissue, synovial fluid, and blood, cover joint surfaces with food wrapping plastic film, apply a thin layer of liquid glue to the plastic film, and place FF into position on top of the plastic film.

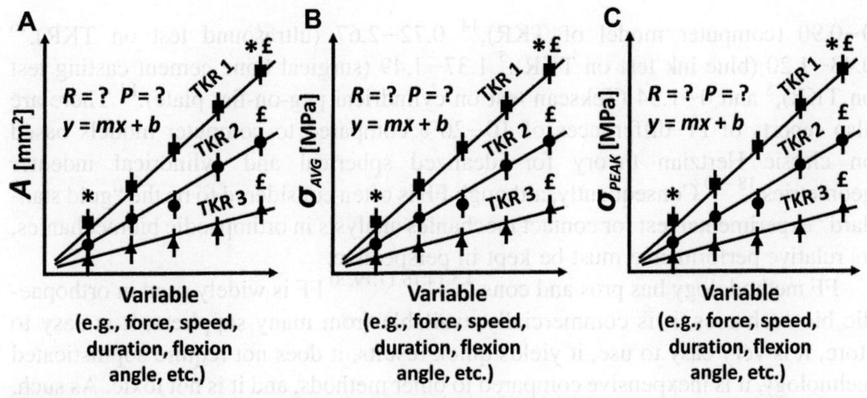

FIGURE 16.8

Main results for FF tests on TKRs. (A) Total contact area, (B) average contact stress, (C) peak contact stress. Each data point is an average ± 1 standard deviation. *P* values are statistical difference results for each pairwise comparison between TKR designs and between variables, which can be indicated by symbols like asterisks (*), pounds (£), etc. Each line of best fit is represented by an equation $y = mx + b$, a correlation coefficient *R*, and also its own statistical *P* value that proves its slope (i.e., slope = *m*) is different than a horizontal line (i.e., slope = 0). Some FF results may instead yield second-order polynomial or exponential lines of best fit, depending on the variables evaluated.

5. DISCUSSION

After completing all FF tests, data collection, data analysis, and data presentation, then final results can be considered and interpreted in the broader context of some important clinical, biomechanical, and/or technological considerations, as follows.

FF results for human and artificial joints depend on the nature of the specimens assessed and the experimental conditions. FF contact area shapes for human joints can be circular (e.g., shoulder),[7] monocentric or bicentric ellipses (e.g., hips),[8,9] and pear-shaped or elliptical (e.g., knees),[13] while artificial joints yield shapes that are circular (e.g., hips)[10] and elliptical or quasirectangular (e.g., knees).[5,15,16] Moreover, FF total contact area and average contact stress, respectively, can range from 58–550 mm^2 and 0.4–5.3 MPa (human shoulder),[6,7] 322–589 mm^2 and 3.6–5.0 MPa (human hip),[8,9] 1010–1120 mm^2 and 1.1–1.2 MPa (human knee),[13] 50–70 mm^2 and 3.4–5.0 MPa (artificial shoulder),[6] 161–319 mm^2 and 9.4–18.6 MPa (artificial hip),[10] and 20–220 mm^2 and 4.0–33.0 MPa (artificial knee).[14–16]

FF results can be directly compared to other methods to identify variations. The ratio of the average contact stress for FF tests vs. other computational and experimental techniques can range from 0.71–0.98 (computer model of human hip),[8]

0–0.90 (computer model of TKR),[14] 0.72–2.67 (ultrasound test on TKR),[16] 0.43–1.20 (blue ink test on TKR),[5] 1.37–1.49 (surgical bone cement casting test on TKR),[5] and 0–1.34 (Tekscan test on cylindrical peg-on-flat plate).[17] There are also reports of FF differences of 10–26% compared to computer models based on classic Hertzian theory for idealized spherical and cylindrical indenter geometries.[18–20] Consequently, although FF is often considered to be the "gold standard" experimental test for contact mechanics analysis in orthopaedic biomechanics, its relative performance must be kept in perspective.

FF methodology has pros and cons.[4,5,14,16,17,19,20] FF is widely used in orthopaedic biomechanics, it is commercially available from many suppliers, it is easy to store, it is very easy to use, it yields quick results, it does not require sophisticated technology, it is inexpensive compared to other methods, and it is not toxic. As such, it is considered to be a kind of "gold standard" approach to measuring contact area and stress in orthopaedic biomechanics. However, FF is only suitable for quasi-static loading, it is limited to *in vitro* laboratory use, it requires a minimum pressure threshold to detect contact, its spatial resolution is restricted, it can be sensitive to shear, it can be affected by temperature and humidity changes, its finite thickness interferes somewhat with actual interfacial contact mechanics, it can crinkle when used for highly curved geometries, and it is sometimes difficult to accurately detect pressure gradients near the edges of the contact area.

6. SUMMARY

- Articulating joints experience complex contact stresses during physiological loading.
- FF is an easy-to-use, inexpensive, nondestructive method for studying contact mechanics.
- FF tests on articulating joints can identify regions at risk of failure due to high contact stress.
- FF results can help optimize the design of TJR shape, texture, and material properties.
- Articulating joint shape, texture, and material properties influence contact area and stress.
- Applied force, loading speed, loading duration, etc., influence contact area and stress.

7. QUIZ QUESTIONS

1. What are the basic operating principles and procedures of the FF technique?
2. What are the general benefits and drawbacks of FF testing in orthopaedics?
3. What is the recommended procedure for doing FF tests on human cadaveric specimens?

4. What are the typical variables that can be evaluated using FF tests on articulating joints?

5. Estimate the peak contact stress σ_{PEAK} in a total hip replacement experiencing a 3 kN load, assuming it can be modeled as a ball-in-cup joint with a metal ball ($D = 30$ mm, $E = 200$ GPa, $\nu = 0.3$) and polymer cup ($D = 40$ mm, $E = 0.9$ GPa, $\nu = 0.4$) (answer: $\sigma_{PEAK} = 56.8$ MPa).

REFERENCES

1. Garino JP, Beredjiklian PK. *Adult reconstruction and arthroplasty: core knowledge in orthopaedics*. Philadelphia (PA, USA): Mosby Elsevier; 2007.
2. Harsha AP, Joyce TJ. Comparative wear tests of ultra-high molecular weight polyethylene and cross-linked polyethylene. *Proceedings of the Institution of Mechanical Engineers (Part H): Journal of Engineering in Medicine* 2013;**227**(5):600–8.
3. Fujifilm Corporation, Tokyo (Japan); www.fujifilm.com/products/prescale. (NJ, USA): Pressure Metrics, Whitehouse Station; www.pressuremetrics.com.
4. Liggins AB, Stranart JCE, Finlay JB, Rorabeck CH. Calibration and manipulation of data from Fuji pressure-sensitive film. In: Little EG, editor. *Experimental mechanics: technology transfer between high tech engineering and biomechanics.* Amsterdam (Netherlands): Elsevier; 1992. pp. 61–70.
5. Liggins AB, Finlay JB. Recording contact areas and pressures in joint interfaces. In: Little EG, editor. *Experimental mechanics: technology transfer between high tech engineering and biomechanics*. Amsterdam (Netherlands): Elsevier; 1992. pp. 71–88.
6. Shapiro TA, McGarry MH, Gupta R, Lee YS, Lee TQ. Biomechanical effects of glenoid retroversion in total shoulder arthroplasty. *Journal of Shoulder and Elbow Surgery* 2007; **16**(3S):90S–5S.
7. Warner JJP, Bowen MK, Deng X-H, Hannafin JA, Arnoczky SP, Warren RF. Articular contact patterns of the normal glenohumeral joint. *Journal of Shoulder and Elbow Surgery* 1998;**7**(4):381–8.
8. Anderson AE, Ellis BJ, Maas SA, Peters CL, Weiss JA. Validation of finite element predictions of cartilage contact pressure in the human hip joint. *Journal of Biomechanical Engineering* 2008;**130**(5):051008-1-10.
9. Konrath GA, Hamel AJ, Guerin J, Olson SA, Bay B, Sharkey NA. Biomechanical evaluation of impaction fractures of the femoral head. *Journal of Orthopaedic Trauma* 1999; **13**(6):407–13.
10. Plank GR, Estok II DM, Muratoglu OK, O'Connor DO, Burroughs BR, Harris WH. Contact stress assessment of conventional and highly crosslinked ultra high molecular weight polyethylene acetabular liners with finite element analysis and pressure sensitive film. *Journal of Biomedical Materials Research Part B: Applied Biomaterials* 2007; **80**(1):1–10.
11. Bougherara H, Zdero R, Mahboob Z, Dubov A, Shah S, Schemitsch EH. The biomechanics of a validated finite element model of stress shielding in a novel hybrid total knee replacement. *Proceedings of the Institution of Mechanical Engineers (Part H): Journal of Engineering in Medicine* 2010;**224**(10):1209–19.
12. Fukubayashi T, Kurosawa H. The contact area and pressure distribution of the knee. *Acta Orthopaedica Scandinavica* 1980;**51**(6):871–9.

13. Ihn JC, Kim SJ, Park IH. In vitro study of contact area and pressure distribution in the human knee after partial and total meniscectomy. *International Orthopaedics* 1993; **17**(4):214–8.
14. Liau JJ, Cheng CK, Huang CH, Lo WH. Effect of Fuji pressure sensitive film on actual contact characteristics of artificial tibiofemoral joint. *Clinical Biomechanics* 2002;**17**(9/10):698–704.
15. Manley MT, Kester M, Averill RG. Femoro-tibial contact area: comparison of contemporary total knee prosthetic systems. Allendale (NJ, USA): Osteonics Corporation; 1990.
16. Zdero R. *A new diagnostic ultrasound technique for studying TKR contact mechanics* [Ph.D. thesis]. Kingston (Canada): Department of Mechanical Engineering, Queen's University; May 1999. http://www.nlc-bnc.ca/obj/s4/f2/dsk1/tape8/PQDD_0001/NQ42992.pdf.
17. Bachus KN, DeMarco AL, Judd KT, Horwitz DS, Brodke DS. Measuring contact area, force, and pressure for bioengineering applications: using Fuji Film and TekScan systems. *Medical Engineering and Physics* 2006;**28**(5):483–8.
18. Johnson KL. *Contact mechanics*. Cambridge (UK): Cambridge University Press; 1985.
19. Hale JE, Brown TD. Contact stress gradient detection limits of Pressensor film. *Journal of Biomechanical Engineering* 1992;**114**(3):352–7.
20. Wu JZ, Herzog W, Epstein M. Effects of inserting a pressensor film into articular joints on the actual contact mechanics. *Journal of Biomechanical Engineering* 1998;**120**(5):655–9.

CHAPTER 17

Tekscan Measurements of Interfacial Contact Area and Stress in Articulating Joints

Tony Chen[1], Hongsheng Wang[1], Bernardo Innocenti[2]
Hospital for Special Surgery, New York, NY, United States[1]; Université Libre de Bruxelles, Brussels, Belgium[2]

1. BACKGROUND

While great progress has been made in understanding the complex kinematic motion of articulating human joints using techniques such as motion capture, less is understood about contact mechanics within the joints. Pressure sensitive film can be used to record joint stresses under static loading conditions, and there are methods to measure cartilage-on-cartilage contact through activities of daily living (ADLs) using open MRI and dual X-ray spectroscopy. While these methods provide some insight into understanding joint contact mechanics, they do not provide direct measurement of the contact mechanics. However, Tekscan sensors permit direct measurement of interfacial contact area and stress, and they are being used for a variety of robotic, automotive, ergonomic, and biomedical applications.[1] In particular, Tekscan sensors can be used to measure changes in joint contact mechanics in healthy, pathologic, and artificial joints under quasi-static and dynamic loading conditions (Fig. 17.1).[2–7] Therefore, this chapter explains how to perform Tekscan testing, as well as how to analyze, present, and interpret results.

2. RESEARCH QUESTIONS

Typical research questions could include one or more of the following:

- What is the typical magnitude of contact stress on the joint surface during ADLs?
- How is load distributed across the surface of an articulating joint?
- What is the effect of knee joint injury on joint contact mechanics?
- How do different repair techniques compare in restoring normal joint stresses?
- What are the stresses seen on total joint replacements?
- etc.

Experimental Methods in Orthopaedic Biomechanics.

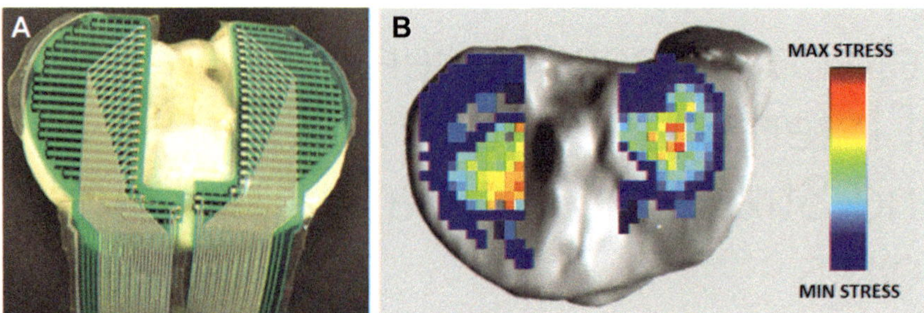

FIGURE 17.1

Cadaveric knee testing using the Tekscan system. (A) Tekscan sensor across the tibial surface, (B) stress map across the tibial surface.

3. METHODOLOGY

3.1 GENERAL STRATEGY

Human or artificial knee joint specimens are tested to serve as an example of how Tekscan sensors can be used to measure contact area and stress for other orthopaedic applications. Knee joints are inspected for visible evidence of chondral defects or other soft tissue damage. For human cadaveric studies, the chances of receiving knees with injuries can be reduced by prescreening donor histories for factors that can indicate poor joint health (e.g., unable to perform ADLs, osteoarthritis, osteosarcoma, knee joint surgery, etc.). After prescreening, specimens deemed suitable for testing are stripped of soft tissue as necessary, and a Tekscan sensor is inserted into the knee joint space. The long bones surrounding the knee joint are cut, leaving an appropriate length above and below the joint line for proper fixation to the testing setup. The long bones are then mounted into their respective fixtures of the test setup with the flexion axis of the knee joint aligned with the flexion–extension axis of the test setup. Suitable quasi-static and/or dynamic loads are applied, and contact areas and stresses at the knee joint are measured. Finally, statistical comparisons are made between test groups.

GLOSSARY

- ✓ **ADL**. Activities of daily living, such as pivoting, stair climbing, walking, etc.
- ✓ **Cell**. One pressure sensing element on the Tekscan sensor array.
- ✓ **I-Scan®**. Computer software distributed with the Tekscan system.
- ✓ **Joint contact mechanics**. The force distribution inside the joint.
- ✓ **K-Scan™**. Hardware that is used for articulating joint tests.
- ✓ **Pressure**. Scalar value measuring normal force per unit area of the body.
- ✓ **Stress**. Tensor value measuring force per cross-sectional area of the body.

SAFETY FIRST

✓ Wear goggles and gloves for protection when testing and handling specimens.
✓ Always cut away from oneself when using a scalpel on specimens.
✓ Never place appendages near the simulator or test jig while the test is running.
✓ Use audible start and stop cues to alert researchers to the test apparatus status.
✓ Clean the work area and all tools with bleach or disinfectant after testing.

3.2 MATERIALS AND TOOLS LIST

- adhesive wound dressing or Teflon tape
- auxiliary materials (i.e., pins, screws, scalpels, sutures)
- bone cement or body filler putty
- human, animal, or artificial knee joints
- joint simulator or test jig
- mechanical tester
- pressure bladder
- saws (i.e., oscillating saw or hand saw)
- Tekscan sensors
- vice or clamp

3.3 SENSOR AND SPECIMEN PREPARATION

Sensor Preparation, Equilibration, and Calibration

Step 1. Obtain Tekscan sensors. Tekscan (i.e., K-Scan™) sensors consist of a matrix of row and column electrodes created by printing conductive silver ink onto a thin polyester film substrate. The electrodes are further coated with a piezo-resistive ink such that a sensing cell is created at each intersection point in the matrix (Fig. 17.2). Changes in the resistance of the cells are measured through scanning electronics that connect to the tail of the sensor. The advantage of resistive technology is that the sensors are thin (100–200 μm thick) to reduce changes in joint kinematics, and they can measure high stress (up to 60 MPa/cell) where average stress in the joint can reach up to 8–11 MPa/cell. Care must be taken when using resistive technology so as to account for issues associated with creep of the resistive ink, sensor shear, and wrinkling of the sensor due to the inherent stiffness of the polyester sheets.

Step 2. Prepare the sensors. Prepare Tekscan sensors so they can be inserted in the joint without wrinkling (i.e., sensors can bend) since wrinkles will be a source of aberrant readings. Wrinkling usually occurs because the sensors are larger than the joint in which they are being inserted, or the radius of curvature of the joint surface is too small. Cut sensors along columns starting from the outer columns to ensure proper fit within the joint space, but do not cut sensors along rows since the connection for the columns begins in the top row (Fig. 17.2A and C). When a sensor is cut,

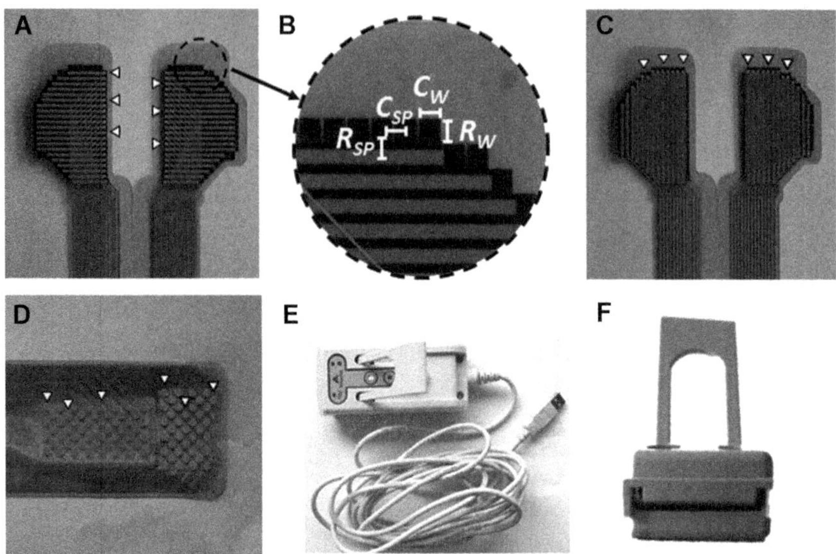

FIGURE 17.2

Tekscan sensor components. (A) Top of the sensor with white arrows denoting the location of the connection between the rows of the sensing elements and the electrical leads to the handle connector, (B) a zoomed-in view of the top of the sensor showing measurements of the column width C_W, row width R_W, column spacing C_{SP}, and row spacing R_{SP}, (C) bottom of the sensor with some of the column sensing element connections indicated with white arrows, (D) the handle connector end of the sensor with white arrows indicating the location of some of the row and column connections for the Tekscan handle, (E) Tekscan handle top view, (F) Tekscan handle side view.

it is no longer watertight and must be sealed. Adhesive wound dressing provides good water resistance while adding very little to the thickness of the sensors. If sensors will be in an environment that consists of high shear forces, covering the sensors with Teflon tape will decrease the rate of sensor degradation. To facilitate anchoring of the sensors to the surrounding soft tissue, plastic tabs supported by a layer of athletic tape under the sensor can be sealed within the adhesive wound dressing to provide suture points without compromising the watertight seal around the electronics.

Step 3. Select the sensor type. Connect the handle with the sensor attached to the computer and then open the I-Scan® software provided by Tekscan. Select the sensor type or model by going to "Options" → "Select Sensor."

Step 4. Select sensitivity. Change the sensitivity of the sensor to suit the mechanical test such that higher sensitivities allow for measurement of smaller changes of stress but have a lower maximum stress value, while lower sensitivities allow for a higher maximum stress but cannot capture small changes in stress. To change the sensitivity, select through the menus in the I-Scan® software as follows: "settings" → "sensitivity" → "sensitivity button." For each new experimental test setup, it is important to determine the proper sensitivity. To determine the best sensitivity for the current test, equilibrate a sensor (see step 5) at a low, medium, and high sensitivity level and run a pilot test on a specimen. The optimal sensitivity will be one that is at 90% of the maximum raw saturation value to account for differences between specimens.

Step 5. Equilibrate the sensors. Equilibration of sensors before calibration is an essential step to account for any discrepancies in the raw output of the cells such that all "hot" and "cold" cells will display a uniform color under the same pressure. Equilibration should be performed at pressures that are near the range of the expected maximum stress. In the case of knee joint testing, place sensors in an air bladder (Fig. 17.3A) and cycle them open/closed about 10–15 times at 90 psi (or 622 kPa) to condition the sensor. Next, place sensors under 45 psi (or 310.3 kPa) under the pressure bladder and allow 30 s for them to relax before initializing the first equilibration (i.e., "Equilibration 1" on the equilibration screen, Fig. 17.3B). After completion of the first equilibration, a second equilibration (i.e., "Equilibration 2" on the equilibration screen) should be performed at 90 psi (or 620.5 kPa).

Step 6. Calibrate the sensors. Calibration should be done at a similar temperature as the test to achieve the best calibration for the testing environment. Perform sensor calibration by applying stress to the sensor using the same materials or other materials that best replicate the test environment. In the case of joint testing, perform calibration using a setup with polyethylene that simulates the cartilage surface and with rubber that provides a surface for uniform load distribution (Fig. 17.4A). Alternatively, calibration can be done directly on the specimen with the caveat that the actual load being transferred to the sensor is not fully known since other soft tissue

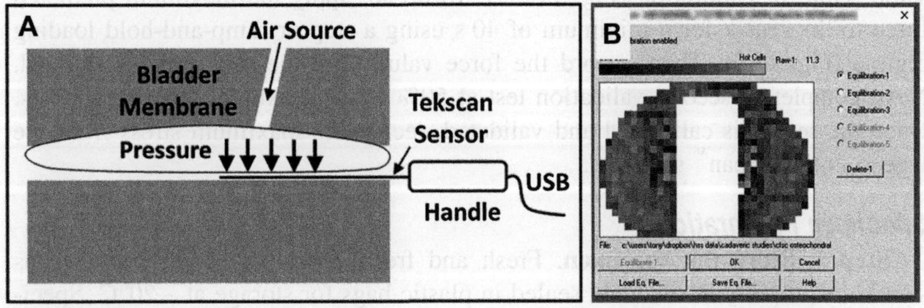

FIGURE 17.3

Equilibration setup. (A) Cross-sectional schematic of the pressure bladder and components, (B) representative equilibration map showing "hot" (very active) and "cold" (less active) cells.

Software window courtesy of www.Tekscan.com.

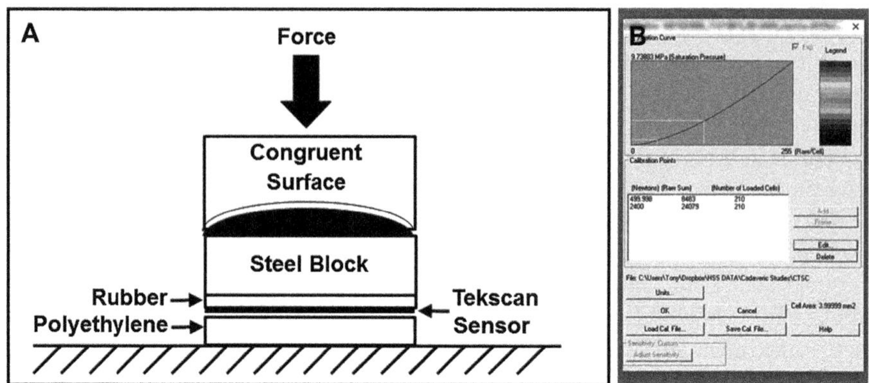

FIGURE 17.4

Calibration setup. (A) Cross-sectional view of the test setup, (B) representative calibration curve.

Software window courtesy of www.Tekscan.com.

structures can redistribute the applied load. The I-Scan® software allows for up to a 10-point sensor calibration with the minimum suggested being a 3-point calibration. To perform a 3-point sensor calibration, place the sensor between the calibration setup on the mechanical tester, and apply three axial loads that result in approximately 20%, 40%, and 80% of the maximum predicted experimental stress. The computer software will then fit the points following a "power calibration" curve provided with the system software (Fig. 17.4B).

Step 7. Validate the sensors. Validation of the calibration should always be done before the sensor is used. Perform the validation on the mechanical tester using the same setup for calibration (Fig. 17.4A). To do so, apply the maximum predicted force to the sensor for a minimum of 40 s using a step or ramp-and-hold loading regime (Fig. 17.5). Then, record the force value once the sensor has stabilized. Next, complete a second validation test at 50% of the maximum predicted stress. Once the sensor is calibrated and validated, record the maximum stress from the legend in the I-Scan® software.

Specimen Preparation

Step 1. Store the specimen. Fresh and fresh–frozen knee joint specimens should be properly labeled and sealed in plastic bags for storage at −20°C. Specimens stripped of soft tissue surrounding the joint should be wrapped in saline-soaked gauze before placing in plastic bag to keep tissues hydrated.

Step 2. Thaw the specimen. If the knee joint specimen was previously frozen, let specimens thaw at room temperature for at least 24 h before performing surgeries. If there is a large quantity of soft tissue, a longer amount of time may be required for complete thawing of specimens.

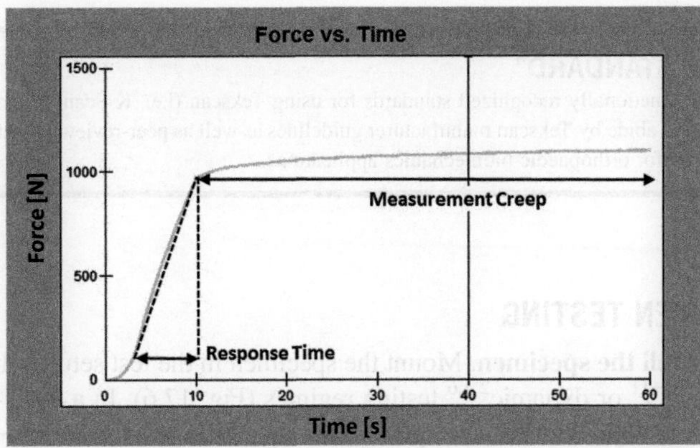

FIGURE 17.5

Validation of Tekscan sensor calibration. In this case, 1000 N is applied to the calibration setup and held for 1 min. Note the response time of the sensor to the ramp-and-hold loading and the subsequent creep of the stress over time.

Software window courtesy of www.Tekscan.com.

Step 3. Examine the specimen. The thawed knee joint specimens should be inspected for visible evidence of chondral defects or soft tissue damage. Keep a record of any chondral defects or soft tissue damage. Specimens should be excluded from the study if chondral and soft tissue damage will confound the results.

Step 4. Strip overlying tissue. If soft tissue was not previously stripped in step 1, use appropriate surgical tools (e.g., scalpel) to strip tissue immediately around the knee joint, being careful to maintain the surrounding soft tissues and capsule.

Step 5. Pot the specimen. Use an oscillating saw or hand saw to remove lengths of long bones until the specimen fits correctly in the test jig. Use a Kirschner wire or pin to define the flexion axis of the knee joint. Use fluoroscopic guidance as an aid to determine if the pin is in the proper orientation. If large torques will be applied to the specimen and potting will be done in bone cement or body filler putty, place crossing screws in locations of the long bones that will be covered in bone cement in order to provide adequate resistance to the applied torque before potting the specimen in the fixtures.

TIPS AND TRICKS

- ✓ When starting a study using a new test setup, always perform a pilot study.
- ✓ Have a clinician perform specimen preparation and surgical procedures.
- ✓ The same researcher or clinician should perform all surgeries for consistency.
- ✓ Record and average 20 loading cycles to account for sensor creep during tests.
- ✓ If using multiple sensors, check sensor alignment with respect to the specimen.

THE "GOLD STANDARD"

There are no internationally recognized standards for using Tekscan (i.e., K-Scan™) sensors. Thus, researchers should abide by Tekscan manufacturer guidelines as well as peer-reviewed journal articles on using Tekscan for orthopaedic biomechanics applications.

3.4 SPECIMEN TESTING

Step 1. Install the specimen. Mount the specimen in the test setup, which can be for quasi-static[8–11] or dynamic[2,4,6] testing regimes (Fig. 17.6). In a quasi-static test, the joint is tested in discrete configurations (e.g., changes in flexion–extension, load, or alignment). In a dynamic test, the joint undergoes continuous loading to simulate an activity (e.g., walking, stair climbing, pivoting). A typical setup involves modification of a wear simulator to apply the loading parameters given in International Organization for Standardization (ISO) 14243-1 to a cadaveric knee specimen.[12]

Step 2. Insert sensor in joint. Place the calibrated sensors between the knee joint surfaces in which the contact stress is to be measured. Once the sensor is in the knee joint space, suture the sensor to the surrounding capsule or soft tissue to maintain sensor placement.

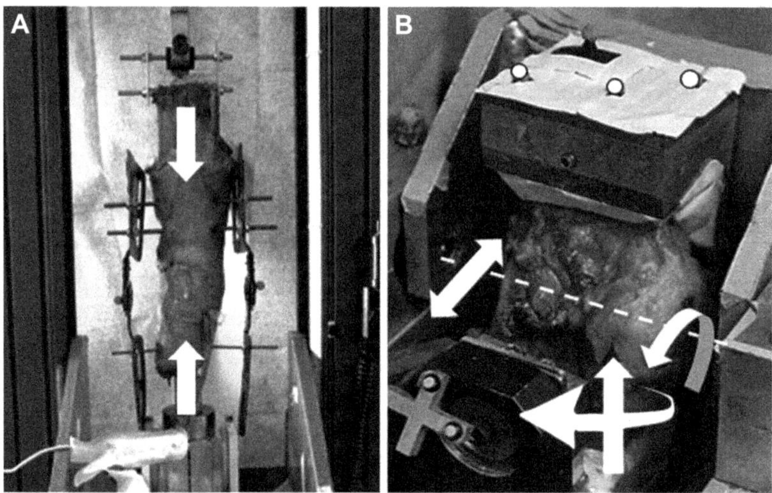

FIGURE 17.6

Tekscan test setup for articulating knee joints. (A) Quasi-static test using external frame fixed to the human knee specimen with metal rods placed perpendicularly to the mechanical axes, (B) dynamic test using multidirectional wear simulator under synchronous inputs (i.e., white arrows) of axial load, flexion–extension angle, internal–external rotation torque, and anterior–posterior force.

Step 3. Apply axial load. Apply a standardized loading regime to validate sensor placement and integrity. For human knee testing, a 1000-N axial load (approximately 50% of maximum load applied to the knee) is applied to check sensor integrity and placement. Adjust or replace the sensor if it is not positioned correctly, or replace the sensor if it is not within 20% of the expected value.

Step 4. Run test of intact specimen. After validation of sensor placement and integrity, run the test using specified loading regime. As given in step 1, the loading parameters provided in the international standard ISO 14243-1 are applied to the cadaveric knee specimen at a cycle frequency of 0.5 Hz for 20 cycles and maximum load of 2600 N.[12] This condition is the control for all subsequent testing of the specimen.

Step 5. Apply axial load. Repeat the standardized loading regime from step 3 to validate sensor placement and integrity.

Step 6. Perform intervention. Perform surgery to simulate injury or repair of the specimen (e.g., anterior cruciate ligament injury, meniscal tear, unicompartmental knee replacement, etc.). Repeat loading of the specimen executed in step 4. Repeat testing until all test conditions are completed.

3.5 RAW DATA COLLECTION

Step 1. Record stress data. For each specimen and test condition, record Tekscan contact stress perpendicular to each cell across the tibial plateau. The frequency at which each test is recorded depends on the rate of each test. The optimal recording frequency can be determined by the Nyquist rate (i.e., sample the data at two times the frequency of the movement to capture the full waveform). For example, in dynamic gait simulations of the knee, the contact stress is collected at 100 Hz for a minimum of 20 gait cycles for each trial to capture all major features during the stance phase, consisting of three peaks, which is approximately equivalent to a 50 Hz waveform.

Step 2. Record specimen information. For each specimen, record pertinent information (e.g., sensor number, max contact stress, contact area, etc.) (Table 17.1). Use the table, since it is important to ensure that the stresses are not dropping below expected levels for each subsequent test.

3.6 RAW DATA ANALYSIS

Step 1. Export representative images. I-Scan® software has a few options to plot stress maps. I-Scan® software can also export the movie to an ASCII file (.csv), which can be imported into data analysis software for further analysis. Before exporting the data, make sure the units are in System Internationale by going to "Options" → "Measurement Units" and change the length, force, and pressure units to mm, N, and MPa, respectively.

Step 2. Calculate contact parameters. The data (i.e., movie) can be analyzed using I-Scan® software to calculate the contact mechanics parameters, such as peak contact stress, contact area, and center of contact. Peak contact stress

Table 17.1 Specimen testing sheet used for recording data during tests.

Specimen ID:		Date Tested:	/ /
Researchers:			
Sensor #:		Sensor Max StressLimit:	
Test 1	Condition:		
	Max Stress [MPa]:		
	Contact Area [mm^2]:		
Test 2	Condition:		
	Max Stress [MPa]:		
	Contact Area [mm^2]:		
Test 3	Condition:		
	Max Stress [MPa]:		
	Contact Area [mm^2]:		
etc.			

Eq. (17.1), total contact area Eq. (17.2), and weighted center of contact stress Eq. (17.3) can be calculated from the raw Tekscan data exported from I-Scan® software. The *WCoCS* is a surrogate for the center of contact that also takes into account the magnitude of contact stress at each cell experiencing stress across the sensor or a portion of the sensor (Fig. 17.7) such that greater weights are assigned to cells with higher contact stresses and smaller weights to peripheral cells with lower contact stresses. The relevant equations are:

$$S_{MAX} = \text{Max}\{S_1, S_2, S_3, \cdots, S_n\} \tag{17.1}$$

$$A = \sum_{i=1}^{n} a \times (S_i > 0) \tag{17.2}$$

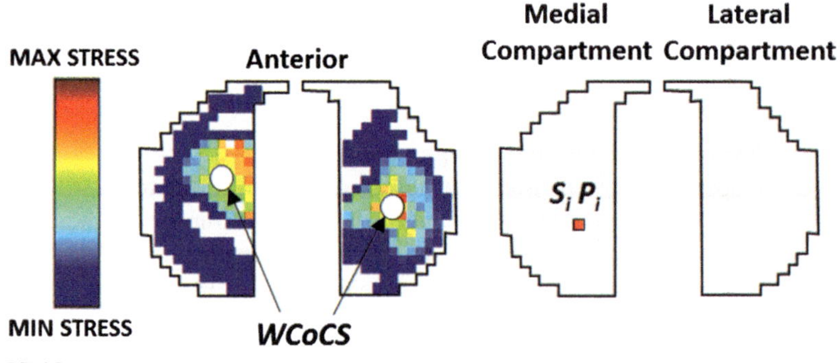

FIGURE 17.7

Calculation of the weight center of contact stress (*WCoCS*) on the medial and lateral tibial plateau. P_i, position of sensing element i; S_i, stress of sensing element i.

$$WCoCS = \sum_{i=1}^{n}(S_i \times P_i) \Bigg/ \sum_{i=1}^{n} S_i \tag{17.3}$$

where n is total number of the cells, S_i is contact stress at the ith cell, P_i is cell position in the local coordinate system, and a is the cell area (row spacing × column spacing; this information can be found in the header information from each exported movie file). For quasi-static and dynamic tests, this method can provide the time history of peak contact stress, contact area, and trajectories of *WCoCS*. Some sensors consist of multiple areas of contact for different joint segments. Try to analyze each region of interest independently instead of extracting global values.

Step 3. Average and normalize the data. Due to differences in specimen size and geometry, data can be normalized for each specimen to a control condition by either dividing or subtracting the control condition from the remaining test conditions. Whether or not normalization is performed, the average $\pm$ 1 standard deviation should be calculated for each test condition. The 95% confidence intervals can also be used in dynamic tests.

Step 4. Alternative analysis. An alternative analysis may also be performed in which a novel algorithm is used to identify characteristic loading patterns from dynamic knee joint testing performed using Tekscan.[13,14] This method uses the time-dependent stress over time at each cell to determine areas on the joint surface that experience similar patterns of stress (i.e., characteristic patterns). By comparing the characteristic patterns found between healthy and pathologic joints, the time-dependent changes in the characteristic patterns and the changes in the pattern locations provides a different perspective on how pathology alters joint mechanics during ADLs.

Step 5. Perform statistical analysis. When comparing test configurations of a quasi-static test, a one-way analysis of variance (ANOVA) should be performed to determine differences between all test conditions (e.g., $P < 0.05$ or <0.01). For a dynamic test, particular points in the loading regime can be analyzed using ANOVA; however, to fully capture differences between the different conditions, principal component analysis should be performed between the conditions. These analyses can be performed using commercially available statistical analysis software. Moreover, to fully understand the underlying relation between specimen age, gender, and geometric parameters, multivariate general linear model can be run against contact mechanics to identify important specimen specific factors.

Step 6. Compute statistical power. Power analysis can be done after the study to ensure there were enough specimens per test group to detect all statistical differences that were actually present (i.e., was type II statistical error avoided?). Statistical power >80% is usually considered to indicate there were enough specimens per test group. Note that if good predictions of averages and standard deviations are available from prior studies, then the number of specimens and/or tests can be chosen before the study begins to ensure a power >80%.

ENGINEER'S TOOLBOX

Early pioneering work by Heinrich Hertz resulted in "Hertzian" theoretical formulas for understanding interfacial contact mechanics of two articulating solid bodies while assuming quasi-static loading, frictionless contact, elastic bodies, small deformations, and symmetry. The formulas can be applied to human or artificial joint biomechanics as shown in the figure by using approximations for geometry and material properties, where F is applied force, D is diameter, L is length, E is elastic modulus, and ν is Poisson's ratio.

Consequently, for ball-in-cup joints (e.g., shoulder and hip), the formulas are average contact stress $\sigma_{AVG} = F/A$, peak contact stress $\sigma_{PEAK} = (3/2)\sigma_{AVG}$, and

$$\text{total contact area } A = \pi\left[\frac{\left(\frac{3F}{8}\right)\left(\frac{1-\nu_1^2}{E_1}+\frac{1-\nu_2^2}{E_2}\right)}{\left(\frac{1}{D_1}-\frac{1}{D_2}\right)}\right]^{\frac{2}{3}}$$

Similarly, for cylinder-in-groove joints (e.g., knee), the formulas are average contact stress $\sigma_{AVG} = F/A$, peak contact stress $\sigma_{PEAK} = (4/\pi)\sigma_{AVG}$, and

$$\text{total contact area } A = \left[\frac{\left(\frac{8FL}{\pi}\right)\left(\frac{1-\nu_1^2}{E_1}+\frac{1-\nu_2^2}{E_2}\right)}{\left(\frac{1}{D_1}-\frac{1}{D_2}\right)}\right]^{\frac{1}{2}}$$

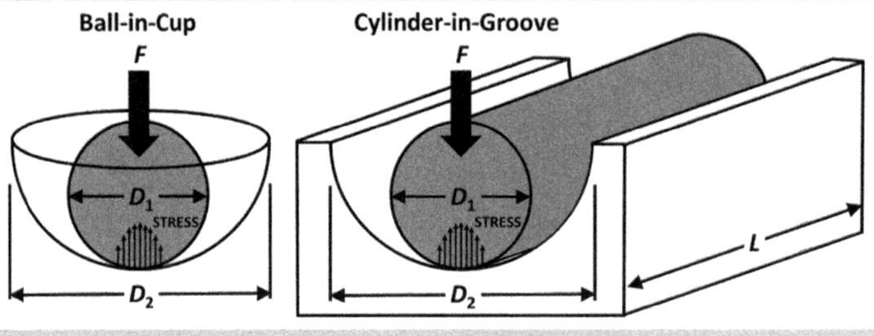

4. RESULTS

Once all Tekscan data collection and analysis have been performed, it is then important to communicate and present the primary results in an understandable and concise manner to the reader of a journal article, conference paper, technical report, or book chapter.

Step 1. Show the pressure sensor at key points. Begin by showing representative stress maps. Orientation of the specimen can be shown by overlaying the stress map over an image of the specimen or shown at key points in the loading regime (Fig. 17.8).

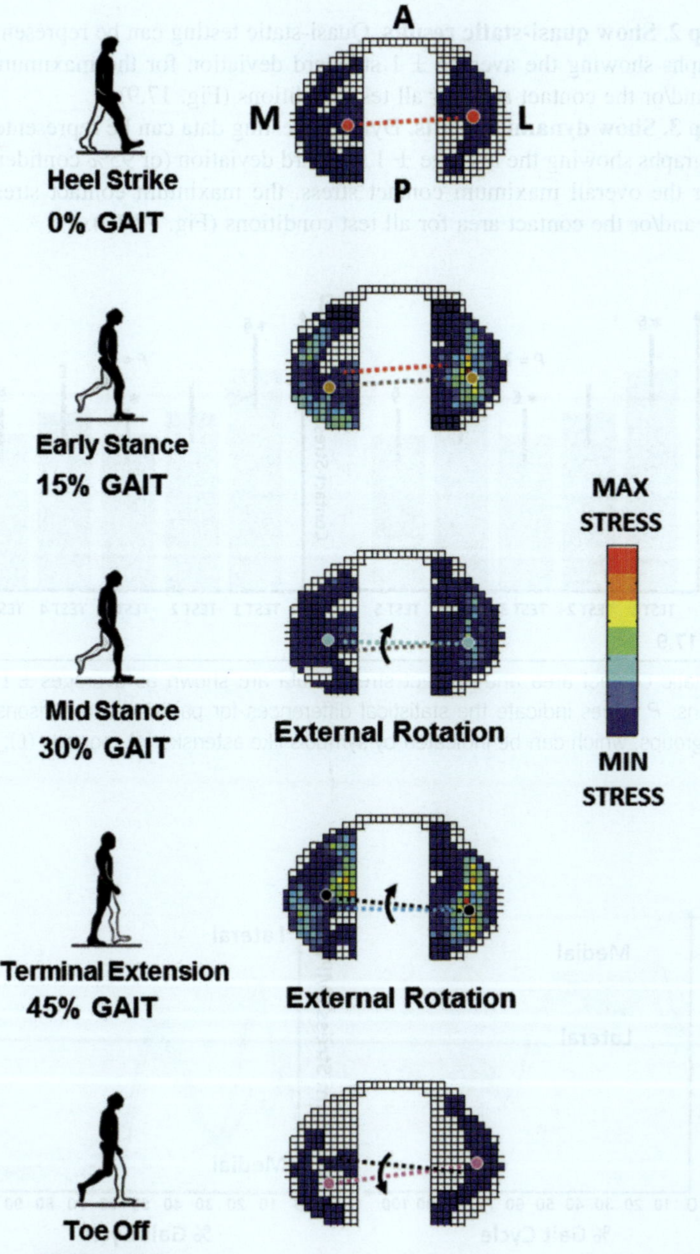

FIGURE 17.8

Representative Tekscan stress maps throughout simulated gait. *A*, Anterior; *L*, lateral; *M*, medial; *P*, posterior.

Step 2. Show quasi-static results. Quasi-static testing can be represented using bar graphs showing the average ± 1 standard deviation for the maximum contact stress and/or the contact area for all test conditions (Fig. 17.9).

Step 3. Show dynamic results. Dynamic testing data can be represented as line or bar graphs showing the average ± 1 standard deviation (or 95% confidence interval) for the overall maximum contact stress, the maximum contact stress at key points, and/or the contact area for all test conditions (Fig. 17.10).

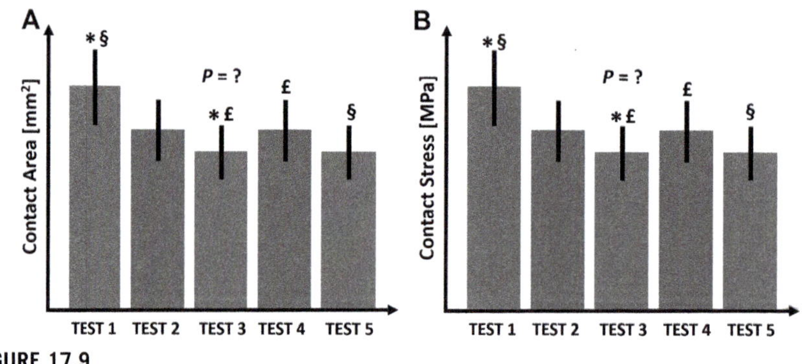

FIGURE 17.9

Quasi-static contact area and contact stress. Data are shown as averages ± 1 standard deviations. *P* values indicate the statistical differences for pairwise comparisons between all test groups, which can be indicated by symbols like asterisks (*), pounds (£), etc.

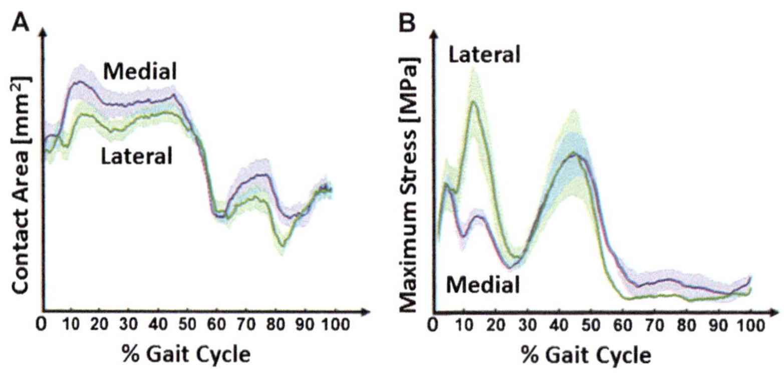

FIGURE 17.10

Dynamic contact area and contact stress variations on the medial and lateral tibial plateau during the gait cycle. Paired testing can be performed in dynamic testing by performing principal component analysis. Solid lines show the average value, while the surrounding envelope indicates ± 1 standard deviation.

ALTERNATIVES AND ADAPTATIONS

✓ **Various joints**. The same calibration and testing protocol can be used for testing biological and artificial ankle, hip, shoulder, spine, and elbow specimens.

✓ **Foot and ankle testing**. A pressure mat can be placed under the foot to record plantar pressures in addition to the stresses within the joint space of interest.

✓ **Arthroscopic implantation of sensors**. The sensor must be carefully rolled so that it can fit through the incision port and run a validation test immediately post-experimentation upon sensor removal.

5. DISCUSSION

After completing all Tekscan tests, data collection, data analysis, and data presentation, then final results can be considered and interpreted in the broader context of some important clinical, biomechanical, and/or technological considerations, as follows.

Tekscan analysis of the effects of articulating joint injury and repair can be compared to prospective, retrospective, and clinical or kinematic observations and, thus, can provide useful evidence in the interpretation of Tekscan results. These findings could be an alteration in the location of stress that corresponds to cartilage damage seen in a majority of patients in a clinical study or relate to changes in *WCoCS* trajectories that correspond to changes seen in kinematic studies. In terms of contact stress and contact area changes with repair techniques, differences can be compared between the "healthy," "injury," and all "repair" conditions. Additionally, a best "repair" by definition restores maximum stress or stress distribution to that of the healthy condition. For example, a "healthy" intact knee can have a maximum contact stress of 4.2 ± 1.2 MPa and a contact area of 546 ± 132 mm^2, a knee with an "injury" (i.e., meniscectomy) can have a maximum contact stress of 6.2 ± 1.0 MPa and a contact area of 192 ± 122 mm^2, and a knee with a bone plug meniscal replacement "repair" can have a maximum contact stress of 5.0 ± 0.7 MPa and a contact area of 398 ± 148 mm^2.[2,6]

Tekscan analysis can be used to compare total joint replacement designs using parameters such as differences in the maximum stress level, changes in the location of the maximum stress, and the total contact area. These findings can be related to clinical observations of implant failures and wear patterns of the particular joint replacement design. For example, typical numerical Tekscan results for total knee replacements can yield a total contact area of 200–800 mm^2 and a contact stress of 5–50 MPa for the tibiofemoral condylar joint,[15] as well as a contact area of 50–250 mm^2 and a contact stress of 5–25 MPa for the patellofemoral joint.[5,16] However, any change in the design of a total knee replacement, the flexion angle, and the load conditions could produce a change in the kinematics and kinetics of the knee joint and, therefore, changes in the contact stress parameters measured.[17]

Tekscan sensors are a valuable tool for measurement of joint contact mechanics during dynamic testing of human, animal, or artificial joints. Due to their thin construction, there is minimal variation in specimen kinematics. Changing the sensor type provides a wide range of spatial resolutions (0.64–3.61 mm^2/cell) and measurable stresses (0.5–60 MPa), and the ability to use these sensors in a variety of specimen types, including human and animal cadaveric knee, ankle, elbow, hip, shoulder, spine, and artificial joint testing. However, because they are based on resistive technology, they are prone to drift and the results are not reproducible. Proper sensor calibration and validation must be done to account for these inaccuracies.

6. SUMMARY

- Tekscan sensors can be used to measure the stress perpendicular to the joint surfaces.
- Tekscan sensors can be used in static, quasi-static, and dynamic testing conditions.
- Tekscan sensors should be conditioned and calibrated after long-time storage.
- Tekscan readings are sensitive to equilibration and calibration processes.
- Wrinkling of sensors can cause aberrant contact area and stress measurements.
- Repeated sensor testing can lead to sensor degradation under shear.

7. QUIZ QUESTIONS

1. What type of sensor technology does Tekscan use?
2. What are the advantages and disadvantages of Tekscan?
3. Why is it important to equilibrate the sensors before calibrating them?
4. How is the correct calibration load determined for a particular application?
5. Estimate the peak contact stress σ_{PEAK} in a total hip replacement experiencing a 3 kN load, assuming it can be modeled as a ball-in-cup joint with a metal ball ($D = 30$ mm, $E = 200$ GPa, $\nu = 0.3$) and polymer cup ($D = 40$ mm, $E = 0.9$ GPa, $\nu = 0.4$) (answer: $\sigma_{PEAK} = 56.8$ MPa).

REFERENCES

1. Tekscan Inc., www.tekscan.com.
2. Bedi A, Chen T, Santner TJ, El-Amin S, Kelly NH, Warren RF, et al. Changes in dynamic medial tibiofemoral contact mechanics and kinematics after injury of the anterior cruciate ligament: a cadaveric model. *Proceedings of the Institution of Mechanical Engineers (Part H): Journal of Engineering in Medicine* 2013;**227**(9):1027–37.
3. Coles LG, Gheduzzi S, Miles AW. In vitro method for assessing the biomechanics of the patellofemoral joint following total knee arthroplasty. *Proceedings of the Institution of Mechanical Engineers (Part H): Journal of Engineering in Medicine* 2014;**228**(12): 1217–26.

4. Wang H, Chen T, Koff MF, Hutchinson ID, Gilbert S, Choi D, et al. Image based weighted center of proximity versus directly measured knee contact location during simulated gait. *Journal of Biomechanics* 2014;**47**(10):2483–9.
5. Luyckx T, Didden K, Vandenneucker H, Labey L, Innocenti B, Bellemans J. Is there a biomechanical explanation for anterior knee pain in patients with patella alta? Influence of patellar height on patellofemoral contact force, contact area and contact pressure. *Bone and Joint Journal* 2009;**91**(3):344–50.
6. Wang H, Gee AO, Hutchinson ID, Stoner K, Warren RF, Chen TO, et al. Bone plug versus suture-only fixation of meniscal grafts: effect on joint contact mechanics during simulated gait. *American Journal of Sports Medicine* 2014;**42**(7):1682–9.
7. Wilharm A, Hurschler C, Dermitas T, Bohnsack M. Use of Tekscan K-scan sensors for retropatellar pressure measurement avoiding errors during implantation and the effects of shear forces on the measurement precision. *BioMed Research International* 2013;**2013**: 829171.
8. Elsner JJ, Portnoy S, Zur G, Guilak F, Shterling A, Linder-Ganz E. Design of a free-floating polycarbonate-urethane meniscal implant using finite element modeling and experimental validation. *Journal of Biomechanical Engineering* 2010;**132**(9):095001.
9. Pianigiani S, Gervasi GL, Antinolfi P, Speziali A, Pascale W, Cerulli GG, et al. Does the chosen experimental test procedure affect the outputs? A pilot study to observe patella-femoral joint mechanics. In: *16th ESSKA (European Society of Sports Traumatology, Knee Surgery, and Arthroscopy) Congress*; 2014. May 14–17, Amsterdam, Netherlands, e-poster #P08-1695.
10. Imhauser CW, Sheikh S, Choi DS, Nguyen JT, Mauro CS, Wickiewicz TL. Novel measure of articular instability based on contact stress confirms that the anterior cruciate ligament is a critical stabilizer of the lateral compartment. *Journal of Orthopaedic Research* 2015;**34**(3):478–88.
11. McCarthy MM, Tucker S, Nguyen JT, Green DW, Imhauser CW, Cordasco FA. Contact stress and kinematic analysis of all-epiphyseal and over-the-top pediatric reconstruction techniques for the anterior cruciate ligament. *American Journal of Sports Medicine* 2013;**41**(6):1330–9.
12. ISO 14243-1. Implants for surgery – wear of total knee-joint prostheses – part 1: loading and displacement parameters for wear-testing machines with load control and corresponding environmental conditions for test. Zurich (Switzerland): International Organization for Standardization (ISO); www.iso.ch.
13. Wang H, Chen T, Gee AO, Hutchinson ID, Stoner K, Warren RF, et al. Altered regional loading patterns on articular cartilage following meniscectomy are not fully restored by autograft meniscal transplantation. *Osteoarthritis and Cartilage* 2015;**23**(3):462–8.
14. Wang H, Chen T, Torzilli P, Warren R, Maher S. Dynamic contact stress patterns on the tibial plateaus during simulated gait: a novel application of normalized cross correlation. *Journal of Biomechanics* 2014;**47**(2):568–74.
15. Shiramizu K, Vizesi F, Bruce W, Herrmann S, Walsh WR. Tibiofemoral contact areas and pressures in six high flexion knees. *International Orthopaedics* 2009;**33**(2):403–6.
16. van Jonbergen HP, Innocenti B, Gervasi GL, Labey L, Verdonschot N. Differences in the stress distribution in the distal femur between patellofemoral joint replacement and total knee replacement: a finite element study. *Journal of Orthopaedic Surgery and Research* 2012;**7**:28.
17. Pianigiani S, Labey L, Pascale W, Innocenti B. Knee kinetics and kinematics: what are the effects of TKA malconfigurations? *Knee Surgery, Sports Traumatology, Arthroscopy* January 24, 2015 [Epub ahead of print].

CHAPTER

Fretting Corrosion Testing of Total Hip Replacements with Modular Heads and Stems

18

Christian Wight[1], Emil H. Schemitsch[2], Radovan Zdero[3]

University of Toronto, Toronto, ON, Canada[1]; London Health Sciences Centre, London, ON, Canada[2]; Western University, London, ON, Canada[3]

1. BACKGROUND

Human hip joints with diseased cartilage and bone are commonly replaced using total hip replacements (THRs), which often have modular head and neck components that connect via matching taper, much like nested cones (Fig. 18.1). Fretting corrosion occurs when subtle relative motion (i.e., micromotion) causes abrasion at the head–neck taper connection (i.e., fretting), thus allowing surrounding biofluids to degrade the material (i.e., corrosion). Corrosion then releases toxic metallic debris that causes local tissue lysis and painful inflammation, ultimately leading to revision surgery. However, concern with long-term systemic effects remains, as the toxic debris diffuses throughout the body.[2] Modular THRs are susceptible to fretting corrosion; however, they are important to allow intraoperative flexibility to reconstruct the joint with optimized biomechanics. Therefore, this chapter describes a procedure for investigating fretting corrosion of THRs, as well as how to analyze, present, and interpret results.

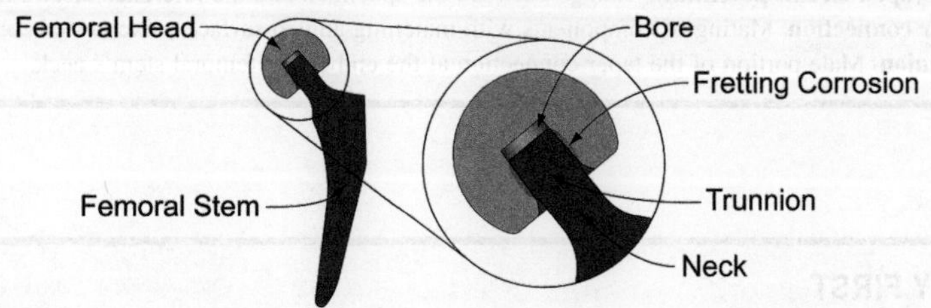

FIGURE 18.1

Modular components of a THR. The femoral head mates with the femoral stem via a taper connection. Fretting corrosion occurs between the contact surfaces of the femoral head's bore and the femoral stem's trunnion at the end of the femoral stem's neck.

Experimental Methods in Orthopaedic Biomechanics.

2. RESEARCH QUESTIONS

Typical research questions might include one or more of the following:

- How does head–neck taper connection geometry affect fretting corrosion?
- What femoral stem material should be used to reduce fretting corrosion?
- Does the severity of corrosion increase with time implanted?
- Are heavier patients at greater risk of fretting corrosion?
- etc.

3. METHODOLOGY

3.1 GENERAL STRATEGY

THR head and stem components are obtained and grouped by material, design, or intended assembly conditions based on the particular research question. Components are oriented, potted, assembled, and left to equilibrate in an environmental chamber filled with electrolyte fluid. Components are instrumented to measure fretting corrosion's electrical activity, after which they are cyclically loaded with a periodically incremented maximum load. Fretting corrosion onset load is determined as the load where the component's open circuit potential (OCP) and corrosion current first increase from the unloaded equilibrium level. To allow a quantifiable amount of surface degradation to take place, components are cyclically loaded long term. Components are then disassembled, and their connection interfaces are qualitatively analyzed for fretting corrosion damage by microscope imaging and semiquantitatively by a scoring rubric. Statistical comparisons are finally made between the various test group results.

GLOSSARY

✓ **Bore.** Female portion of the taper connection in the femoral head.
✓ **Corrosion.** Degradation of a metal by chemical reaction with its environment.
✓ **Fretting.** Cyclic relative motion between contacting surfaces causing gradual abrasive wear.
✓ **OCP (open circuit potential).** Voltage between the specimen and the reference electrode.
✓ **Taper connection.** Mating of components with matching angled surfaces, like nested cones.
✓ **Trunnion.** Male portion of the taper connection at the end of the femoral stem's neck.

SAFETY FIRST

✓ Pot the stem in a well-ventilated area per the potting medium manufacturer's instructions.
✓ Position the test apparatus to avoid impingement with the mechanical tester's actuator.
✓ Set actuator limits so that the mechanical tester powers off if the specimen breaks.
✓ Place a protective shield around the specimen when the mechanical tester is running.
✓ Design and verify that the test apparatus is sufficiently strong to withstand the loading.

3.2 MATERIALS AND TOOLS LIST

- electrometer
- head disassembly jig
- mechanical tester
- optical microscope
- poly(methyl methacrylate) (PMMA)
- potting jig
- scanning electron microscope (SEM)
- total hip replacement (THR)
- watertight test chamber
- zero resistance ammeter

3.3 SPECIMEN PREPARATION

Step 1. Orient the femoral stem. Using a ruler and marker, mark the THR's distal stem axis as the center line in the anterior–posterior projection of the most distal 50 mm of the stem. The distal stem axis would coincide with the patient's femoral canal axis if the THR were implanted. Grip the stem's neck with a ring clamp and test tube holder. Orient the stem such that the distal stem axis is $10 \pm 1°$ from the vertical axis (which will coincide with the load application axis) in the anterior–posterior plane (Fig. 18.2). Use a protractor or similar device to ensure the accuracy of the stem's orientation. Once the stem is oriented, place the stem into the potting jig with care not to affect the orientation. Ensure there is a minimum of a 10-mm gap between the stem and the potting jig. Screw a bolt through the potting jig until contact is made with the distal stem.

Step 2. Pot the femoral stem. Connect the assembly to an angled vice such that it may be oriented in the anterior–posterior plane. Tilt the assembly until the stem's

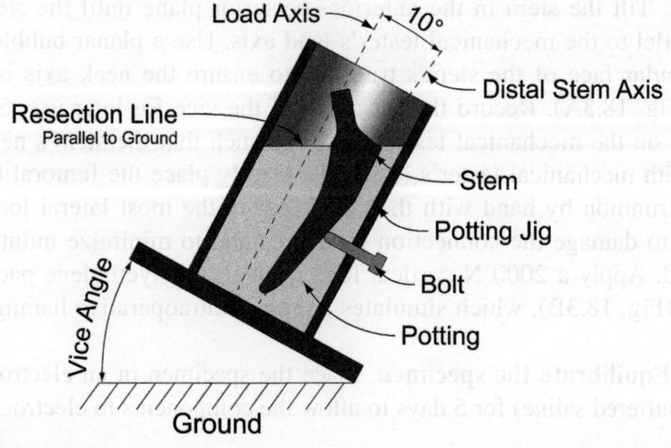

FIGURE 18.2

Femoral stem potting in cement.

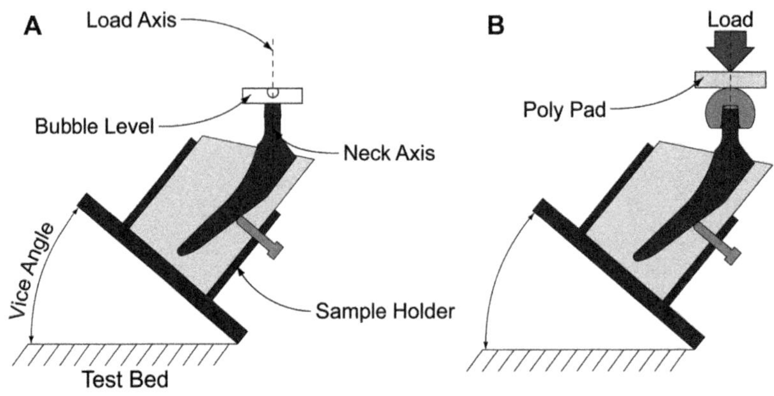

FIGURE 18.3

Femoral stem preparation. (A) Orienting the stem, (B) mounting the head.

resection line is level with the ground. The resection line is the level where the patient's bone would be if the stem were implanted. PMMA is recommended as the potting medium because it represents bone cement; however, a similar acrylic of equivalent strength may be used. Prepare the potting per the manufacturer's instructions; then, pour it into the potting jig around the stem until the stem is encapsulated up to the resection line. Allow the potting to set overnight.

Step 3. Mark the femoral head. The femoral head is typically axisymmetric; therefore, it must be marked to allow rotational orientation during later analysis. Place a small epoxy dot just outside the head's bore using a small pin. Leave the epoxy to set overnight.

Step 4. Assemble the femoral head and stem. Grip the potted stem in the angled vice. Tilt the stem in the anterior–posterior plane until the stem's neck axis is parallel to the mechanical tester's load axis. Use a planar bubble level on the top circular face of the stem's trunnion to ensure the neck axis is oriented correctly (Fig. 18.3A). Record the tilt angle of the vice for later use. Secure the angled vice on the mechanical tester's test bed such that the stem's neck axis is collinear with mechanical tester's load axis. Gently place the femoral head onto the stem's trunnion by hand with the epoxy dot in the most lateral location. Be careful not to damage the connection interfaces and to minimize unintended assembly load. Apply a 2000-N vertical load through a polyethylene pad at a rate of 500 N/s (Fig. 18.3B), which simulates a typical intraoperative hammer impact assembly.

Step 5. Equilibrate the specimen. Place the specimen in an electrolyte (e.g., phosphate buffered saline) for 5 days to allow the components to electrochemically stabilize.

TIPS AND TRICKS

- ✓ A minimum of five specimens per test group is recommended.
- ✓ Make the counter electrode by cutting the distal stem from the specimen prior to potting.
- ✓ Isolate all electrochemically active surfaces from the electrolyte with sealant.
- ✓ Immersing too little of the counter electrode in the electrolyte may reduce fretting corrosion.
- ✓ 10 million cycles at 5 Hz lasts 3 weeks, so adjust the number of cycles if required.
- ✓ Image processing software can be used to aid fretting and corrosion scoring.

THE "GOLD STANDARD"

The American Society for Testing and Materials (ASTM) provides guidelines in its document ASTM F1875 (Standard practice for fretting corrosion testing of modular implant interfaces: hip femoral head-bore and cone taper interface). However, this procedure depends on complex chemical analysis to allow comparison between components of different material.

3.4 SPECIMEN TESTING

Step 1. Assemble the test apparatus. Connect a load bearing polyethylene pad to the mechanical tester's actuator via a low friction X–Y linear bearing to minimize extraneous loading. Secure the specimen in the holder to the base of the mechanical tester such that the head's center is aligned vertically with the mechanical tester's actuator (Fig. 18.4). Assemble the test chamber onto the holder so the entirety of the specimen outside the potting is within the volume of the test chamber. Connect the specimen to a reference electrode (i.e., Ag/AgCl) via a high-impedance potentiostat (i.e., electrometer with impedance $>10^{10}$ Ω) by connecting a wire from the bolt contacting the distal stem to the electrometer's positive terminal, and the wire from the reference electrode to the electrometer's negative terminal. The reference electrode provides a constant potential against which the specimen's OCP may be measured. Connect the specimen to a counter electrode via a zero resistance ammeter by connecting a wire from the bolt to the ammeter's positive terminal, and a wire from the counter electrode to the ammeter's negative terminal. The counter electrode acts as an electron sink for the corrosion reactions and serves to represent the distal femoral stem. Therefore, the counter electrode should be manufactured from the same material as the distal femoral stem, and a 400-cm^2 surface area is recommended.[3] Place the reference and counter electrodes in the environmental chamber. The ammeter and electrometer must be connected to a data acquisition device, such as a computer, so that the data may be stored for later analysis.

Step 2. Fill the test chamber. Prepare the electrolyte by dissolving salt tablets in deionized water per the manufacturer's recommendations. Pour the electrolyte

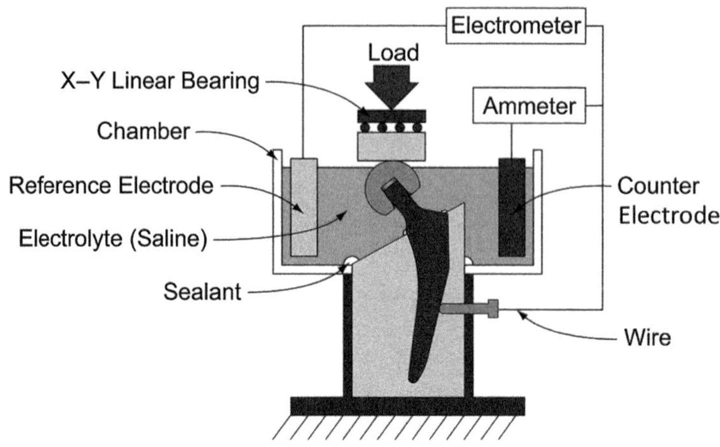

FIGURE 18.4

Fretting corrosion test setup.

into the test chamber so that the active areas of the counter and reference electrolyte are engulfed, as well as the entirety of the specimen's modular connection of interest. The electrolyte must not engulf the articulating surface of the head and polyethylene load application pad. The volume of electrolyte, as well as the areas of the test specimen and electrodes engulfed, must be consistent between specimens.

Step 3. Conduct the corrosion test. Execute an automatic test sequence that applies a cyclic load with max/min load ratio of 10, with the load incrementally increasing after a period of cycling. For example, the cyclic load begins with a sinusoidal curve from 10–100 N at 2 Hz for 600 cycles, then increments to 20–200, 30–300, 40–400 N, etc., every 600 cycles until the desired maximum load is reached (e.g., 300–3300 N). The load magnitude and frequency are recommended to represent gait. The period of cycling at each load level is selected to allow transient effects to pass. Upon reaching the maximum load level, cyclic loading is left to continue at 5 Hz for 10 million cycles to allow the long-term effects of corrosion to take place.

Step 4. Disassemble the specimen. After the test sequence is complete, jog the actuator up to allow access to the specimen. Remove the specimen from the mechanical tester and test chamber. Connect the disassembly jig to the mechanical tester's actuator. Load the specimen into the angled vice, and tilt it to an angle so the stem's neck is parallel with the mechanical tester's load axis (Fig. 18.3B). Mount the angled vice to the mechanical tester's base with the stem's neck aligned with the mechanical tester's load direction. Connect a suitable grip to the specimen's femoral head. Apply a tensile load to the femoral head at 0.008 mm/s until the head is pulled from the stem. Rinse the connection interfaces of head and stem with deionized water to remove any debris.

3.5 RAW DATA COLLECTION

Step 1. Photograph the specimens. Prior to testing, photograph the connection surfaces of a representative specimen of the femoral heads and stems. The surface of final product quality implants should be consistent. Thus, only a selection of the specimens require SEM or optical microscope imaging; however, all specimens must be inspected to ensure there are no defects.

Step 2. Record electrical values. After the 5-day unloaded equilibration period, record OCP and current at 200 Hz for 1 s using the electrometer and ammeter, respectively, to provide a baseline (Table 18.1). At the end of each loading increment, OCP and corrosion current are again sampled at 200 Hz for 1 s (Table 18.1).

Step 3. Record load levels. Using data processing software, calculate average OCP and corrosion current for each incremental load level. Then, determine the fretting corrosion onset load as the load at which there is a marked shift in OCP and corrosion current magnitude (Table 18.1).

Step 4. Record imaging scores. Image each quadrant of the head and stem taper interface with optical microscopy, making sure to identify each image with specimen number, quadrant, and image number (Table 18.2). Be sure to capture any irregularities relative to the pretest imagery. Corrosion may present as pitting or intergranular corrosion. Fretting is often noted as wear scars perpendicular to the longitudinal machine lines typically found on the components. A variable pressure SEM may be necessary to image nonconductive specimens. Be sure to capture images where the entirety of the taper interface quadrant is visible for use during semiquantitative analysis of the taper interfaces. Specimens will need to be carefully angled to allow imaging of the depths of the taper interface into the femoral head's bore. Ensure specimens are oriented and lit consistently to improve repeatability between imaging. Finally, using optical microscopy in combination with an established rubric developed previously (Table 18.3),[4] enter the score for the degree of fretting and corrosion on the head's bore and stem's trunnion (Table 18.1).

3.6 RAW DATA ANALYSIS

Step 1. Perform statistical analysis. Statistical analysis techniques are dependent on the investigation's hypothesis. The criterion for statistical difference first needs to be chosen (e.g., $P < 0.01$ or < 0.05). Then, data can be used to compare different test groups investigated, such as those differing by design (e.g., short vs. long head offset), material (e.g., cobalt−chrome alloy vs. titanium alloy stems), or assembly technique (e.g., wet vs. dry assembled components). The Mann–Whitney U test is recommended for comparing fretting corrosion onset loads, fretting scores, or corrosion scores between test groups due to the potentially nonnormal behavior of discrete variables.

Step 2. Compute statistical power. Power analysis can be done after the study to ensure there were enough specimens per test group to detect all statistical differences that were actually present (i.e., was type II statistical error avoided?).

Table 18.1 Electrical, mechanical, and imaging scores for individual specimens.

Specimen	OCP [mV]	Current [μA]	Onset Load [N]	Head Taper Fretting Score	Head Taper Corrosion Score	Stem Taper Fretting Score	Stem Taper Corrosion Score	Head Taper Microscopy Observations	Stem Taper Microscopy Observations
1									
2									
3									
etc.									

Table 18.2 Optical microscopy and/or SEM identification table.

Specimen	Image Number	Quadrant	Microscopy Type	Remarks
1				
2				
3				
etc.				

Table 18.3 Semiquantitative scoring metric for fretting and corrosion damage.[4]

Severity	Score	Criteria
None	1	No sign of corrosion or fretting.
Mild	2	<30% of taper surface discolored. Single band of fretting scars across less than three machine lines.
Moderate	3	>30% of taper surface discolored, or <10% of taper surface with black debris, pits, or etch marks. Several fretting scar bands or single band across more than three machine lines.
Severe	4	>10% of taper surface with black debris, pits, or etch marks. Several fretting scar bands across several adjacent machine lines, or flattened areas with nearby fretting scars.

Statistical power >80% is usually considered to indicate there were enough specimens per test group. Note that if good predictions of averages and standard deviations are available from prior studies, then the number of specimens and/or tests can be chosen before the study begins to ensure a power >80%.

ENGINEER'S TOOLBOX

THR head offset may be used to more accurately reconstruct the patient's natural hip joint by increasing the lateral distance of the femoral head center to the taper engagement level (TEL). However, these THRs are more susceptible to fretting corrosion because they induce torsion at the TEL. The torsion induced by a lateral femoral head offset is $T = Fd$, where F is force applied to the femoral head by patient activity and weight and d is perpendicular distance from the head center to the TEL.

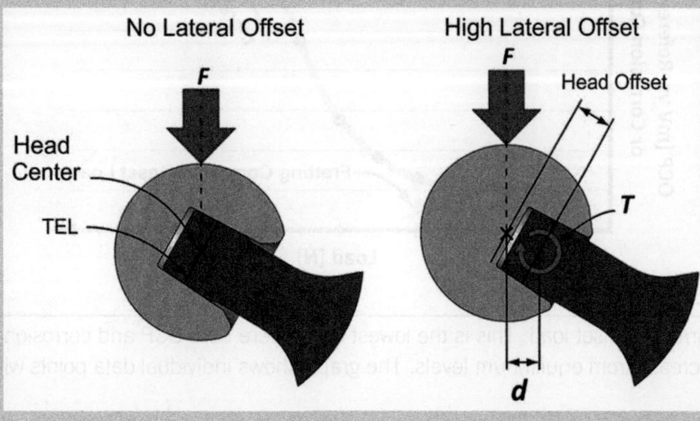

4. RESULTS

Once all fretting corrosion data collection and analysis have been performed, it is then important to communicate and present the primary results in an understandable and concise manner to the reader of a journal article, conference paper, technical report, or book chapter.

Step 1. Show raw OCP and current data. Begin with a typical profile of OCP or corrosion current vs. applied load. Initially, at low loads, OCP and corrosion current remain at unloaded equilibrium levels. As load increases, a threshold is passed when fretting corrosion begins (Fig. 18.5). OCP and corrosion current will then increase with load as the load-induced micromotion abrades a greater area of the connection surfaces. This plot is influenced by material combination, assembly procedure, test medium, connection design, and loading regimen.

Step 2. Show main results. Present the main numerical findings for fretting corrosion onset load, fretting scores, and corrosion scores (Table 18.4). Identify statistically significant differences (or nondifferences) using the *P* values for all relevant pairwise comparisons.

Step 3. Show SEM or optical microscope images. Present SEM or optical microscope images of pretest and posttest specimens. Identify the THR component (i.e., head or stem), the surface location (i.e., medial, lateral, anterior, or posterior; and proximal or distal), and the wear mechanism (i.e., fretting, pitting, intergranular corrosion, etc.). Representative images of wear mechanisms common between all specimens are presented first, followed by those representative of a particular test group, and finally those unique to only a few specimens or less, so that images

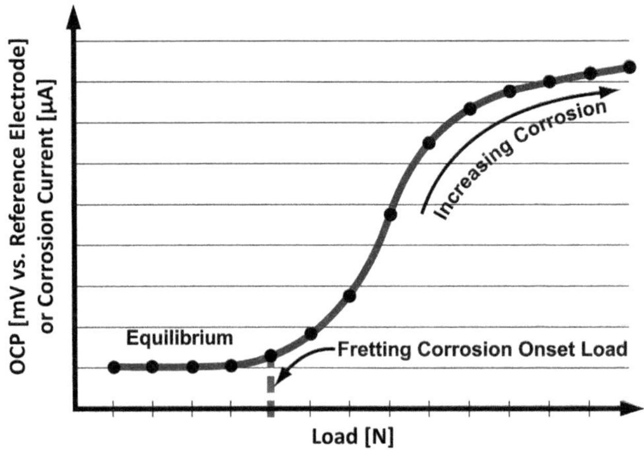

FIGURE 18.5

Fretting corrosion onset load. This is the lowest load where both OCP and corrosion current begin to increase from equilibrium levels. The graph shows individual data points with a line of best fit.

Table 18.4 Fretting corrosions results.

Test Group	Fretting Corrosion Onset Load [N]	Head Taper Fretting Score	Head Taper Corrosion Score	Stem Taper Fretting Score	Stem Taper Corrosion Score
1					
2					
3					
etc.					

Provide numerical entries using average ± 1 standard deviation. Show statistical P *values for intergroup comparisons and visually indicate differences (or nondifferences) using symbols (e.g., asterisks).*

are arranged in order of increasing specificity. For example, typical pretest SEM images of unused specimens show no damage (Fig. 18.6A), while fretting wear scars indicate the direction of micromotion (Fig. 18.6B), widespread pitting corrosion may indicate greater corrosive activity (Fig. 18.6C), and intergranular corrosion may indicate poor material metallurgy (Fig. 18.6D).

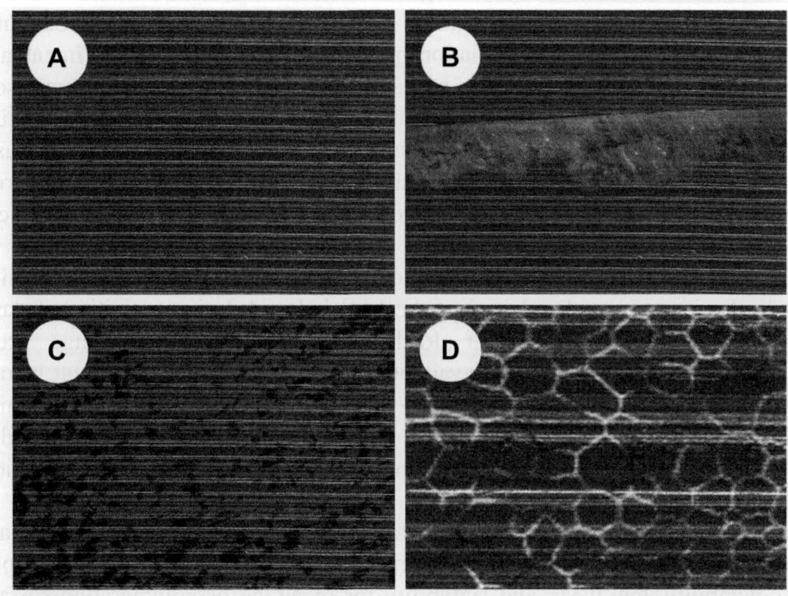

FIGURE 18.6

SEM images of cobalt–chrome alloy femoral head taper surface. (A) No damage (pretest), (B) fretting (posttest), (C) pitting (posttest), (D) intergranular corrosion (posttest).

ALTERNATIVES AND ADAPTATIONS

- ✓ **Trunnion test coupons.** These can be used as a less expensive alternative to an entire femoral stem implant. The trunnion test coupon must be identical in design, manufacture, and material to the implant quality of the trunnion.
- ✓ **Proteinaceous solution.** A solution of 10% calf serum in 0.9% NaCl in distilled water may be used instead of a saline electrolyte to more accurately represent the *in vivo* environment. Manual maintenance of pH and addition of an antimicrobial may be required.
- ✓ **Chemical analysis.** Electrolyte analysis may be performed to quantify fretting corrosion debris, but requires meticulous component preparation, setup, and retrieval. Total debris release = concentration × fluid volume, while total weight loss = sum of each element's total debris.

5. DISCUSSION

After completing all fretting corrosion tests, data collection, data analysis, and data presentation, then final results can be considered and interpreted in the broader context of some important clinical, biomechanical, and/or technological considerations, as follows.

OCP and corrosion current are sensitive to both specimen material and test setup. Regarding materials, each metal undergoes a unique corrosion reaction, such that $M \rightarrow M^{n+} + ne^-$, where M is the original metal molecule, M^{n+} is the resulting metal ion, e^- is the electron that produces corrosion current, and n is a whole number. Regarding test setup, any deviation in the relative position of specimens and electrodes, as well as their coverage in electrolyte, will confound the comparison. Consequently, OCP and corrosion current can only be used to compare fretting corrosion between specimens of the same material within the same meticulously controlled study. However, fretting corrosion onset load, fretting score, and corrosion score are more robust and allow a degree of comparison between specimens of different material.

To determine the likelihood of fretting corrosion clinically, the fretting corrosion onset load may be considered against the load induced during activities of daily living. Fretting corrosion onset load has been found to range from 200–2500 N[5–7] for commercially available THRs. A typical 70 kg patient who is walking will apply a load of approximately 1650 N through the hip joint, which can double or triple while jogging or stumbling, respectively.[8,9] It is clear that with the current suboptimal implant design, fretting corrosion will remain a clinical concern.

Fretting corrosion would ideally be eliminated through use of inert biomaterials. However, due to the many advantages of metallic components, such as biocompatibility, durability, low wear, and suitability for receiving osseointegration surface treatments, it is unlikely that they will be entirely replaced in the short term.

ASTM provides guidelines on fretting corrosion testing of modular hip femoral head–neck taper connection.[3] However, this procedure does not allow comparison between materials in a short-term electrochemical study, but it entirely depends on the quantification of corrosion products following long-term loading.

6. SUMMARY

- THR metallic modular connections are inherently susceptible to fretting corrosion.
- Surface abrasion (i.e., fretting) exposes underlying metal to chemical attack (i.e., corrosion).
- Fretting corrosion gradually releases toxic debris over a long period of time.
- Corrosion reactions may be measured using both the OCP and the corrosion current.
- Careful consideration is required when comparing specimens of different materials.
- Improved materials and THR design could potentially reduce fretting corrosion.

7. QUIZ QUESTIONS

1. Is onset load or corrosion current better for comparing specimens of different materials?
2. Why is a student's *t*-test not used to statistically compare fretting corrosion scores?
3. How may the existing modular connection be redesigned to reduce fretting corrosion?
4. Why might different test setups yield different corrosion currents for the same specimen?
5. A patient undergoes bilateral THR surgery, consisting of a +12 mm femoral head offset in their right leg and a +4 mm femoral head offset in their left leg, each paired with an identical femoral stem with a 135° neck-stem angle. At mid-stance during gait, when the femur is approximately vertical, each THR is subject to 2300 N. Compute the torque induced by the head offset in each leg, and indicate which THR is more susceptible to head–neck taper connection fretting corrosion (answer: 19.5 N·m (right), 6.5 N·m (left), right).

REFERENCES

1. Bosker BH, Ettema HB, Boomsma MF, Kollen BJ, Maas M, Verheyen CCPM. High incidence of pseudotumour formation after large-diameter metal-on-metal total hip replacement: a prospective cohort study. *Journal of Bone and Joint Surgery (British)* 2012; **94**(6):755–61.

2. Jacobs JJ, Urban RM, Gilbert JL, Skipor AK, Black J, Jasty M, et al. Local and distant products from modularity. *Clinical Orthopaedics and Related Research* 1995;**319**: 94–105.
3. ASTM F1875. Standard practice for fretting corrosion testing of modular implant interfaces: hip femoral head-bore and cone taper interface. West Conshohocken (PA, USA): American Society for Testing and Materials (ASTM). Available at: http://www.astm.org.
4. Goldberg JR, Gilbert JL, Jacobs JJ, Bauer TW, Paprosky W, Leurgans S. A multicenter retrieval study of the taper interfaces of modular hip prostheses. *Clinical Orthopaedics and Related Research* 2002;**401**:149–61.
5. Gilbert JL, Mehta M, Pinder B. Fretting crevice corrosion of stainless steel stem-CoCr femoral head connections: comparisons of materials, initial moisture, and offset length. *Journal of Biomedical Materials (Part B): Applied Biomaterials* 2009;**88**:162–73.
6. Goldberg JR, Gilbert JL. In vitro corrosion testing of modular hip tapers. *Journal of Biomedical Materials (Part B): Applied Biomaterials* 2003;**64**:78–93.
7. Mroczkowski ML, Hertzler JS, Humphrey SM, Johnson T, Blanchard CR. Effect of impact assembly on the fretting corrosion of modular hip tapers. *Journal of Orthopaedic Research* 2006;**24**:271–9.
8. Bergmann G, Graichen F, Rohlmann A. Hip joint loading during walking and running, measured in two patients. *Journal of Biomechanics* 1993;**26**(8):969–90.
9. Bergmann G, Graichen F, Rohlmann A, Bender A, Heinlein B, Duda GN, et al. Realistic loads for testing hip implants. *Biomedical Materials and Engineering* 2010;**20**:65–75.

CHAPTER

Pin-on-Disk Wear Testing of Biomaterials Used for Total Joint Replacements

19

Radovan Zdero[1], Leah E. Guenther[2], Trevor C. Gascoyne[2]

Western University, London, ON, Canada[1]; Orthopaedic Innovation Centre, Winnipeg, MB, Canada[2]

1. BACKGROUND

Osteoarthritis is a joint disease caused by cartilage deterioration in the shoulder, hip, and knee, thereby resulting in bone-on-bone grinding, functional impairment, and pain.[1] Every year, about one million people worldwide are implanted with a total joint replacement (TJR) to restore function and reduce pain (Fig. 19.1A).[2] TJRs are made from biomaterial combinations like metal-on-metal, ceramic-on-ceramic, and ceramic-on-polymer, but most commonly, metal-on-polymer.[1,3] Metal components

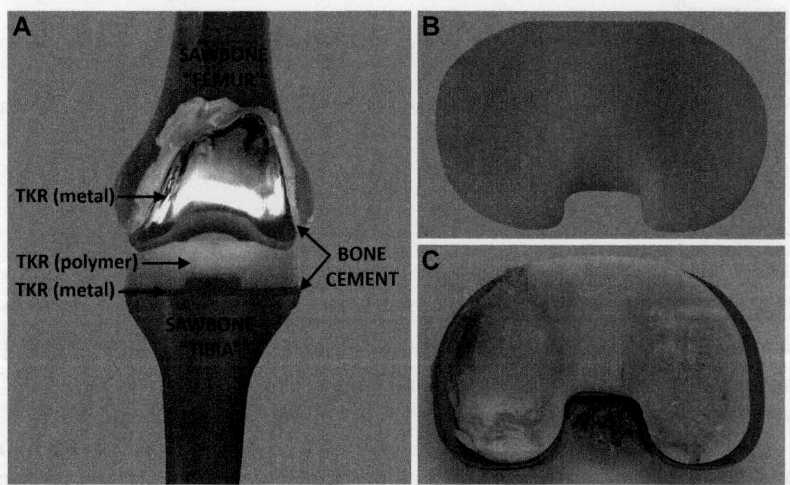

FIGURE 19.1

Total knee replacement (TKR). (A) Main components of a TKR implanted in artificial sawbones, (B) unused polymer component of a TKR, (C) explanted polymer component of a TKR with substantial wear.

Images courtesy of Orthopaedic Innovation Centre, www.OrthoInno.com.

Experimental Methods in Orthopaedic Biomechanics.

may be fabricated from cobalt–chrome alloy, stainless steel, or titanium alloy. Ceramic parts are usually manufactured from alumina. The polymer mostly employed is ultra-high molecular weight polyethylene (UHMWPE). Once implanted, TJRs experience forces and motions that create wear debris, which may cause osteolysis, device loosening, and revision surgery as early as 6–8 years post-arthroplasty (Fig. 19.1B and C).[1] Implant manufacturers use commercially available joint simulators that mimic clinical conditions to optimize TJR wear performance, but these machines are expensive to set up and maintain. In response, researchers often use "classic" pin-on-disk (POD) wear testers, which are easy to use, relatively inexpensive, and can reasonably mimic clinical wear conditions.[2,4–12] Therefore, this chapter explains how to perform POD wear testing of TJR biomaterials, as well as how to analyze, report, and interpret results.

2. RESEARCH QUESTIONS

Typical research questions might include one or more of the following:

- How does force, lubricant, motion, surface roughness, etc., affect a given biomaterial's wear?
- Do biomaterials 1 vs. 2 vs. 3, etc., have a different wear performance for the same test conditions?
- Do biomaterials 1 vs. 2 vs. 3, etc., produce wear debris that has the same visual appearance?
- Do different POD research laboratories and wear testers produce the same wear results?
- Which POD wear testing parameters generate data most similar to clinically explanted TJRs?
- etc.

3. METHODOLOGY

3.1 GENERAL STRATEGY

POD wear testing is done using established protocols for the most common TJR biomaterials, namely, UHMWPE articulating against cobalt–chrome (CoCr).[13–15] UHMWPE cylindrical pins and CoCr circular disks are fabricated, UHMWPE pins are presoaked in test lubricant, and both the pins and disks are mounted in a POD wear tester. Contact stress is applied to the POD, cyclic interfacial motion is employed for a certain number of cycles, and then pins are removed for weight change analysis (i.e., gravimetry). Both the pins and disks may then undergo surface roughness analysis (i.e., profilometry). Statistical comparisons of wear parameters (e.g., cumulative weight loss, average surface roughness, and wear rate) are made between test groups. Finally, correlations are made for wear outcomes vs. test groups (e.g., pin size, contact stress, number of cycles, etc.) to determine vital factors.

GLOSSARY

✓ **CoCr**. Cobalt–chrome alloy that is commonly used to construct components of a TJR.
✓ **Lubricant**. A fluid medium intended to mimic synovial fluid in human synovial joints.
✓ **POD**. Pin-on-disk configuration, which simulates TJR interfacial articulation.
✓ **TJR**. Total joint replacement, such as those used for the shoulder, hip, or knee.
✓ **UHMWPE**. Ultra-high molecular weight polyethylene used to fabricate TJRs.
✓ **Wear**. Degradation of surface quality and material due to interfacial articulation.

SAFETY FIRST

✓ Read the material safety data sheet for chemicals before use and exposure.
✓ Sodium azide is an extremely hazardous poison that should be used with caution.
✓ Handle test lubricants only while wearing goggles, gloves, a mask, and a lab coat.
✓ Clean test lubricant spills with soapy water, and then rinse with alcohol and air dry.
✓ Keep clothes, long hair, and fingers away from the POD tester's moving parts during use.

3.2 MATERIALS AND TOOLS LIST

- fixtures (mounting plate, acrylic parts, O-rings, etc.)
- fume hood
- microbalance
- POD wear tester and computer software
- profilometer (noncontact or contact type)
- scanning electron microscope (SEM)
- silicone sealant
- temperature control unit and/or hot plate
- test lubricant (e.g., bovine blood serum–based)
- test pins and disks

3.3 SPECIMEN PREPARATION

Step 1. Determine the number of groups and specimens. Determine how many test groups, as well as how many specimens (i.e., pins and disks) per group are required. This is important in order to answer the research question in a statistically rigorous way. Consequently, a typical POD wear study can use anywhere from 3—25 specimens per test group.[2,5,6,10,13]

Step 2. Fabricate the pins. Prepare a polymer pin to mimic the articulating UHMWPE surface of a TJR (Fig. 19.2A).[2,5,6,8,10,11,13,14] Obtain a long, cylindrical rod of medical grade UHMWPE, which may or may not be gamma irradiated with 2.5–5 Mrad. Use a band saw to cut the rod into preliminary segments that are about 20 mm long so the ends can be gripped during the fabrication process. Then, use a

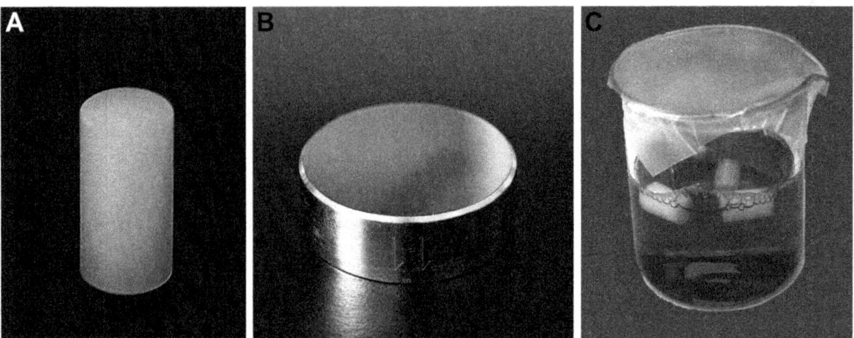

FIGURE 19.2

Specimen preparation for POD wear tests. (A) UHMWPE pin, (B) CoCr disk, (C) UHMWPE pins presoaking in bovine blood serum–based lubricant.

Images courtesy of Orthopaedic Innovation Centre, www.OrthoInno.com.

lathe to reduce the outer surface to a 2–10 mm diameter, use a band saw to cut the rod into segments that are 12–13 mm long, and use a milling machine or lathe to make the ends flat with an average surface roughness of $Ra \leq 0.8$ μm.

Step 3. Fabricate the disks. Prepare a metal disk to mimic the articulating CoCr surface of a TJR (Fig. 19.2B).[2,6,8,13] Obtain a medical grade CoCr plate with a 2–10 mm thickness. Cut out a circular disk using an industrial metal band saw or milling machine. Reduce the disk's outer edge roughness using a grinding wheel or milling machine. Machine the disk's outer edge to a 30–100 mm diameter using a lathe. Work the main surface by dry grinding, then wet grinding with #1200 and #2400 grit paper, and then cloth wheel polishing using 6-μm (and subsequently 1-μm) diamond paste. The resulting surface should have an average roughness of $Ra \leq 0.8$ μm.

Step 4. Prepare the lubricant. In a fume hood, prepare a lubricant simulating human joint synovial fluid.[5–11,14,15] Obtain commercially available, filtered, sterilized, non-iron-supplemented bovine blood serum. Then, dilute it with deionized water or phosphate buffered saline solution to a chosen bovine blood serum protein concentration of 20–72 g/L. To inhibit microbial growth, add an amount of sodium azide that is 0.2–0.3% of lubricant total mass. The lubricant can now be frozen at −15 to −20°C for up to 5 years or refrigerated at 2–8°C for up to 30 days prior to use.

Step 5. Presoak the pins. Fully immerse each pin into a glass beaker containing about 150 mL of lubricant (Fig. 19.2C).[2,5,6,11,14,16] Keep the lubricant at 37 ± 3°C using a temperature control unit or hot plate for a minimum of 5 weeks to reproduce *in vivo* human body conditions. Use a watch glass to weigh down the pins. While most pins eventually will be wear tested, several pins can act as "soak controls" by permanently remaining in lubricant. This minimizes lubricant fluid uptake by pins during wear tests, as well as eliminating weight change measurement artifacts.

Step 6. Clean the pins. Remove each pin from the lubricant bath in which it was presoaked and clean it, as follows: rinse it in warm water with a small amount of laboratory liquid detergent, wipe off any lubricant residue using a nonabrasive cloth, rinse it with deionized water, dry it with nitrogen gas, soak it in 95% methyl alcohol for 5 min, and dry it again with nitrogen gas.[16]

Step 7. Measure the pins. Use a precision microbalance with a 10-μg resolution to weigh each pin to the nearest 0.1 mg. Then, measure the surface roughness of each pin surface using a noncontact profilometer. Do this three times in order to get an average weight and an average surface roughness for each pin. Return the "soak control" pins to their lubricant baths, whereas the other pins are now ready to be wear tested.

TIPS AND TRICKS

- ✓ The same researcher(s) should perform all POD wear tests for consistency in results.
- ✓ Metal disks can be successfully polished using a spinning cloth wheel and diamond paste.
- ✓ Apply silicone sealant along the inside edges of test chambers to prevent lubricant leaks.

THE "GOLD STANDARD"

Researchers should consult the American Society for Testing and Materials (ASTM) documents ASTM G99 (Standard test method for wear testing with a pin-on-disk apparatus), ASTM F732 (Standard test method for wear testing of polymeric materials used in total joint prostheses), and ASTM F2025 (Standard practice for gravimetric measurement of polymeric components for wear assessment). Previously published peer-reviewed journal articles may also be useful.

3.4 SPECIMEN TESTING

Step 1. Choose a wear tester. The POD wear tester should produce relative motion between the pin and disk that generates a change in shear direction, which mimics *in vivo* TJR conditions and wear results (Fig. 19.3A).[2,4,7,8,10,13,14] Specifically, the motion should be "multidirectional" or "circularly translating" between a stationary (or moving) pin and a moving (or stationary) disk. This is in contrast to some wear testers that only produce linear reciprocation or unidirectional rotation, which do not mimic TJR conditions. Also, the POD wear tester should be able to precisely control or monitor the force, speed, frequency, and lubricant temperature.

Step 2. Secure the pins. Insert each pin perpendicular to the plane of motion (i.e., ± 1°) into its own overhead pin holder in the POD wear tester (Fig. 19.3B).[13,14] A typical pin holder is composed of a hollow shaft made of metal or plastic into which the pin is press-fitted by hand or by a special hand tool and then brought down into direct contact with the disk.

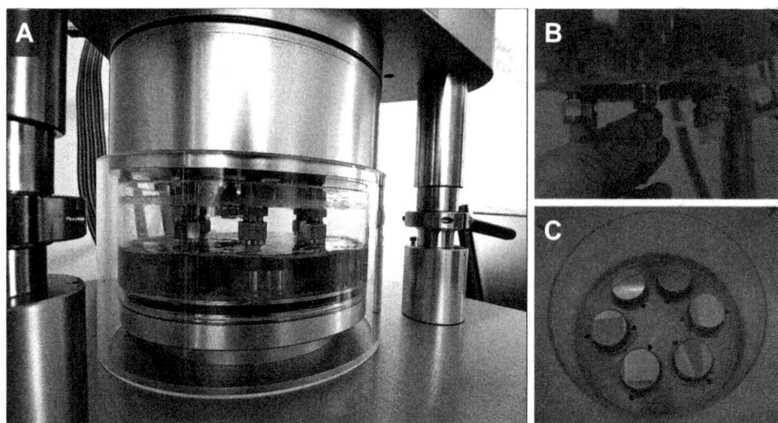

FIGURE 19.3

Typical POD wear tester. (A) Assembled POD wear tester, (B) UHMWPE pins secured in the holder, (C) CoCr disks secured in the test chamber.

Images courtesy of Orthopaedic Innovation Centre, www.OrthoInno.com.

Step 3. Secure the disks. Insert each disk perpendicular to the plane of motion (i.e., $\pm 1°$) into its own designated station in the test chamber of the POD wear tester (Fig. 19.3C).[13,14] A typical test chamber is composed of an acrylic circular tube surrounding a stainless steel plate with or without through-holes and screws to accommodate different disk diameters. Place plastic plugs with O-rings into any empty through-holes to prevent lubricant from leaking out during wear testing. Apply a line of silicone sealant along the inside edges of the test chamber to reduce leakage.

Step 4. Configure the tester. The POD wear tester should be configured manually or by computer control for each test group by choosing parameters that represent a particular TJR *in vivo* clinical condition, such as POD contact stress (=applied force/cross-sectional area = 1.1−3.5 MPa),[2,4,7,11,14] sliding frequency (1 Hz),[7,14] sliding speed (24−50 mm/s),[7,14] sliding distance (25−35 mm/cycle),[2,14] total number of cycles (1−7 Mc or million cycles),[2,7,9,10] and lubricant temperature ($37 \pm 3°C$).[7,14] Of course, different nonclinical levels may be used, depending on the research question.

Step 5. Run the test. Begin wear testing with pins and disks in contact. Stop the test after every "interval" (i.e., every 250,000−500,000 cycles),[2,5−7,10] and then perform these tasks: disassemble test chambers, dispose old lubricant, clean the pins, measure pin weight using a microbalance, measure pin surface roughness using a noncontact profilometer, photograph any interesting features on the disk's wear path (e.g., cracks, discoloration, protrusions, etc.), add new lubricant to the test chamber, resecure the same pins, and continue wear testing under the same test conditions. Repeat this after every interval until the total number of wear cycles is completed.

3.5 RAW DATA COLLECTION

Step 1. Record specimen characteristics. Before each wear test, ensure that important information is documented on pins and disks, such as physical dimensions, material name, and supplier. Lubricant name, composition, and bovine serum supplier should also be noted.

Step 2. Record test groups. Each test group represents a change in some numerical or categorical factor that is controlled by the researcher (e.g., contact stress, sliding frequency, sliding speed, sliding distance, pin density, lubricant protein concentration, pin material, etc.) in order to assess its influence on wear performance (Table 19.1).

Step 3. Record wear outcomes. Wear outcomes are quantities that are measured by the researcher (i.e., pin cumulative weight loss ΔW and pin average surface roughness *Ra*), which represent wear performance of the pin material (Table 19.1). Keep several things in mind. First, the weight loss of the UHMWPE wear pin is adjusted to account for fluid absorption during testing. At a certain test interval, the net gravimetric wear ΔW (i.e., net weight loss) of the UHMWPE wear pin is the apparent weight loss of the individual wear pin (W_{loss}) plus the average weight gain of the soak control pins (S_{gain}). Therefore, at a certain test interval, $\Delta W = W_{loss} + S_{gain} = (W_i - W_f) + (S_f - S_i)$, where i is initial weight and f is final weight. Second, if time is of the essence, *Ra* can be measured before the study begins and after the very last interval of the study, rather than at each interval, since it can be a time-consuming process. Third, the initial phase of a wear test is called "running-in wear," which is a nonsteady-state phenomenon that occurs for many materials starting from 0 cycles to as high as 1 Mc. Data within this range may be collected since they provide some information about the material's behavior, but they are ignored later when analyzing and presenting the study's main results for steady-state wear.

Step 4. Record wear surface features. Before the wear study begins and once all tests are done at the end of the last interval, obtain SEM images and/or

Table 19.1 Raw POD wear test data.

		"Running-in Wear" Intervals				Steady-State Intervals			
Test Group	**Wear Outcome**	**1**	**2**	**3**	**etc.**	**1**	**2**	**3**	**etc.**
1	ΔW [mg]								
	Ra [μm]								
2	ΔW [mg]								
	Ra [μm]								
3	ΔW [mg]								
	Ra [μm]								
etc.									

Ra, *pin average surface roughness;* ΔW, *pin cumulative weight loss. Each numerical entry of* ΔW *and* Ra *measurements has an average* ± *1 standard deviation or* ±*95% confidence interval.*

photographs of the pin's wear surface to detect any changes or damage (e.g., cracks, discoloration, protrusions, etc.).

3.6 RAW DATA ANALYSIS

Step 1. Calculate correlation coefficients. Graphically plot all wear outcomes vs. test groups to visualize their interrelationship; then, calculate the correlation coefficient R for each line of best fit to determine the strength of the interrelationship. $R > 0.8$ is considered to indicate a strong correlation and that steady-state wear was achieved.

Step 2. Perform statistical comparisons. Determine the criterion for statistical difference (e.g., $P < 0.01$ or < 0.05). Choose from among the various software programs for comparing two test groups (e.g., unpaired t-test) or two or more test groups influenced by multiple factors (e.g., analysis of variance, ANOVA). Parametric or nonparametric statistical tests are chosen, depending on whether or not results have a normal distribution.

Step 3. Compute statistical power. Power analysis can be done after the study to ensure there were enough specimens per test group to detect all statistical differences that were actually present (i.e., was type II statistical error avoided?). Statistical power >80% is usually considered to indicate there were enough specimens per test group. Note that if good predictions of averages and standard deviations are available from prior studies, then the number of specimens and/or tests can be chosen before the study begins to ensure a power >80%.

ENGINEER'S TOOLBOX

Wear factor is a traditional way to report the pin material's performance during POD wear testing. Wear factor is based on Archard's Law, which states that the amount of wear debris from the pin is proportional to the applied force and sliding distance. This is expressed mathematically as $K = \dot{W}/(\rho F D)$, where K is wear factor [$mm^3/(N \cdot m)$], $\dot{W}$ is wear rate of removed material [mg/Mc], ρ is material density [mg/mm^3], F is applied force [N], and D is sliding distance [m/Mc]. (Note: The unit Mc means million cycles.)

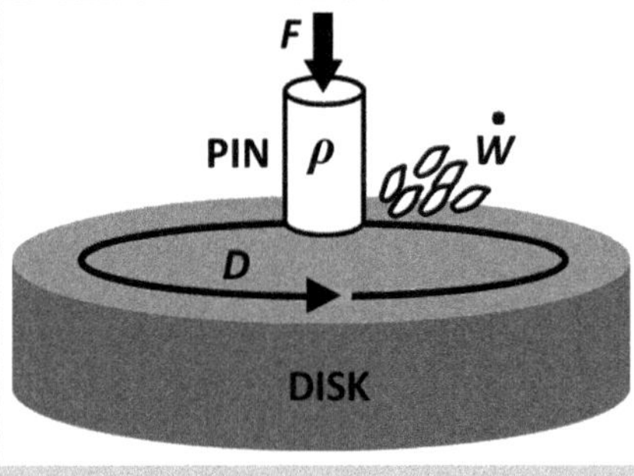

4. RESULTS

Once all POD wear test data collection and analysis have been performed, it is then important to communicate and present the primary results in an understandable and concise manner to the reader of a journal article, conference paper, technical report, or book chapter.

Step 1. Show the main steady-state results. Cumulative weight loss ΔW and average surface roughness Ra vs. number of cycles should be given (Fig. 19.4A and B). Statistical P values should be stated for pairwise comparisons between wear outcomes at each interval (e.g., ΔW for test group 1 vs. 2 vs. 3 at interval 1, etc.) and between wear outcomes for each test group (e.g., ΔW at interval 1 vs. 2 vs. 3 for test group 2, etc.). For each line of best fit, several items can be presented: an equation $y = mx + b$ with slope m and intercept b, a linear correlation coefficient R, and its own P value to ensure that the slope of each line of best fit (i.e., slope $= m$) is statistically different than a horizontal line (i.e., slope $= 0$). (Note: "Running-in wear" data do not represent steady-state wear; thus, they are excluded from the lines of best fit. Also, data from soak control pins are not shown since they do not undergo wear testing and are only used to correct wear data for fluid absorption.)

Step 2. Present steady-state wear rates. The slope m for each ΔW line of best fit is known as the "gravimetric wear rate" $\dot{W}$ in milligrams per million cycles

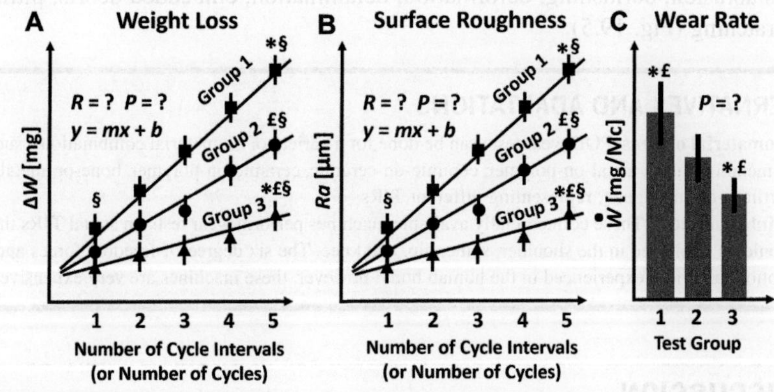

FIGURE 19.4

Steady-state wear outcomes. (A) Cumulative weight loss, (B) average surface roughness, (C) average wear rate. Each data point is an average $\pm$ 1 standard deviation or $\pm$ 95% confidence interval. P indicates the many statistical difference values obtained for each pairwise comparison between test groups and between intervals which can be indicated by symbols like asterisks (*), pounds (£), etc. Each line of best fit is represented by an equation $y = mx + b$, a linear correlation coefficient R, and its own statistical P value to ensure that its slope (i.e., slope $= m$) is statistically different than a horizontal line (i.e., slope $= 0$).

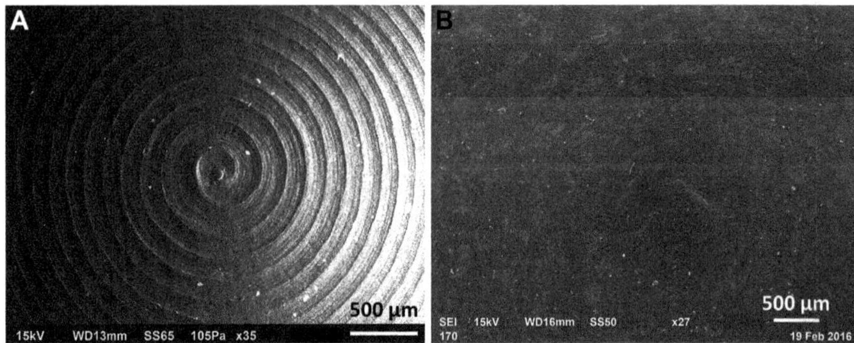

FIGURE 19.5

Polymer pin surfaces. (A) SEM image of pin surface showing concentric machining marks (prestudy), (B) SEM of pin surface showing wear (poststudy).

Images courtesy of Orthopaedic Innovation Centre, www.OrthoInno.com.

[mg/Mc], which should also be statistically compared between test groups (Fig. 19.4C). The material with the lower wear rate is considered to have superior resistance to wear.

Step 3. Illustrate wear patterns. Provide SEM images and/or photos of prestudy and poststudy pin wear surfaces to identify wear mechanisms of interest, such as abrasion, burnishing, deformation, delamination, embedded debris, pitting, and scratching (Fig. 19.5).

ALTERNATIVES AND ADAPTATIONS

- ✓ **Biomaterial options**. POD wear tests can be done for a variety of biomaterial combinations, such as metal-on-metal, metal-on-polymer, ceramic-on-ceramic, ceramic-on-polymer, bone-on-metal, cartilage-on-metal, etc., representing different TJRs.
- ✓ **Joint simulators**. These commercially available machines perform wear tests on actual TJRs that would be implanted in the shoulder, spine, hip, and knee. The six degree-of-freedom forces and motions are those experienced in the human body; however, these machines are very expensive.

5. DISCUSSION

After completing all POD wear tests, data collection, data analysis, and data presentation, then final results can be considered and interpreted in the broader context of some important clinical, biomechanical, and/or technological considerations, as follows.

Osteoarthritis of weight-bearing synovial joints (e.g., shoulder, hip, knee) causes cartilage wear, bone-on-bone grinding, functional impairment, and pain.[1] Annually, one million people worldwide receive a TJR for this condition.[2] Most TJRs work via metal-on-polymer articulation at the joint interface, but *in vivo* forces and motions create polymer wear debris through wear mechanisms like abrasion, burnishing, deformation, delamination, embedded debris, pitting, and scratching.[3,17] Subsequently, polymer wear debris launches a cascade of important events: auto-immune response → osteolysis (i.e., bone resorption) → implant loosening → implant failure → revision surgery.[1]

Standardized POD wear test protocols have been developed in order to predict and reproduce *in vivo* clinical wear of TJRs.[13,14] However, the studies that have been particularly successful at doing this have incorporated several key factors, such as a contact stress of 1.1–3.5 MPa.[2,4,7,11,14] Multidirectional or circularly translating motion that changes the cross-shear direction,[2,4,7,8,10,13,14] and bovine blood serum–based lubricant, which permits some polymer surface scratching via third body wear.[4–11,14] Even so, the standardized test protocols do not always fully incorporate the latest research findings and, thus, may need to be occasionally updated to reflect clinical conditions.

POD wear testing of UHMWPE can produce a range of numerical results, depending on the type of UHMWPE used and the experimental conditions. For example, average wear rate W can range from 1.69—13.55 mg/Mc (standard UHMWPE),[2,5,8,9] 0.41–2.06 mg/Mc (generic cross-linked UHMWPE),[2] 0.15–6.43 mg/Mc (commercial cross-linked UHMWPE),[5,6] and 4.28–36.68 mg/Mc (explanted UHMWPE from TJR patients).[7] Similarly, average surface roughness Ra can reach values of 0.88 μm (standard UHMWPE),[2] 0.68 μm (generic cross-linked UHMWPE),[2] and 0.007–0.012 μm (commercial cross-linked UHMWPE).[6]

6. SUMMARY

- Osteoarthritis is a joint disease that can be addressed by surgical implantation of a TJR.
- TJRs can experience different types of wear mechanisms leading to component damage.
- POD wear testers have been developed that reasonably reproduce *in vivo* TJR conditions.
- POD wear tests assess the effects of contact stress, material, sliding type, and lubrication.
- POD wear tests can provide pin cumulative weight loss, surface roughness, and wear rate.
- POD wear tests on UHMWPE pins can yield a wide range of numerical results.

7. QUIZ QUESTIONS

1. Why does osteoarthritis often lead to surgical implantation of a TJR?
2. What are three wear mechanisms that are commonly experienced by an implanted TJR?

3. What are the different types of motion that can be created by various POD wear testers?
4. What are the main factors that researchers can control during a typical POD wear test?
5. Calculate the wear factor of a UHMWPE pin during a POD wear test with these pin characteristics: gravimetric wear rate (10 mg/Mc), applied force (157 N), density (0.95 mg/mm^3), and sliding distance (30 mm/cycle) (answer: 2.2×10^{-6} mm^3/(N·m)).

REFERENCES

1. Garino JP, Beredjiklian PK. *Adult reconstruction and arthroplasty: core knowledge in orthopaedics*. Philadelphia (PA, USA): Mosby Elsevier; 2007.
2. Harsha AP, Joyce TJ. Comparative wear tests of ultra-high molecular weight polyethylene and cross-linked polyethylene. *Proceedings of the Institution of Mechanical Engineers (Part H): Journal of Engineering in Medicine* 2013;**227**(5):600–8.
3. Hosseinzadeh HRS, Eajazi A, Shahi AS. "The bearing surfaces in total hip arthroplasty: options, material characteristics, and selection" (Chapter 10). In: Fokter S, editor. *Recent advances in arthroplasty*. Rijeka (Croatia): InTech; 2012. Available free online at: http://cdn.intechopen.com/pdfs-wm/26863.pdf.
4. Baykal D, Siskey RS, Haider H, Saikko V, Ahlroos T, Kurtz SM. Advances in tribological testing of artificial joint biomaterials using multidirectional pin-on-disk testers. *Journal of the Mechanical Behavior of Biomedical Materials* 2014;**31**:117–34.
5. Brandt JM, Vecherya A, Guenther LE, Koval SF, Petrak MJ, Bohm ER, et al. Wear testing of crosslinked polyethylene: wear rate variability and microbial contamination. *Journal of the Mechanical Behavior of Biomedical Materials* 2014; **34**:208–16.
6. Guenther LE, Turgeon TR, Bohm ER, Brandt JM. The biochemical characteristics of wear testing lubricants affect polyethylene wear in orthopaedic pin-on-disc testing. *Proceedings of the Institution of Mechanical Engineers (Part H): Journal of Engineering in Medicine* 2015;**229**(1):77–90.
7. Kurtz SM, MacDonald DW, Kocagoz S, Tohfafarosh M, Baykal D. Can pin-on-disk testing be used to assess the wear performance of retrieved UHMWPE components for total joint arthroplasty? *Biomed Research International* 2014:1–6, 581812.
8. Saikko V. A multidirectional motion pin-on-disk wear test method for prosthetic joint materials. *Journal of Biomedical Materials Research* 1998;**41**(1):58–64.
9. Saikko V. A hip wear simulator with 100 test stations. *Proceedings of the Institution of Mechanical Engineers (Part H): Journal of Engineering in Medicine* 2005;**219**(5): 309–18.
10. Saikko V. Effect of contact pressure on wear and friction of ultra-high molecular weight polyethylene in multidirectional sliding. *Proceedings of the Institution of Mechanical Engineers (Part H): Journal of Engineering in Medicine* 2006;**220**(7):723–31.
11. Saikko V, Ahlroos T. Type of motion and lubricant in wear simulation of polyethylene acetabular cup. *Proceedings of the Institution of Mechanical Engineers (Part H): Journal of Engineering in Medicine* 1999;**213**(4):301–10.

12. Wright KMJ, Dobbs HS, Scales JT. Wear studies on prosthetic materials using the pin-on-disc machine. *Biomaterials* 1982;**3**:41–8.
13. ASTM G99. Standard test method for wear testing with a pin-on-disk apparatus. West Conshohocken (PA, USA): American Society for Testing and Materials (ASTM); www.astm.org.
14. ASTM F732. Standard test method for wear testing of polymeric materials used in total joint prostheses. West Conshohocken (PA, USA): American Society for Testing and Materials (ASTM); www.astm.org.
15. Serum Source International Inc., Charlotte (NC, USA), www.serumsourceintl.com.
16. ASTM F2025. Standard practice for gravimetric measurement of polymeric components for wear assessment. West Conshohocken (PA, USA): American Society for Testing and Materials (ASTM); www.astm.org.
17. Brown TD, Bartel DL. What design factors influence wear behavior at the bearing surfaces in total joint replacements? *Journal of the American Academy of Orthopaedic Surgeons* 2008;**16**(Suppl. 1):S101–6.

CHAPTER

Vibration Analysis of the Biomechanical Stability of Total Hip Replacements

20

Kathleen Denis, Leonard C. Pastrav, Steven Leuridan
KU Leuven, Leuven, Belgium

1. BACKGROUND

In cases of severe rheumatoid arthritis or osteoarthritis, the hip joint is substituted by an artificial joint composed of a femoral stem fitted with a spherical head that can rotate inside a cup inserted in the acetabulum (Fig. 20.1A). This procedure is called total hip replacement (THR) and is one of the most frequently performed

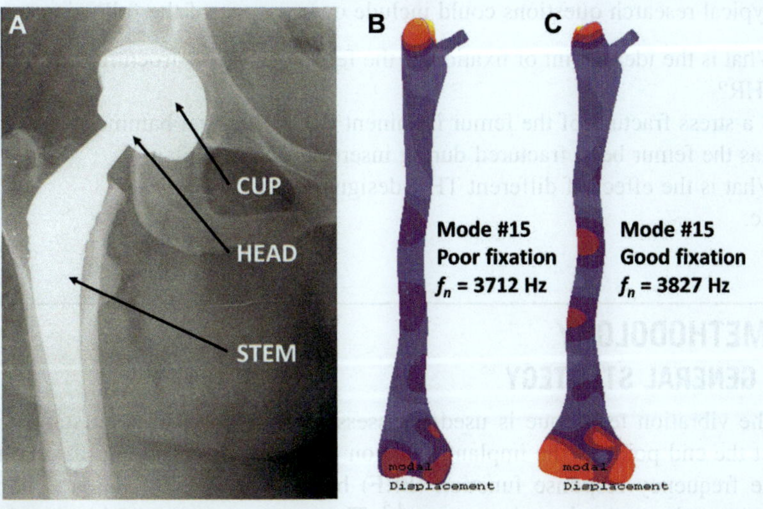

FIGURE 20.1

Total hip replacement. (A) Radiograph showing the femoral stem, spherical head, and acetabular cup, (B) vibration mode #15 of a femur–implant structure showing poor fixation with only 15% contact at the femur–implant interface, (C) vibration mode #15 of a femur–implant structure showing good fixation with 98% contact at the femur–implant interface. The natural frequency f_n increases from 3712 to 3827 Hz, according to finite element modal analysis.

Experimental Methods in Orthopaedic Biomechanics.

orthopaedic surgeries. For a cementless femoral implant, the fixation is achieved by preparing a slightly undersized bone bed, and the implant is forcefully hammered into the bone. The initial stability at the time of surgery is one of the most important factors to establish long-term survival of the implant.[1] With each surgical hammer blow, the fixation of the implant in the bone increases. However, introducing an implant with a diameter wider than the bone's inner canal contour (i.e., press fit) introduces stresses in the cortical bone, which can cause femoral fracture.[2] In response to mechanical excitations (i.e., hammer blows or external vibrations), a femur–implant structure will display vibration modes and frequencies, just like any other mechanical structure. Changes in material properties or boundary conditions of a femur–implant structure will change its vibration modes and frequencies, which can be obtained numerically and experimentally (Fig. 20.1B and C).[3] Therefore, this chapter explains how to perform vibration analysis on a THR component (i.e., the femoral implant) in order to assess femur–implant stability, as well as how to analyze, present, and interpret results.

2. RESEARCH QUESTIONS

Typical research questions could include one or more of the following:

- What is the ideal point of fixation of the femur–implant structure in cementless THR?
- Is a stress fracture of the femur imminent due to surgical hammering?
- Has the femur been fractured during insertion of the implant?
- What is the effect of different THR designs on stability?
- etc.

3. METHODOLOGY

3.1 GENERAL STRATEGY

The vibration technique is used to assess THR femoral implant stability and detect the end point of the implant insertion process by interpreting the evolution of the frequency response function (FRF) between different fixation conditions (i.e., two subsequent insertion steps).[4,5] The surgeon or researcher inserts the implant in the femoral canal through successive controlled hammer blows. After each hammer blow that is intended to insert the implant more deeply into the femoral cavity, the FRF of the femur–implant structure is measured directly on the implant neck in the range of 100–10,000 Hz. The FRF changes are used as indicators of the evolution of the stiffness of the femur–implant structure and the evolution of implant fixation. When the FRF graph does not change noticeably, the fixation is considered stable and the hammering is stopped. Extra blows would not improve the fixation of the implant, but would increase fracture risk. The similarity of two successive FRF graphs is evaluated by using Pearson's correlation coefficient *R*. A correlation between the FRFs of successive stages of $R = 0.99 \pm 0.01$ over the range 100–10,000 Hz is proposed as an endpoint criterion. Finally, statistical analysis is performed to compare data from various test groups representing different THRs (e.g., implant 1 vs. 2 vs. 3, etc.), bone densities (e.g., normal vs. osteopenic vs. osteoporotic), bone types (e.g., human femur vs. artificial femur), etc.

GLOSSARY

- ✓ **Implant stability.** Ability to support physiological loads without changing position in bone.
- ✓ **Micromotion.** Cyclic relative movements at femur–implant interface induced by cyclic loads.
- ✓ **Migration.** Irreversible displacement of the implant inside the bone under physiological load.
- ✓ **Vibration analysis.** Assessment of a mechanical structure by studying vibration behavior.
- ✓ **Vibration modes.** Inherent properties of mechanical structures defined by three parameters.
- ✓ **Vibration mode parameters.** Mode shape, natural (resonant) frequency, and damping.

SAFETY FIRST

- ✓ Remember to always wear a lab coat, safety glasses, and gloves for protection.
- ✓ Secure the bone during the reaming and insertion procedures to avoid injuries.
- ✓ Take specific safety measures against biological hazards when using human bones.
- ✓ Vaccination against hepatitis and tetanus is a necessary precaution.
- ✓ Take specific safety measures while using electrical equipment.

3.2 MATERIALS AND TOOLS LIST

- accelerometer and force cell
- computer and software for vibration signal processing
- data acquisition and vibration analysis system (e.g., hardware)
- excitation hammer or shaker
- human or artificial femora
- impedance head
- supporting structure and elastic bands
- THR femoral cementless implants
- tools for implant insertion (e.g., hammer, inserter, etc.)

3.3 SPECIMEN PREPARATION

Step 1. Store the femora. Fresh or fresh–frozen cadaveric femora should be sealed in plastic wrappings and properly stored in a freezer before the tests. The temperature inside the freezer should be lower than −20°C. Dried/dehydrated cadaveric, embalmed cadaveric, or artificial femora can be stored in an appropriate package at room temperature.

Step 2. Thaw the femora. After removal from the freezer, the cadaveric femora should be left to thaw in their wrappings at ambient room temperature or in a warm water bath for at least 12 h. Then, to prevent dehydration after removing the wrappings, the femora should be soaked or sprayed with saline water solution.

Step 3. Ream the femora. The femora must be prepared by an orthopaedic surgeon for insertion of a cementless implant. A standard surgical reaming protocol should be used unless the vibration testing method is intended to assess the implant stability produced by applying a new reaming method. The geometry of the reamed medullary canal should be in agreement with the bone and implant sizes and geometries unless the vibration testing method is intended to assess the implant stability in other situations, such as over- or under-reaming, local bone resorption, etc.

TIPS AND TRICKS

✓ At least one research team member should be familiar with THR surgery.
✓ At least one research team member should be familiar with vibration analysis.
✓ Before testing, calibrate the accelerometer, force cell, and impedance head.
✓ Ensure no external sources exist that cause mechanical or electrical interference.
✓ The shaker excitation technique is less influenced by the operator's skills.

THE "GOLD STANDARD"

No known international standards exist specifically for using vibration analysis on bones, implants, or bone–implant constructs. Thus, researchers should consult peer-reviewed journal articles that have used vibration analysis for similar orthopaedic biomechanics applications.

3.4 SPECIMEN TESTING

Step 1. Assemble the test setup. In the case of the hammer impulse excitation method, attach an accelerometer using bee wax to the implant neck (Fig. 20.2). Attach supplementary accelerometers to the femur in order to obtain more information related to the vibration behavior of the femur–implant structure. In the case of the shaker excitation method, connections between the implant and the shaker should be as rigid as possible in the direction of excitation, and their properties must not change during the measuring procedure in order to avoid errors (Fig. 20.3).

Step 2. Apply hammer impulse excitation. If using the hammer excitation method, the FRF of the femur–implant system is measured using five to eight averages by impulse excitation on the implant neck. The input force is measured by a force cell included in an instrumented hammer, while the response of the analyzed structure is measured by an accelerometer attached opposite to the impact site (Fig. 20.2B).

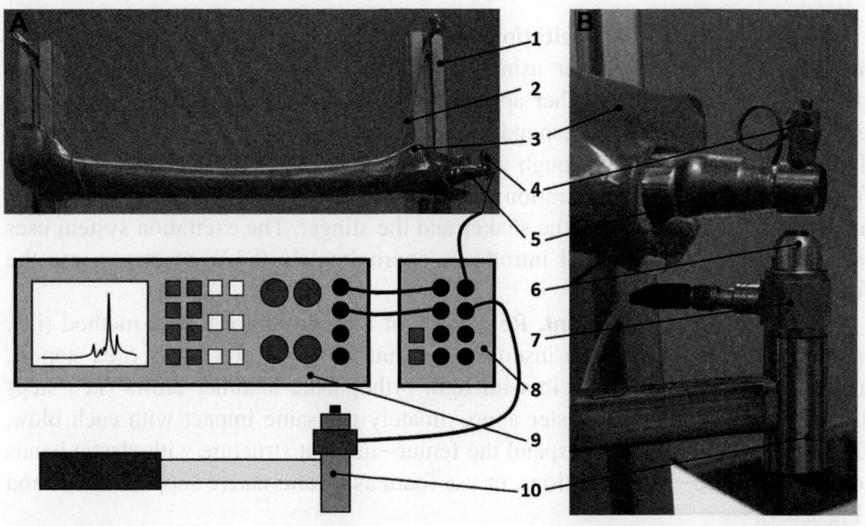

FIGURE 20.2

Experimental setup for hammer impulse excitation. (A) General view, (B) detail of impulse excitation setup showing (1) support, (2) elastic bands, (3) femur specimen, (4) accelerometer, (5) hip stem, (6) hammer tip, (7) force cell, (8) input signal amplifier, (9) spectrum analyzer, and (10) excitation hammer.

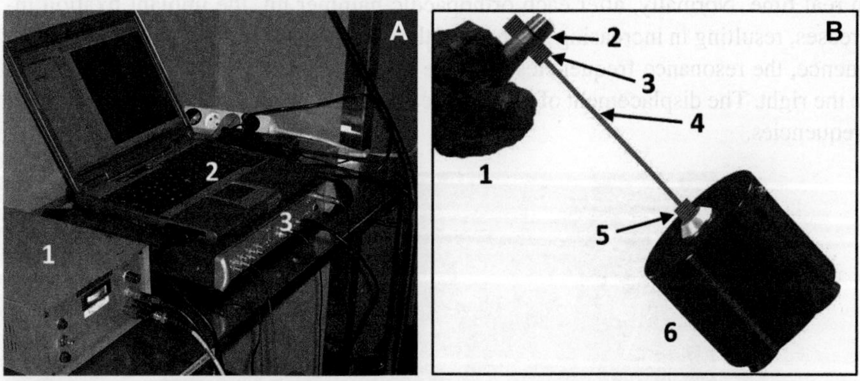

FIGURE 20.3

Experimental setup for shaker excitation. (A) Electronic components showing (1) power amplifier, (2) laptop, and (3) vibration analyzer, (B) mechanical components showing (1) femur, (2) implant, (3) clamping system, (4) stinger, (5) impedance head, and (6) shaker.

Step 3. Apply shaker excitation. If using the shaker excitation method, attach the implant neck to a shaker using a stinger provided with a clamping system (Fig. 20.3). The stinger is either approximately perpendicular to the plane defined by the longitudinal axes corresponding to the stem and to the neck or in that plane. The excitation is realized through white noise in the range 0–12.5 kHz. The input force and the response acceleration are measured at the same point with an impedance head mounted between the shaker and the stinger. The excitation system uses low-amplitude vibrations and introduces approximately 0.5 W of power into the femur–implant system.

Step 4. Insert the implant. Regardless of the chosen excitation method (i.e., hammer or shaker), manually insert the implant loosely in the cavity (i.e., step 0), and subsequently hammer it in with light orthopaedic hammer blows (i.e., steps 1...*n*), taking care to administer approximately the same impact with each blow. During the measurements, suspend the femur–implant structure with elastic bands to reproduce free–free conditions, or use foam as an alternative supporting method (Fig. 20.4).

Step 5. Measure FRF data. After each insertion step, the femur–implant structure is excited either by impulses using instrumented hammer hits or by white noise provided by a shaker. The signals corresponding to the input force and response acceleration are amplified and then processed by a vibration analyzer (Figs. 20.2A and 20.3A). Calculate and record FRF by a computer connected to the vibration analyzer and provided with the appropriate software. In addition, the vibration analyzer generates the excitation signal, which is amplified and sent to the shaker. Set up the vibration analysis software in such a way that the FRF graph evolution can be followed in real time. Normally, after each orthopaedic hammer hit, the implant fixation increases, resulting in increasing stiffness of the femur–implant structure. As a consequence, the resonance frequencies increase as well; therefore, the FRF graph shifts to the right. The displacement of the FRF peaks is more visible in the range of higher frequencies.

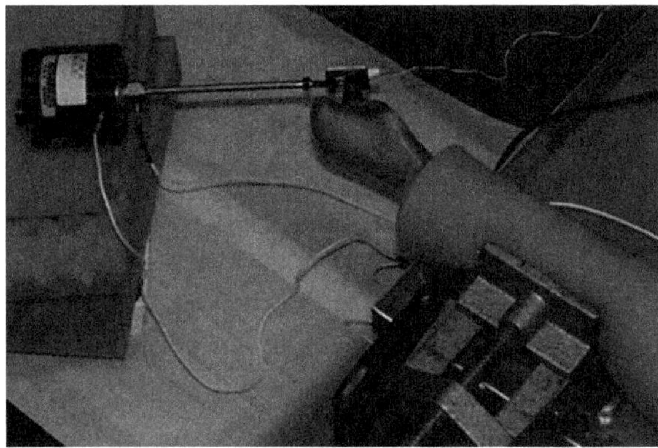

FIGURE 20.4

Experimental setup for shaker excitation. Alternative supporting method on foam.

Step 6. Stop the insertion. Stop the insertion process when the FRF peak positions do not change noticeably anymore (Fig. 20.5). At this moment, the implant fixation is stable and the insertion process has reached the end point. Hammering should also stop when an abnormal evolution of the FRF graph is observed, such as the FRF graph shifts to the left or the FRF graph drops in amplitude. The stable position indicated by vibration analysis could be validated by measuring the implant migration and the relative femur–implant micromotion under physiological load (e.g., micromotion below 150 μm and migration below 1 mm correspond to a stable implant).[6,7]

3.5 RAW DATA COLLECTION

Step 1. Record femur related data. Note the femur type (i.e., frozen, embalmed, dried), quality (i.e., normal, osteopenic, osteoporotic), and its physical

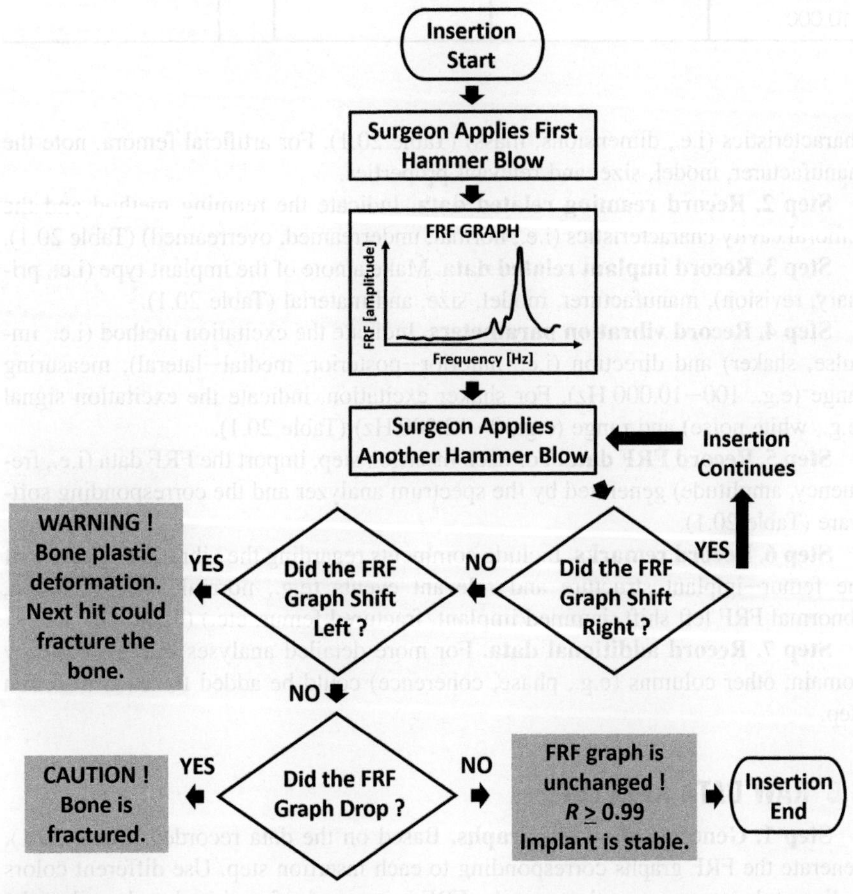

FIGURE 20.5

Implant insertion decision chart based on FRF analysis.

Table 20.1 FRF data summary table.

Experiment ID:				
Femur:				
Reaming:				
Implant:				
Vibration parameters:				
Remarks:				
Frequency [Hz]	**Insertion Step 0 [FRF Amplitude]**	**Insertion Step 1 [FRF Amplitude]**	...	**Insertion Step *n* [FRF Amplitude]**
100				
...				
10,000				

characteristics (i.e., dimensions, mass) (Table 20.1). For artificial femora, note the manufacturer, model, size, and relevant properties.

Step 2. Record reaming related data. Indicate the reaming method and the femoral cavity characteristics (i.e., normal, underreamed, overreamed) (Table 20.1).

Step 3. Record implant related data. Make a note of the implant type (i.e., primary, revision), manufacturer, model, size, and material (Table 20.1).

Step 4. Record vibration parameters. Indicate the excitation method (i.e., impulse, shaker) and direction (i.e., anterior—posterior, medial—lateral), measuring range (e.g., 100—10,000 Hz). For shaker excitation, indicate the excitation signal (e.g., white noise) and range (e.g., 0—12,500 Hz) (Table 20.1).

Step 5. Record FRF data. For each insertion step, import the FRF data (i.e., frequency, amplitude) generated by the spectrum analyzer and the corresponding software (Table 20.1).

Step 6. Record remarks. Include comments regarding the vibration behavior of the femur—implant structure and relevant events (e.g., normal FRF evolution, abnormal FRF left shift, jammed implant, fractured femur, etc.) (Table 20.1).

Step 7. Record additional data. For more detailed analyses in the frequency domain, other columns (e.g., phase, coherence) could be added for each insertion step.

3.6 RAW DATA ANALYSIS

Step 1. Generate the FRF graphs. Based on the data recorded (Table 20.1), generate the FRF graphs corresponding to each insertion step. Use different colors or line styles to compare them easily. FRF is a transfer function that describes the input—output relationship between two points on a mechanical system as a function of frequency. FRF is a measure of the dynamic properties of a structure by indicating

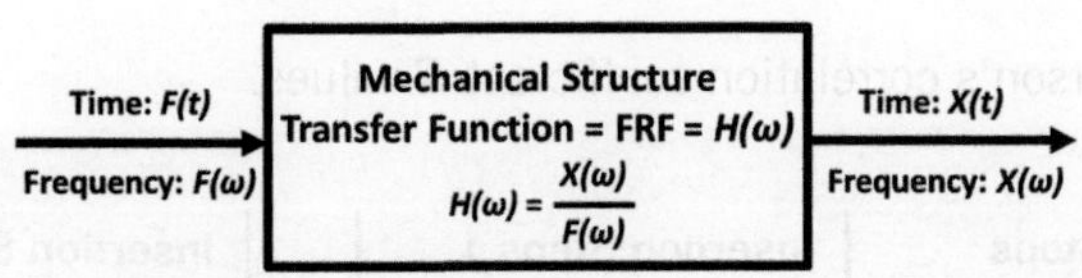

FIGURE 20.6

Block diagram of frequency response function (FRF).

the displacement, velocity, or acceleration response at an output point per unit of excitation force at an input point. The FRF is also defined as the ratio of the Fourier transform of an output response $X(\omega)$ divided by the Fourier transform of the input force $F(\omega)$ that caused the output (Fig. 20.6).[3]

Step 2. Superpose the FRF graphs. Superpose the FRF graphs corresponding to subsequent insertion step pairs to visualize the FRF graph evolution (Fig. 20.7).

Step 3. Calculate the Pearson's correlation coefficient. Pearson's correlation coefficient R is a dimensionless index that ranges from -1.0 to 1.0 inclusive and reflects the extent of a linear relationship between two data sets (i.e., two variables x_i and y_i, where $i = 0, 1, 2 \ldots N-1$). The equation for the correlation coefficient is:

$$R = \frac{\sum (x - \overline{x})(y - \overline{y})}{\sqrt{\sum (x - \overline{x})^2 \sum (y - \overline{y})^2}} \tag{20.1}$$

where x and y are the two variables, and N is the number of (x_i, y_i) pairs.[8] Calculate R between each FRF pair to determine the degree of similarity. R should be calculated

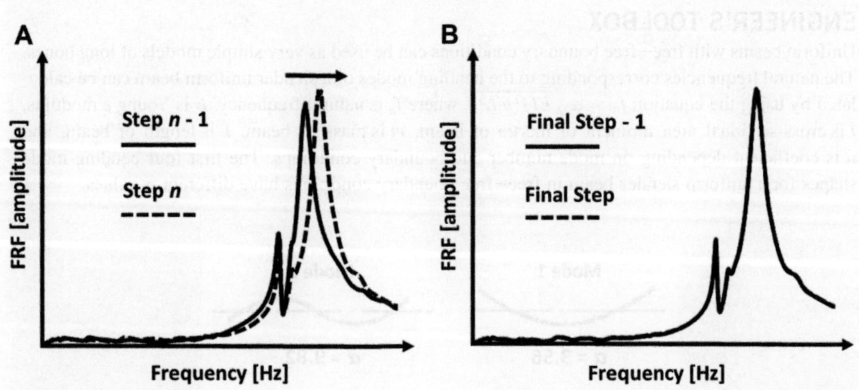

FIGURE 20.7

Typical FRF graph pairs obtained during an insertion process. (A) FRF graphs of two subsequent intermediary insertion steps, where the graph shifts to the right, indicating a stiffness increase that corresponds to a better fixation, (B) FRF graphs of the two final insertion steps, where the graphs substantially overlap each other without any left or right shift, thus indicating a stable implant and the endpoint of insertion.

Table 20.2 Pearson's correlation coefficient R values.

Experiment ID:				
	Insertion Steps 0 and 1	Insertion Steps 1 and 2	...	Insertion Steps (n-1) and n
R				

over the whole measuring range (e.g., 100–10,000 Hz). Since there is no linear dependence of one FRF with respect to the other, two FRF graphs are identical if $R = 1$. Collect R values as indicated (Table 20.2).

Step 4. Perform statistical comparisons. Determine the criterion for statistical difference (e.g., P <0.01 or <0.05) to compare the various test groups assessed. Choose from among the various software programs for comparing two test groups (e.g., unpaired t-test) or two or more test groups influenced by multiple factors (e.g., analysis of variance, ANOVA). Parametric or nonparametric statistical tests are chosen, depending on whether or not results have a normal distribution.

Step 5. Compute statistical power. Power analysis can be done after the study to ensure there were enough specimens per test group to detect all statistical differences that were actually present (i.e., was type II statistical error avoided?). Statistical power >80% is usually considered to indicate there were enough specimens per test group. Note that if good predictions of averages and standard deviations are available from prior studies, then the number of specimens and/or tests can be chosen before the study begins to ensure a power >80%.

ENGINEER'S TOOLBOX

Uniform beams with free–free boundary conditions can be used as very simple models of long bones. The natural frequencies corresponding to the bending modes of a slender uniform beam can be calculated by using the equation $f_n = \alpha\sqrt{EI/(mL^3)}$, where f_n is natural frequency, E is Young's modulus, I is cross-sectional area moment of inertia of beam, m is mass of beam, L is length of beam, and α is coefficient depending on mode number and boundary conditions. The first four bending mode shapes for a uniform slender beam in free–free boundary conditions have different α values.

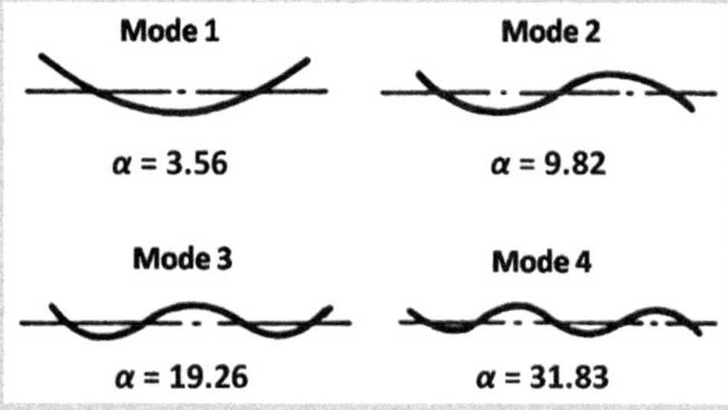

4. RESULTS

Once all vibration method data collection and analysis have been performed, it is then important to communicate and present the primary results in an understandable and concise manner to the reader of a journal article, conference paper, technical report, or book chapter.

Step 1. Show the main FRF graphs. Show the qualitative interpretation according to the decisional chart (Fig. 20.5) and the corresponding FRF graphs in pairs (Fig. 20.7).

Step 2. Show correlation coefficient data. The detection of the optimal implant stability, normally corresponding to the end point of insertion, should not be based only on visual analysis of the final FRF graphs. The degree of similarity should be based on the Pearson's correlation coefficient *R* (Fig. 20.8).

Step 3. Show unusual FRF graphs. Unusual FRF graph evolutions corresponding to singular situations encountered during the insertion process should also be presented and analyzed (Fig. 20.9). For example, FRF graphs slightly shifted to the left indicate a potentially jammed implant or a risk for bone cracking, while abnormal FRF graph changes occur after a femur fracture.

Step 4. Show cemented implant FRF graphs. The same vibration method used for uncemented implants (Figs. 20.5–20.9) can also be applied to assess the fixation of cemented implants (Fig. 20.10A) or to follow the cement curing process (Fig. 20.10B).

Step 5. Show comparisons between test groups. If relevant, present a bar graph that statistically compares any parameter(s) measured while achieving maximum implant stability (e.g., final Pearson correlation coefficient *R*, final most sensitive

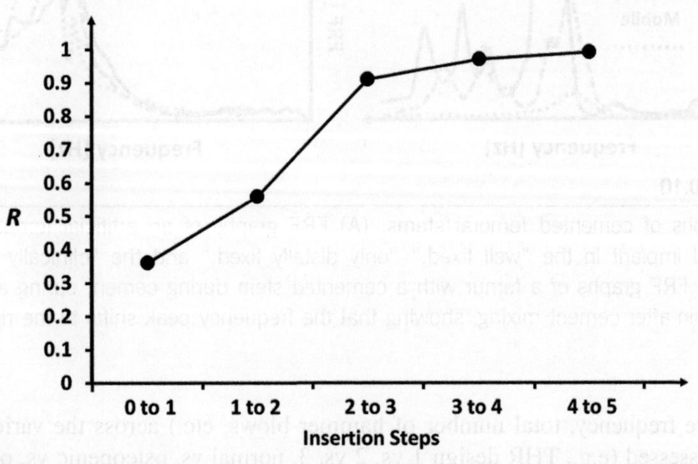

FIGURE 20.8

Typical Pearson's correlation coefficient *R* graph. At the end of a normal insertion process, the value of the coefficient *R* should be at least 0.99.

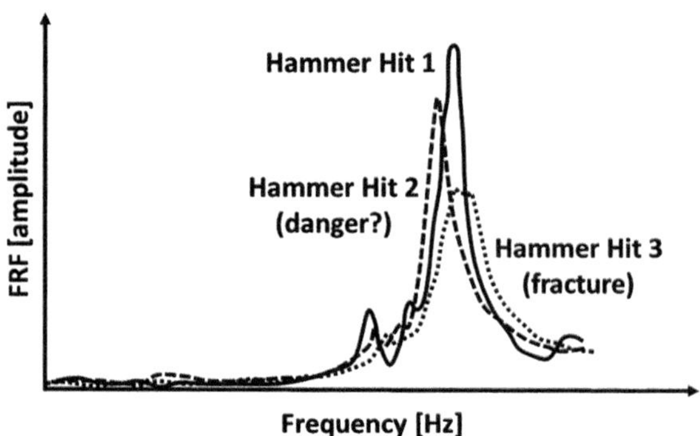

FIGURE 20.9

Abnormal FRF graph behavior. For hammer hit 1, the FRF graph appears normal. For hammer hit 2, the FRF graph shifts to the left, indicating a potentially dangerous situation in which the structural stiffness decreases, possibly due to bone plastic deformation. For hammer hit 3, the FRF graph substantially drops in amplitude, indicating femur fracture.

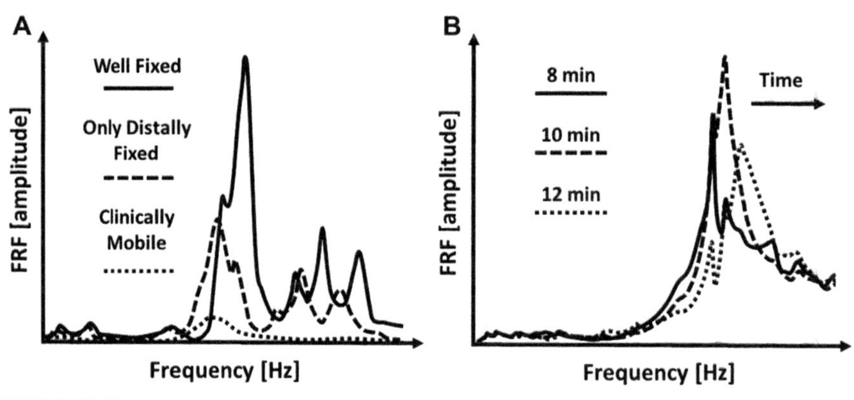

FIGURE 20.10

FRF graphs of cemented femoral stems. (A) FRF graphs of an artificial femur with a cemented implant in the "well fixed," "only distally fixed," and the "clinically mobile" state, (B) FRF graphs of a femur with a cemented stem during cement curing at 8, 10, and 12 min after cement mixing, showing that the frequency peak shifts to the right over time.

resonance frequency, total number of hammer blows, etc.) across the various test groups assessed (e.g., THR design 1 vs. 2 vs. 3, normal vs. osteopenic vs. osteoporotic bone, underreamed vs. ideally-reamed vs. overreamed, etc.) (Fig. 20.11).

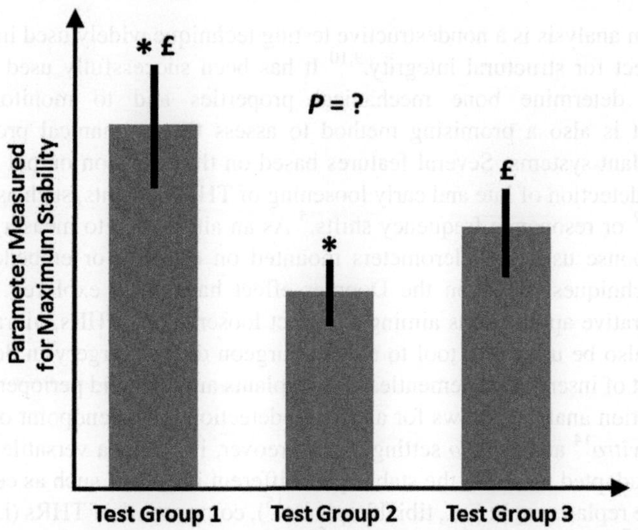

FIGURE 20.11

Bar graph for parameters measured at maximum implant stability. Each bar represents an average ± 1 standard deviation. Multiple P values indicate the various statistical differences obtained (e.g., $P<0.05$) for each pairwise comparison between test groups, which can be indicated by symbols like asterisks (*), pounds (£), etc.

ALTERNATIVES AND ADAPTATIONS

- ✓ **Clinical applications.** The end point of insertion is accurately detected, avoiding excessive press fit and preventing intraoperative fractures. The best initial stability of the femoral implant is achieved, and the micromotions are reduced to a minimum, thus preventing loosening. Consequently, vibration analysis can be successfully applied *in vivo* on cementless and cemented THR femoral implants.
- ✓ **Various implants.** Vibration analysis is a promising and versatile technique that can be adapted to assess the stability of different implants, such as components of total knee replacements (i.e., tibial implants), components of THRs (i.e., acetabular cups), glenoidal components used in reverse shoulder arthroplasty, dental and cranio−facial implants, etc.
- ✓ **Learning tool.** The vibration analysis method can be used as an *in vitro* training tool for junior orthopaedic surgeons to learn THR surgical procedures.

5. DISCUSSION

After completing all vibration analysis tests, data collection, data analysis, and data presentation, then final results can be considered and interpreted in the broader context of some important clinical, biomechanical, and/or technological considerations, as follows.

Vibration analysis is a nondestructive testing technique widely used in engineering to inspect for structural integrity.[9,10] It has been successfully used in biomechanics to determine bone mechanical properties and to monitor fracture healing.[11] It is also a promising method to assess the mechanical properties of femur–implant systems. Several features based on the vibration output have been used in the detection of late and early loosening of THR implants, such as harmonic distortions[12] or resonance frequency shifts.[5] As an alternative to measuring the vibration response using accelerometers mounted on the skin or embedded in the implant, techniques based on the Doppler effect have been explored.[13] Besides the postoperative applications aiming to detect loosening of THRs, vibration analysis could also be used as a tool to aid the surgeon during surgery in determining the endpoint of insertion of cementless hip implants and to avoid perioperative fractures. Vibration analysis allows for a reliable detection of the endpoint of insertion both for *in vitro*[14] and *in vivo* settings.[4,5] Moreover, it is also a versatile technique that can be adapted to assess the stability of different implants, such as components of total knee replacements (i.e., tibial implants[15]), components of THRs (i.e., acetabular cups[16,17]), glenoidal components used in reverse shoulder arthroplasty,[6] dental and cranio–facial implants,[18] etc.

Vibration analysis has several pros and cons. Being noninvasive and nondestructive, the assessment of femoral implant stability by vibration analysis could be used intraoperatively to assist the surgeon in taking optimal decisions. Implant stability is tested by using normal orthopaedic hammer blows that generally result in increasing the degree of fixation. Since the load induced by an average hammer blow has at least a magnitude of 7000 N[19] and the physiological load has a magnitude lower than 3000 N,[20] it is logical to assume that an implant that remains stable after hammer blows will also be stable to physiological load as well. Although it is able to discriminate quite accurately between well-fixated implant and quasi-well-fixated implant situations, this method can only qualitatively assess the stability for a specific bone–implant structure; therefore, it is difficult to compare the stabilities of different bone–implant structures. This limitation could be overcome by running calibration tests to correlate the magnitude of implant micromotions to FRF shifts or to resonance frequency values.

6. SUMMARY

- Initial implant stability achieved during surgery is crucial for the success of THR.
- It is important to accurately detect the end point of implant insertion to avoid fractures.
- Vibration analysis can assess the stability of femoral implants during insertion.
- FRF evolution indicates the implant stability change between subsequent insertion steps.
- If the FRF graph does not change significantly, then the end point of insertion is reached.
- Pearson's correlation coefficient is used to check the similarity of two FRF graphs.

7. QUIZ QUESTIONS

1. Why is it important to accurately detect the end point of implant insertion during surgery?
2. What is the normal evolution of the FRF graph during implant insertion?
3. How can the similarity of two FRF graphs be quantitatively determined?
4. What are three vibration methods that could be applied in orthopaedics?
5. A cylindrical straight rod made of polymer reinforced with glass fibers (Young's modulus $E = 15$ GPa, material density $\rho = 1.6$ g/cm^3, outer diameter $D = 25$ mm, inner diameter $d = 15$ mm, and length $L = 0.45$ m) is used as a rough model of a femur. Calculate the resonance frequencies f_n corresponding to the first three bending modes in free–free conditions (answer: 392.3, 1082.3, 2122.6 Hz).

REFERENCES

1. Viceconti M, Brusi G, Pancanti A, Cristofolini L. Primary stability of an anatomical cementless hip stem: a statistical analysis. *Journal of Biomechanics* 2006;**39**(7): 1169–79.
2. Davidson D, Pike J, Garbuz D, Duncan C, Masri B. Intraoperative periprosthetic fractures during total hip arthroplasty. *Journal of Bone and Joint Surgery (American)* 2008;**90**(9):2000–12.
3. Thorby D. *Structural dynamics and vibration in practice*. Oxford (UK): Butterworth Heinemann,; 2008.
4. Pastrav LC, Jaecques SVN, Jonkers I, Van der Perre G, Mulier M. In vivo evaluation of a vibration analysis technique for the per-operative monitoring of the fixation of hip prostheses. *Journal of Orthopaedic Surgery and Research* 2009;**4**, article ID #10.
5. Pastrav LC. *Monitoring of the fixation of orthopaedic implants by vibration analysis* [Ph.D. thesis]. KU Leuven, Belgium: Arenberg Doctoral School; 2010.
6. Pilliar RM, Lee JM, Maniatopoulos C. Observations on the effect of movement on bone ingrowth into porous-surfaced implants. *Clinical Orthopaedics and Related Research* 1986;**208**:108–13.
7. Kärrholm J, Borssen B, Lowenhielm G, Snorrason F. Does early micromotion of femoral stem prostheses matter? 4-7-year stereoradiographic follow-up of 84 cemented prostheses. *Journal of Bone and Joint Surgery (British)* 1994;**76**(6):912–7.
8. Spiegel MR. *Theory and problems of statistics*. New York, NY, USA: McGraw-Hill Publishing Co. Ltd; 1972.
9. Broch JT. *Mechanical vibration and shock measurements*. Naerum, Denmark: Bruel & Kjaer; 1984.
10. Doebling SW, Farrar CR, Prime MB, Shevitz DW. *Damage identification and health monitoring of structural and mechanical systems from changes in their vibration characteristics: a literature review*. 1996. Los Alamos National Laboratory report LA-13070-MS.
11. Van der Perre G, Lowet G. In vivo assessment of bone mechanical properties by vibration and ultrasonic wave propagation analysis. *Bone* 1996;**18**:29S–35S.
12. Georgiou AP, Cunningham JL. Accurate diagnosis of hip prosthesis loosening using a vibrational technique. *Clinical Biomechanics* 2001;**16**(4):315–23.

13. Rowlands A, Duck FA, Cunningham JL. Bone vibration measurement using ultrasound: Application to detection of hip prosthesis loosening. *Medical Engineering and Physics* 2008;**30**(3):278–84.
14. Cristofolini L, Varini E, Pelgreffi I, Cappello A, Toni A. Device to measure intra-operatively the primary stability of cementless hip stems. *Medical Engineering and Physics* 2006;**28**(5):475–82.
15. Denis K, Pastrav L, Leuridan S, Goethals P, Bellemans J, Desmet W, Vander Sloten J. Assessing the stability of tibial implants: an experimental and numerical structural health monitoring approach. *The 3rd International Symposium on Engineering, Energy, and Environment,* Bangkok, Thailand. November 17–20, 2013. Paper #71.
16. Pastrav C, Leuridan S, Denis K, Delport H, Debeer P, De Wilde L, et al. Vibrational techniques to assess the stability of spherical press-fitted implants: preliminary results. *ISMA 2010*, Leuven, Belgium. September 20–22, 2010. Paper #597.
17. Leuridan S, Pastrav L, Denis K, Mulier M, Desmet W, Vander Sloten J. Assessing the initial stability of the acetabular implant: development of an intraoperative tool. *Journal of Biomechanics* 2012;**45**(Suppl. 1):S360.
18. Meredith N, Book K, Friberg B, Jemt T, Sennerby L. Resonance frequency measurements of implant stability in vivo. A cross-sectional and longitudinal study of resonance frequency measurements on implants in the edentulous and partially dentate maxilla. *Clinical Oral Implants Research* 1997;**8**(3):226–33.
19. Crisman A, McCuskey M, Yoder N, Meneghini RM, Cornwell PJ. Femoral component insertion monitoring using human cadaveric specimens. *Proceedings of the 25th IMAC conference on structural dynamics,* Orlando *(FL, USA)*. 2007. Paper #109.
20. Bergmann G, Deuretzbacher G, Heller M, Graichen F, Rohlmann A, Strauss J, et al. Hip contact forces and gait patterns from routine activities. *Journal of Biomechanics* 2001; **34**(7):859–71.

3 Soft Tissues

CHAPTER

In Situ and Ex Vivo Biomechanical Testing of Articular Cartilage

21

Joanna F. Weber, Stephen D. Waldman

Ryerson University, Toronto, ON, Canada

1. BACKGROUND

Articular cartilage is a viscoelastic tissue, which provides a low friction surface for articulating joints (i.e., shoulder, hip, knee, etc.) and allows load transmission to underlying bones (Fig. 21.1).[1] Its strength and viscoelastic properties are determined by the density and structural organization of its extracellular matrix constituents. As a result, cartilage mechanical properties are reflective of tissue health.[2] It is important to map spatial variation in mechanical properties over the joint articular surface for various applications, such as osteochondral transfer (i.e., mosaic arthroplasty), computer-augmented surgical procedures, and the development of computational biomechanical models.[3,4] *In situ* articular cartilage tests may also be useful for determining the state of osteoarthritic degeneration and for assessing the quality of tissue

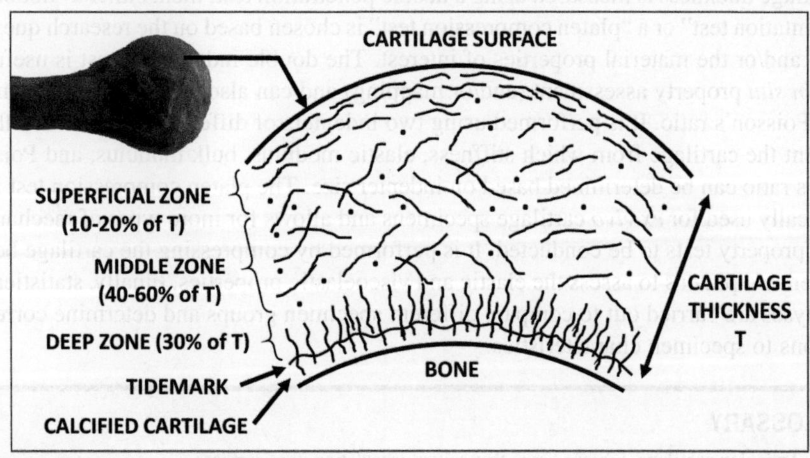

FIGURE 21.1

A human humeral head covered with articular cartilage, which has a structural organization of collagen fibers immersed in proteoglycan gel.

repair surgery.[5] Therefore, this chapter explains two common methods for articular cartilage mechanical testing (i.e., *in situ* and *ex vivo*), as well as describing how to collect, analyze, present, and interpret the results.

2. RESEARCH QUESTIONS

Typical research questions might include one or more of the following:

- What are the typical mechanical properties of macroscopically healthy articular cartilage?
- Do the mechanical properties differ between macroscopically healthy vs. damaged cartilage?
- Do the mechanical properties differ over the entire surface of the articulating joint?
- How does the biochemical composition of cartilage influence its mechanical properties?
- etc.

3. METHODOLOGY

3.1 GENERAL STRATEGY

Human or animal articulating joints are used fresh, ideally less than 24 h after collection and stored at 4°C prior to testing. Joints are then grouped based on the exact research question (e.g., age, disease state, treatment group, etc.). Articular cartilage thickness is measured using a needle penetration test; then, either a "double indentation test" or a "platen compression test" is chosen based on the research question and/or the material properties of interest. The double indentation test is useful for *in situ* property assessment (and/or mapping) and can also be used to determine the Poisson's ratio. It is performed using two indenters of differing radii that briefly indent the cartilage from which stiffness, elastic modulus, bulk modulus, and Poisson's ratio can be determined based on indenter size. The platen compression test is typically used for *ex vivo* cartilage specimens and allows for more types of mechanical property tests to be conducted. It is performed by compressing the cartilage between two platens to assess the elastic and viscoelastic properties. Finally, statistical analyses are carried out to compare different specimen groups and determine correlations to specimen characteristics.

GLOSSARY

- ✓ **Articular cartilage**. Fibrous tissue that covers articulating joint surfaces.
- ✓ **Biphasic**. A tissue that is made from two distinct phases: solid and liquid.
- ✓ **Extracellular matrix**. Noncellular components of cartilage (i.e., collagens, proteoglycans).
- ✓ **Osteochondral**. A combined sample of articular cartilage and its underlying bone.
- ✓ **Strain**. Applied deformation that is normalized by initial length.
- ✓ **Stress**. Applied load that is normalized by cross-sectional area.

SAFETY FIRST

✓ Wear adequate respiratory and eye protection when drilling core samples.
✓ Always follow recommended safety protocols for the mechanical tester used.
✓ Clean the work area and all tools with ethanol and/or disinfectant after testing.
✓ Be sure to discard sharps and used specimens appropriately.

3.2 MATERIALS AND TOOLS LIST

In Situ (i.e., Double Indentation) Tests

- animal or human articular joints from tissue banks (fresh or up to 1 day old at 4°C)
- fine-gage needle
- fine-toothed saw
- isotonic saline solution
- mechanical tester
- specimen holder with adjustable stage
- two indenters of different sizes

Ex Vivo (i.e., Platen Compression) Tests

- animal or human articular joints from tissue banks (fresh or up to 1 day old at 4°C)
- biopsy punch
- calipers
- diamond-tipped coring bit
- fine gage needle
- isotonic saline solution
- mechanical tester
- specimen holder with adjustable stage

3.3 SPECIMEN PREPARATION

Whole Joint

Step 1. Remove the limb. Detach the limb from the animal, and expose the articular surface of the joint by dissecting through skin and muscle (Fig. 21.2A). When opening the synovial joint capsule, be extremely careful to avoid damaging the articular surface by cutting away from the joint surface. Once exposed, ensure cartilage is constantly kept hydrated with an isotonic saline solution.

Step 2. Remove excess bone. Cut off any excess bone such that the joint fits into the specimen holder (Fig. 21.2B). At this time, remove excess soft tissue, which would sit under the joint when in the specimen holder, as this may cause the specimen to sit loosely and affect testing accuracy.

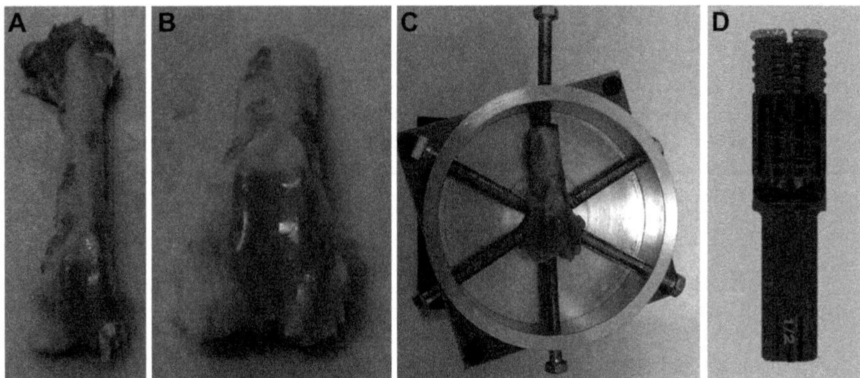

FIGURE 21.2

Specimen preparation for mechanical tests. (A) Rabbit femur, (B) distal end of rabbit femur (knee joint), (C) holder used for larger limb segments or smaller core specimens, (D) diamond-tipped coring bit for cutting osteochondral cores.

Step 3. Mount the specimen. Mount the specimen in the holder by tightening the screws and allowing them to secure into the surrounding bone of the specimen (Fig. 21.2C). If properly mounted, the specimen will not move or rotate within the holder. Then, mount the specimen holder into the mechanical tester. For small animal joints, identify the largest and flattest region for the indentations. Using the adjustable base, rotate the specimen such that the region of interest is normal (i.e., perpendicular) to the testing axis. For large animal joints where surface location is of interest, anatomical reference points should be used to define the testing location.

Osteochondral Core

Step 1. Remove the limb. Detach the limb from the animal, and expose the articular surface of the joint by dissecting through skin and muscle (Fig. 21.2A). When opening the synovial joint capsule, be extremely careful to avoid damaging the articular surface by cutting away from the joint surface. Once exposed, ensure cartilage is constantly kept hydrated with an isotonic saline solution.

Step 2. Remove a core specimen. Using a biopsy punch, prepunch the area of cartilage to be cored. Next, use a drill press and the same diameter coring bit (Fig. 21.2D) to drill into the joint by following the scoring line from the punch. When drilling, keep a constant stream of isotonic saline solution on the site to reduce friction-generated heat, which would compromise the specimen. Ensure that the core is taken perpendicular to the joint surface. Prior to compression testing, measure the diameter of the core using calipers.

Step 3. Mount the specimen. Mount the core into the specimen holder by tightening the screws and allowing them to secure into the surrounding bone of the specimen (Fig. 21.2C). Make sure the core specimen is completely vertical while the

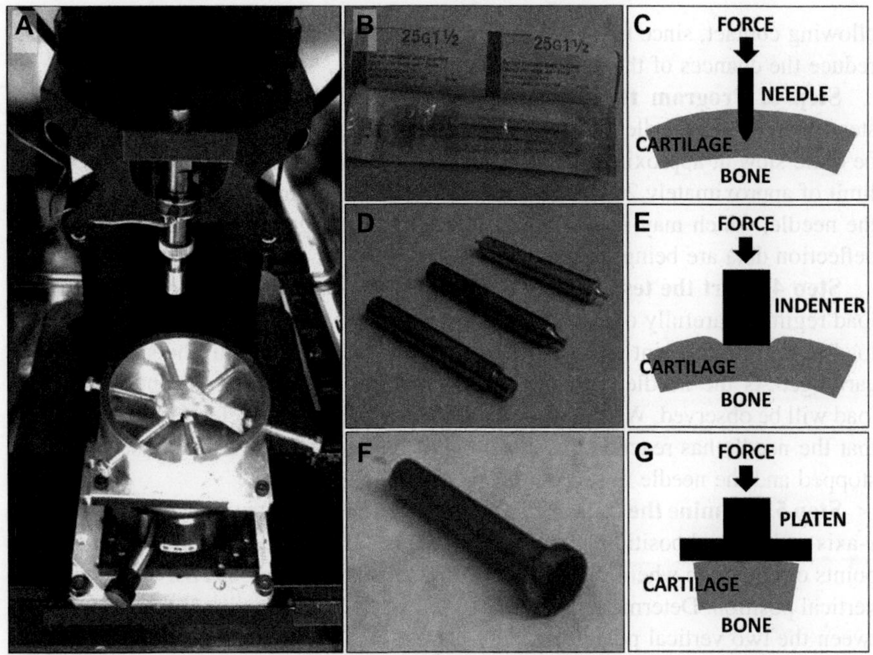

FIGURE 21.3

Typical hardware used for testing. (A) Mechanical tester, (B) steel fine-gage needle, (C) needle penetration test, (D) steel flat-end indenters, (E) double indentation test, (F) steel flat-end platen with waterproof sandpaper covering its end, (G) platen compression test.

cartilage surface of the specimen is horizontal. Mount the specimen holder into the mechanical tester.

3.4 SPECIMEN TESTING

Needle Penetration Tests

Step 1. Mount the needle. Once the mechanical tester has been prepared (Fig. 21.3A), mount a short, steel, small gage needle (e.g., 3/4 inch 25G) (Fig. 21.3B and C). Ensure the needle is secure and aligned with the vertical test axis. A crooked needle (bent or improperly aligned) will yield inaccurate results. Position the needle above a region adjacent to the indentation site since needle penetration damages the cartilage. This will preserve the actual test site and will yield representative thicknesses. Ensure the cartilage is hydrated with isotonic saline.

Step 2. Lower the needle. In manual control mode for the mechanical tester, slowly lower the actuator (which is holding the needle) toward the cartilage surface. Position the tip of the needle as close as possible to the cartilage surface without allowing contact, since this will shorten the amount of time needed for the test and reduce the chances of the specimen drying out.

Step 3. Program the tester. Program the mechanical tester to slowly and steadily lower the needle toward and into the cartilage. The rate of movement should be quite slow at approximately 0.005–0.010 mm/s. As a safety feature, add a load limit of approximately 20 g to halt the testing device to avoid bending or damaging the needle, which may inadvertently affect the cartilage. Ensure that the load and deflection data are being recorded and saved.

Step 4. Start the test. Commence the mechanical tester's computer-controlled load regime. Carefully observe the load cell readout. The beginning of the nonzero load indicates the point at which the needle makes contact with the surface of the cartilage. As the needle penetrates the cartilage, a slow but constant increase in load will be observed. When the increase in load suddenly accelerates, this indicates that the needle has reached the subchondral bone. At this time, the test should be stopped and the needle raised out of the cartilage.

Step 5. Examine the data. Plot load vs. position data with vertical load along the y-axis and vertical position along the x-axis (Fig. 21.4A). Identify the two inflection points of the graph where the rate of loading changes, and record the corresponding vertical position. Determine the cartilage thickness by computing the difference between the two vertical positions.

Double Indentation Tests

Step 1. Choose the indenters. Select two indenters of different sizes (Fig. 21.3D and E). The large-diameter indenter should be selected based on the thickness of the cartilage (as measured above) and the size of the largest flat area available on the joint. The small-diameter indenter should be the smallest diameter possible, based on what is commercially available and what can be mounted properly in the mechanical tester. The ratio of the large indenter diameter to cartilage thickness should ideally be less than 4. Typical indenter diameters are 2–6 mm (large indenter) and 1–2 mm (small indenter).[6–8] This makes computations later with the Hayes method more reliable.[9]

Step 2. Position the indenter. Install the large indenter into the mechanical tester. Carefully align the indenter with the cartilage surface to be tested. Using manual control, slowly bring the actuator (holding the large indenter) down toward the specimen, such that the indenter is only 1 mm or so away from the cartilage surface. Ensure that the cartilage is kept hydrated with isotonic saline (or submerged) before conducting this test.

Step 3. Program the mechanical tester. Program the mechanical tester to slowly move the indenter toward the specimen to find contact with the cartilage surface. Apply a small preload to the cartilage to ensure contact, such that the preload is proportional to the surface area of the indenter (i.e., the larger diameter indenter will have a higher preload than the smaller diameter indenter). The recommended ratio of preload-to-indenter surface area is about 0.001–0.003 N/mm^2; however, the lower limit for preload will be limited by the noise and load resolution of the mechanical

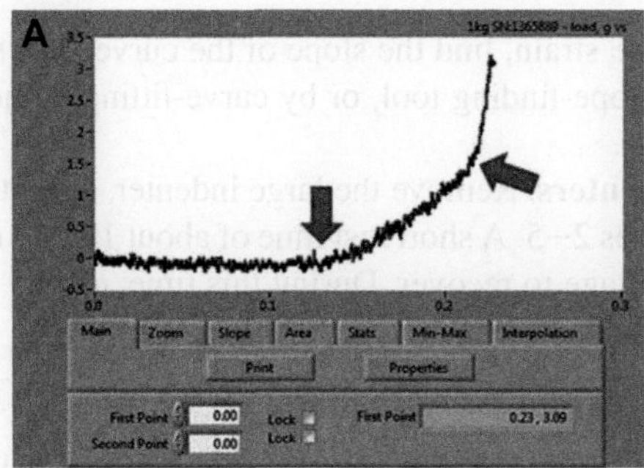

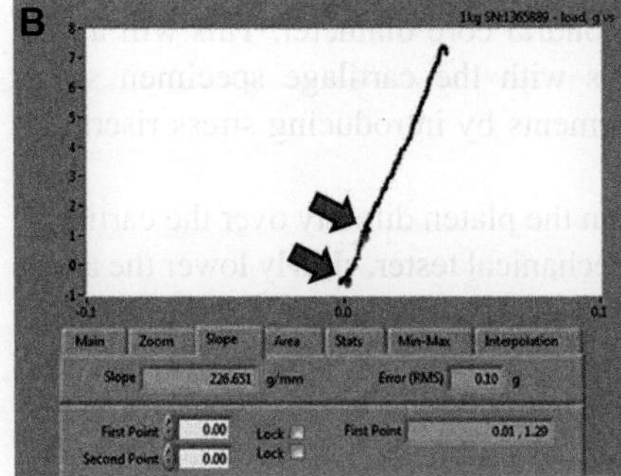

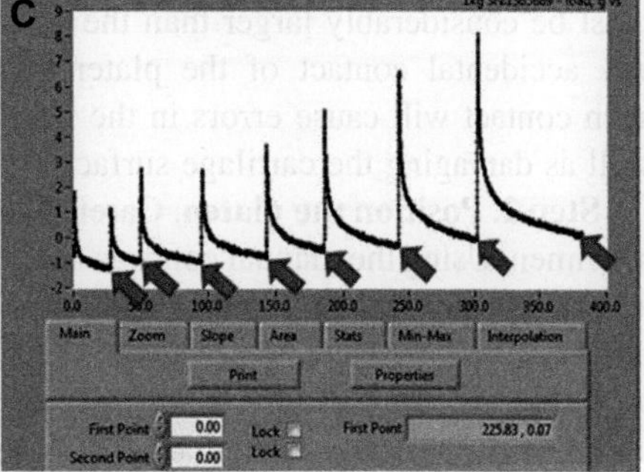

FIGURE 21.4

Raw data profiles during mechanical tests. (A) Needle penetration test readout showing vertical load vs. position data for calculating cartilage thickness using inflection points, (B) double indentation test readout showing vertical load vs. vertical deflection data for measuring slope (i.e., elastic modulus), (C) platen compression test readout showing vertical load vs. time data for measuring equilibrium loads.

Software images courtesy of Biomomentum Inc., www.biomomentum.com.

tester. Add a ramp function to apply a physiologic strain (5–10%) at a rate of about 10% strain/s.[10] Faster or slower strain rates may be required, depending on the capture rate of the load cell and the speed and accuracy of the position actuator. Immediately remove the indenter from the specimen, either manually or by programming the motion control system of the mechanical tester. Ensure that load and deflection data will be recorded and saved.

Step 4. Start the test. Commence the mechanical tester's computer-controlled load regime. Carefully observe the load cell readout. A small increase in load will be observed as the preload is applied when the indenter is making contact with the cartilage surface. A larger increase in load will be observed as the linear ramp strain is applied; then, a decrease in load back to zero will be observed as the indenter is removed from the specimen.

Step 5. Examine the data. Plot the load vs. deflection data with the vertical load along the *y*-axis and the vertical deflection along the *x*-axis (Fig. 21.4B). Between the deflections of 1–5% strain, find the slope of the curve. The slope may be found either with a built-in slope-finding tool, or by curve-fitting a linear equation to that section of the graph.

Step 6. Switch indenters. Remove the large indenter, mount the small-diameter indenter, and repeat steps 2–5. A short rest time of about 10–20 min between indentation tests allows cartilage to recover. During this time, ensure that the cartilage is kept hydrated with isotonic saline.

Platen Compression Tests

Step 1. Install the compression platen. Mount the platen to be used for compression tests in the mechanical tester (Fig. 21.3F and G). The platen's diameter must be considerably larger than the osteochondral core diameter. This will avoid any accidental contact of the platen's edges with the cartilage specimen since such contact will cause errors in the measurements by introducing stress risers, as well as damaging the cartilage surface.

Step 2. Position the platen. Carefully align the platen directly over the cartilage specimen. Using the manual controls of the mechanical tester, slowly lower the actuator (holding the platen) toward the cartilage surface, such that the gap between the platen and the cartilage is only about 1 mm. Ensure that the cartilage is kept hydrated with isotonic saline before conducting this test.

Step 3. Program the mechanical tester. Program the mechanical tester to slowly move the platen toward the cartilage until there is contact. Apply a small preload of about 50 mN to the cartilage surface to ensure full contact across the surface. Add a series of step strains to a maximum strain of 20–25%.[11,12] The magnitude of each step strain should be chosen to maximize the total number of step strains applied, but also take into account the resolution of the position control of the mechanical tester and the time required to complete the test. Consequently, about five to eight step strains should be sufficient. The ramp rate of each step should ideally be as fast as possible; however, setting ramp strain rate (1/s) to three times the step strain should be sufficient (i.e., each step should be completed in approximately 1/3 s). Then, program the mechanical tester with a function that will immediately remove the platen from the specimen after maximum strain relaxation is completed. Ensure that load and deflection data will be recorded and saved.

Step 4. Start the test. Commence the mechanical tester's computer-controlled load regime. Carefully observe the load cell readout. A small increase in load will be observed as the preload is applied when the platen is making contact with the cartilage surface. A larger increase in load will be observed as the first step strain is applied. The load will decrease as the cartilage relaxes and will eventually reach equilibrium (i.e., no further changes in load). After equilibrium is reached for the first step strain, the next step strain will be applied causing the load to increase again, etc.

Step 5. Examine the data. Plot the data with vertical load along the y-axis and time along the x-axis. Identify and record each equilibrium load for each applied step strain (Fig. 21.4C). Averaging the last few equilibrium load values at each step strain will improve the accuracy of this method.

TIPS AND TRICKS

- ✓ For large intact joints, use a mapping system to evaluate differences between locations.
- ✓ Use a new needle for every two penetration tests to ensure the needle tip remains sharp.
- ✓ Choose indenters with the largest possible difference in radii for the most accurate results.
- ✓ Allow equilibration for a few minutes after installing hardware into the mechanical tester.

THE "GOLD STANDARD"

No known international standards exist for *in vitro* mechanical testing of articular cartilage. Researchers should use peer-reviewed journal articles for precedents and guidelines.

3.5 RAW DATA COLLECTION

Double Indentation Tests

Step 1. Record demographics. For each individual specimen, enter demographic data (i.e., age, sex, limb side, etc.), as well as quantitative and qualitative macroscopic score (Table 21.1). For large joints, record testing region or limb. From needle penetration tests, enter cartilage thickness.

Step 2. Record test results. From the double indentation tests, record indenter radii. Also, find the slope corresponding to the strains of 1–5%, but keep slope in g/mm or N/mm units (i.e., do not convert to stresses and strains).

Table 21.1 Raw data for articular cartilage double indentation tests.

	Test			
Parameter	**1**	**2**	**3**	**etc.**
Specimen number				
Age or time postop				
Sex				
Region or limb side				
Macro score				
Cartilage thickness h [mm]				
Radius of large indenter a_1 [mm]				
Radius of small indenter a_2 [mm]				
Slope of large indenter test S_1 [g/mm]				
Slope of small indenter test S_2 [g/mm]				
Remarks				

Step 3. Record extra remarks. Document any extra observations for later review about unusual cartilage behavior (e.g., damage, deformation, etc.) during testing.

Platen Compression Tests

Step 1. Record demographics. For each individual specimen, enter demographic data (i.e., age, sex, limb side, etc.), as well as quantitative and qualitative macroscopic score (Table 21.2). From earlier needle penetration tests, enter computed cartilage thickness. Record the location where osteochondral core was obtained, as well as the core diameter.

Step 2. Record test results. From the platen compression tests, at each step strain record the corresponding equilibrium load in g or N units.

Step 3. Record extra remarks. Any extra observations about unusual cartilage behavior (e.g., damage, deformation, etc.) during testing should be recorded as well.

3.6 RAW DATA ANALYSIS

Double Indentation Tests

Step 1. Compute ratios. Calculate the ratios of the large indenter radius a_1 and small indenter radius a_2 to cartilage thickness h, respectively, as a_1/h and a_2/h.

Step 2. Convert slopes. If necessary, convert large and small indenter slopes from g/mm to N/mm units by multiplying by 9.81×10^{-3} m/s^2. Thus, large and small slopes, respectively, will now be expressed as $S_1 = (F/d)_1$ and $S_2 = (F/d)_2$, where F is vertical indentation force and d is vertical indentation deflection. Also, calculate the ratio of the slopes as $(F/d)_1/(F/d)_2$.

Step 3. Interpolate values. Use Table 1 data from the classic paper by Hayes et al.[9] to interpolate values for the factor κ, which is a function of Poisson's ratio

Table 21.2 Raw data for articular cartilage platen compression tests.

	Test			
Parameter	**1**	**2**	**3**	**etc.**
Specimen number				
Age or time postop				
Sex				
Region or limb side				
Macro score				
Core location				
Core diameter [mm]				
Cartilage thickness h [mm]				
Equilibrium force at step strain 1 [g or N]				
Equilibrium force at step strain 2 [g or N]				
Equilibrium force at step strain 3 [g or N]				
etc.				
Remarks				

ν, for each a/h ratio. Plot κ vs. the corresponding ν, then curve fit a third-order polynomial equation for each a/h graph. Begin by estimating that $\nu = 0.5$, but change ν to satisfy the equation $(a_1/a_2) \times (\kappa_1/\kappa_2) = (F/d)_1/(F/d)_2$.

Step 4. Compute moduli. The mechanical properties of the cartilage can now be determined. Compute cartilage bulk modulus $B = [(F/d)_1(1 - \nu)]/(4a_1\kappa_1)$. Calculate cartilage elastic modulus $E = 2B(1 + \nu)$.

Platen Compression Tests

Step 1. Compute stresses. Calculate the applied equilibrium surface stress at each step strain during platen compression tests such that $\sigma = F/A$, where σ is applied surface stress [MPa], F is equilibrium force [N], and A is surface area of the osteochondral core [mm^2].

Step 2. Plot the data. Graph the equilibrium surface stresses calculated above along the y-axis vs. the step strains along the x-axis. Then, find the equation of the line of best fit through the data, which will likely be either a first- or second-order polynomial line.

Step 3. Find equilibrium modulus. Take the derivative of the equation of the line of best fit from above, which will provide the equilibrium modulus Q as a function of applied strain. Remember that from a first-order polynomial, a derivative will result in a single value, while from a second-order polynomial, a derivative will result in a first-order polynomial.

ENGINEER'S TOOLBOX

Force vs. deflection (and stress vs. strain) relationships of biological materials differ from elastic materials. The toe region shows a small increase in force (and stress) with a large increase in deflection (and strain). The linear region is where the time-dependent viscoelastic material behaves as an elastic material. The physiological range extends slightly beyond the linear region; thus, force (and stress) or deflection (and strain) past this point result in permanent cartilage damage. Double indentation tests should be used to obtain force vs. deflection curves in the linear region such that slope $S = \Delta F/\Delta d$, where F is vertical force and d is vertical deflection.

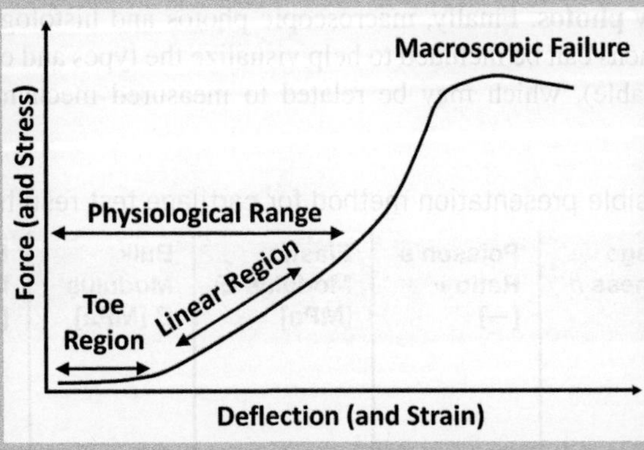

4. RESULTS

Once all articular cartilage data collection and analysis have been performed, it is then important to communicate and present the primary results in an understandable and concise manner to the reader of a journal article, conference paper, technical report, or book chapter.

Step 1. Give platen compression equations. Generally, equilibrium stress vs. step strain curves do not need to be shown. However, it may be beneficial to report the equation for the line of best fit if the relationship is not first-order. In this case, present the equation in the form $y = ax^2 + bx + c$, as well as the quality of the data fit via the coefficient of determination R^2. Within the physiological strain range, equilibrium stress should increase with step strain; however, past this point and when the tissue is mechanically damaged, this trend does not necessarily continue.

Step 2. Provide numerical results. Measured values for cartilage thickness, Poisson's ratio, elastic modulus, bulk modulus, and equilibrium modulus can be presented in tables (Table 21.3), scatter plots with data fitting (Fig. 21.5A), or bar charts (Fig. 21.5B). For each measured value presented, all appropriate statistical pairwise comparisons should be made, averages ± 1 standard deviation or 1 standard error of the mean should be shown, and statistical significance (e.g., $P < 0.05$) should be indicated on the tables, graphs, and legends. If appropriate, measurements can also be assessed for the effect of various factors (e.g., age, sex, limb side, etc.). Where relevant, remember to include line-of-best-fit equations in the form of $y = mx + b$ and associated R^2 values.

Step 3. Present statistical power. If appropriate, statistical power analysis results can be stated to indicate if there were enough specimens per test group to detect all statistical differences that were actually present (i.e., was type II statistical error avoided?). Statistical power >80% is usually considered to indicate there were enough specimens per test group. Note that if good predictions of averages and standard deviations were available from prior studies, then the number of specimens and/or tests could have been chosen before the study began to ensure a power >80%.

Step 4. Show photos. Finally, macroscopic photos and histological images of particular specimens can be included to help visualize the types and extent of degradation (if applicable), which may be related to measured mechanical properties (Fig. 21.6).

Table 21.3 Possible presentation method for cartilage test results.

Test Group	Cartilage Thickness h [mm]	Poisson's Ratio ν [–]	Elastic Modulus E [MPa]	Bulk Modulus B [MPa]	Equilibrium Modulus Q [MPa]
1					
2					
3					
etc.					

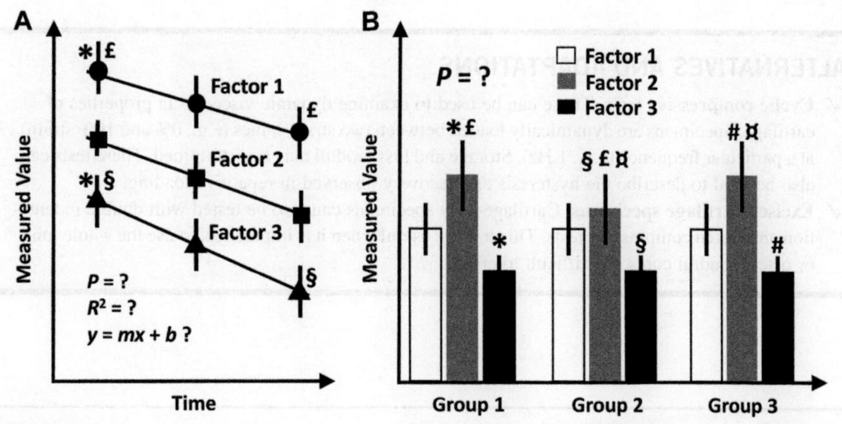

FIGURE 21.5

Measured values from mechanical tests on cartilage. (A) Scatter plot for time-related studies, (B) bar graph for highlighting differences between test groups. Error bars indicate $\pm$ 1 standard deviation or 1 standard error of the mean. *P* values indicate statistical differences (e.g., $P < 0.05$) between test groups and/or factors, which can be indicated by symbols like asterisks (*), pounds (£), etc. R^2 shows the quality of data fit for the lines of best fit represented by equations in the form $y = mx + b$.

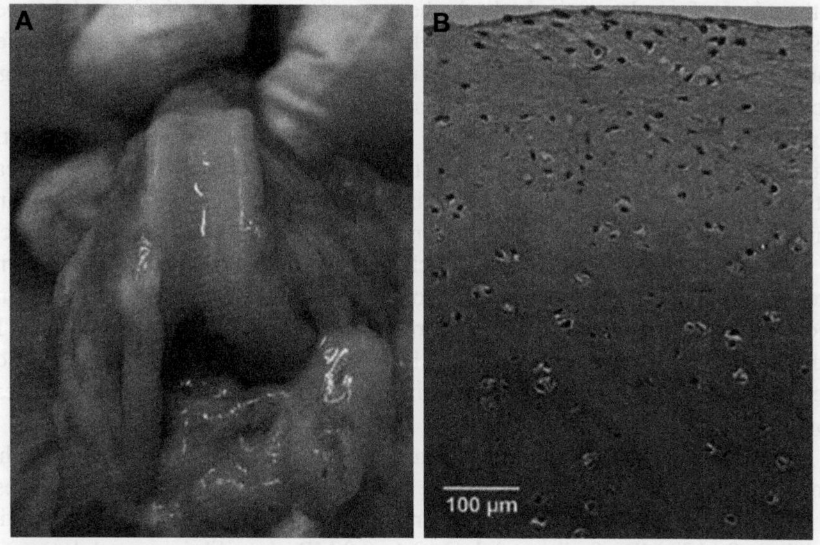

FIGURE 21.6

Images of representative cartilage specimens. (A) Macroscopic photos of the joint surface, (B) Safranin O staining of histological samples.

ALTERNATIVES AND ADAPTATIONS

- ✓ **Cyclic compressive tests.** These can be used to examine dynamic viscoelastic properties of cartilage. Specimens are dynamically loaded between two strain values (e.g., 0% and 10% strain) at a particular frequency (e.g., 1 Hz). Storage and loss moduli may be determined. These tests can also be used to describe the hysteresis and recovery observed in repetitive loading.
- ✓ **Excised cartilage specimens.** Cartilage-only specimens can also be tested with double indentation and platen compression tests. This may be useful when it is impractical to use the whole joint or osteochondral cores are difficult to create.

5. DISCUSSION

After completing all articular cartilage tests, data collection, data analysis, and data presentation, then final results can be considered and interpreted in the broader context of some important clinical, biomechanical, and/or technological considerations, as follows.

Cartilage thickness can be obtained using needle penetration tests. This direct method is particularly useful when the joint surface needs to be kept intact. While somewhat of a destructive method, its impact is minimized if testing locations are taken immediately adjacent to the double indentation and plate compression test sites. As expected, cartilage thickness can vary greatly, depending on the location on the joint surface, the particular limb, the age or time postoperatively, and the species. For example, human adult cartilage thickness can range from 1.5–4.5 mm (femur), 2–7 mm (tibia), and 1–2 mm (humerus),[13–15] while rabbit adult cartilage thickness can range from 0.5–5 mm (femur) and 0.45–0.84 mm (tibia).[13,16,17]

Mechanical tests on articular cartilage can be done in various ways due to cartilage's biphasic, anisotropic, and viscoelastic properties.[18] The choice of mechanical test will depend on several factors. The size and surface characteristics of the joint limits the indenter and platen size(s); therefore, indentation tests may be indicated for small joints from small animals. Mechanical properties of interest also inform potential testing options, since instantaneous properties are determined from indentation tests and viscoelastic properties are determined from compression tests. If both types of tests can be done, it may be beneficial to pilot test both on a few specimens to find out which test is more sensitive in detecting differences between test groups.

Mechanical properties can vary substantially, depending on the location on the joint surface, the particular limb, the age or time postoperatively, and the species. For instance, human adult cartilage properties can range from 40–160 MPa (elastic modulus), 9–170 MPa (bulk modulus), 0.4–0.9 MPa (equilibrium modulus), and 0–0.15 (Poisson's ratio).[13,19,20] Similarly, rabbit adult cartilage properties can range from 0.53—0.81 MPa (elastic modulus), 0.83–1.35 MPa (bulk modulus), 0.45–1.5 MPa (equilibrium modulus), and 0.1–0.4 (Poisson's ratio).[13,16,17] It is worth noting that because human cartilage is sometimes legally, ethically, and financially more difficult to obtain for research, animal models are often used as a surrogate to study cartilage properties.

Clinically, articular cartilage relies on matrix constituents and structural organization for strength and load-bearing capability.[1,18] Changes in these can cause a decrease in mechanical properties, leaving the cartilage susceptible to mechanical injury and degradation. Mechanical testing of cartilage, therefore, provides an indicator of tissue health. Cartilage thickness and mechanical properties vary from location to location within the same joint surface; therefore, testing of several locations for these large joints can be useful for both computational modeling[4] as well as surgical repair procedures (e.g., osteochondral transfer).

6. SUMMARY

- Cartilage thickness can be measured *in situ* using needle penetration tests.
- Elastic modulus, bulk modulus, and Poisson's ratio can be measured using indentation tests.
- Equilibrium modulus can be measured by the use of platen compression tests.
- Indentation and compression tests are usually nondestructive; thus, specimens can be reused.
- Large joints may benefit from mapping to examine variations across a joint surface.

7. QUIZ QUESTIONS

1. What is the difference between elastic, bulk, and equilibrium modulus?
2. How will cartilage thickness and mechanical properties vary across large joints?
3. How will the loss of matrix constituents affect cartilage mechanical properties?
4. Why should mechanical tests be performed within physiological strain ranges?
5. Calculate the slope of the linear region of a force vs. deflection curve obtained from a platen compression test on cartilage. Data points from far ends of the linear region yield forces of 1 and 5 mN with corresponding deflections 0.010 and 0.015 mm (answer: 0.8 N/mm).

REFERENCES

1. Mankin HJ, Grodzinsky AJ, Buckwalter JA. Articular cartilage and osteoarthritis. In: Einhorn TA, O'Keefe RJ, Buckwalter JA, editors. *Orthopaedic basic science: foundations of clinical practice*. 3rd ed. Rosemount (IL, USA): American Academy of Orthopaedic Surgeons; 2006. pp. 161–174.
2. Armstrong CG, Mow VC. Variations in the intrinsic mechanical properties of human articular cartilage with age, degeneration, and water content. *Journal of Bone and Joint Surgery (American)* 1982;**64**(1):88–94.

3. Changoor A, Hurtig MB, Runciman RJ, Quesnel AJ, Dickey JP, Lowerison M. Mapping of donor and recipient site properties for osteochondral graft reconstruction of subchondral cystic lesions in the equine stifle joint. *Equine Veterinary Journal* 2006;**38**(4):330–6.
4. Seifzadeh A, Oguamanam DC, Trutiak N, Hurtig M, Papini M. Determination of nonlinear fibre-reinforced biphasic poroviscoelastic constitutive parameters of articular cartilage using stress relaxation indentation testing and an optimizing finite element analysis. *Computer Methods and Programs in Biomedicine* 2012;**107**(2):315–26.
5. Obeid EM, Adams MA, Newman JH. Mechanical properties of articular cartilage in knees with unicompartmental osteoarthritis. *Journal of Bone and Joint Surgery (British)* 1994;**76**(2):315–9.
6. Brenner JM, Kunz M, Tse MY, Winterborn A, Bardana DD, Pang SC, et al. Development of large engineered cartilage constructs from a small population of cells. *Biotechnology Progress* 2013;**29**(1):213–21.
7. Rosa RG, Joazeiro PP, Bianco J, Kunz M, Weber JF, Waldman SD. Growth factor stimulation improves the structure and properties of scaffold-free engineered auricular cartilage constructs. *PLoS One* 2014;**9**(8):e105170. 12 pages.
8. Giardini-Rosa R, Joazeiro PP, Thomas K, Collavino K, Weber J, Waldman SD. Development of scaffold-free elastic cartilaginous constructs with structural similarities to auricular cartilage. *Tissue Engineering (Part A)* 2014;**20**(5–6):1012–26.
9. Hayes WCC, Keer LMM, Herrmann G, Mockros LFF. A mathematical analysis for indentation tests of articular cartilage. *Journal of Biomechanics* 1972;**5**(5):541–51.
10. Lötjönen P, Julkunen P, Tiitu V, Jurvelin JS, Töyräs J. Ultrasound speed varies in articular cartilage under indentation loading. *IEEE Transactions on Ultrasonics, Ferroelectrics, and Frequency Control* 2011;**58**(12):2772–80.
11. Korhonen RK, Laasanen MS, Töyräs J, Rieppo J, Hirvonen J, Helminen HJ, et al. Comparison of the equilibrium response of articular cartilage in unconfined compression, confined compression and indentation. *Journal of Biomechanics* 2002;**35**(7):903–9.
12. Jurvelin JS, Buschmann MD, Hunziker EB. Mechanical anisotropy of the human knee articular cartilage in compression. *Proceedings of the Institution of Mechanical Engineers (Part H): Journal of Engineering in Medicine* 2003;**217**(3):215–9.
13. Athanasiou KA, Rosenwasser MP, Buckwalter JA, Malinin TI, Mow VC. Interspecies comparisons of in situ intrinsic mechanical properties of distal femoral cartilage. *Journal of Orthopaedic Research* 1991;**9**(3):330–40.
14. Ateshian GA, Soslowsky LJ, Mow VC. Quantitation of articular surface topography and cartilage thickness in knee joints using stereophotogrammetry. *Journal of Biomechanics* 1991;**24**(8):761–76.
15. Graichen H, Jakob J, von Eisenhart-Rothe R, Englmeier K-H, Reiser M, Eckstein F. Validation of cartilage volume and thickness measurements in the human shoulder with quantitative magnetic resonance imaging. *Osteoarthritis and Cartilage* 2003;**11**(7):475–82.
16. Roemhildt ML, Coughlin KM, Peura GD, Fleming BC, Beynnon BD. Material properties of articular cartilage in the rabbit tibial plateau. *Journal of Biomechanics* 2006;**39**(12):2331–7.
17. Hoch DH, Grodzinsky AJ, Koob TJ, Albert ML, Eyre DR. Early changes in material properties of rabbit articular cartilage after meniscectomy. *Journal of Orthopaedic Research* 1983;**1**(1):4–12.

18. Lu XL, Mow VC. Biomechanics of articular cartilage and determination of material properties. *Medicine and Science in Sports and Exercise* 2008;**40**(2):193–9.
19. Ding M, Dalstra M, Linde F, Hvid I. Mechanical properties of the normal human tibial cartilage-bone complex in relation to age. *Clinical Biomechanics* 1998;**13**(4–5):351–8.
20. Hori RY, Mockros LF. Indentation tests of human articular cartilage. *Journal of Biomechanics* 1976;**9**(4):259–68.

CHAPTER 22

Uniaxial Biomechanical Testing of Ligaments and Tendons

Karen Gordon[1], **Allan Brett**[2], **Joanna F. Weber**[3]

University of Guelph, Guelph, ON, Canada[1]; *Running Injury Clinic, Calgary, AB, Canada*[2]; *Ryerson University, Toronto, ON, Canada*[3]

1. BACKGROUND

Ligaments and tendons are tightly packed collagen fiber bundles that connect, respectively, bone to bone and muscle to bone in order to facilitate the loads and motions of limbs and articulating joints (Fig. 22.1). As such, understanding the mechanical properties of soft tissues like ligaments and tendons is an integral

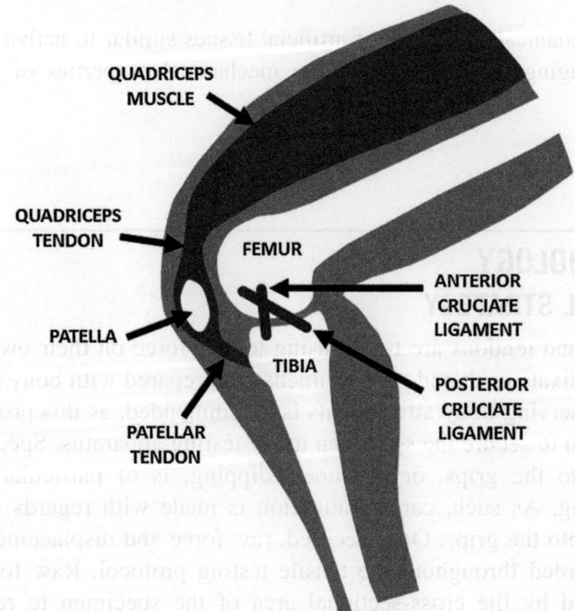

FIGURE 22.1
Schematic of knee joint anatomy showing various ligaments and tendons. Diagram is not to scale.

concept in the study of orthopaedic biomechanics. Specific applications include, but are not limited to, computational modeling, comparing treatment groups in a clinical study, verifying the use of an animal model for a human surrogate, evaluating surgical and clinical treatment options, and inferring a structure–function relationship for the soft tissue structure in question. Therefore, this chapter explains and compares how to effectively perform ligament and tendon tensile testing in a uniaxial configuration, as well as how to analyze, present, and interpret results.

2. RESEARCH QUESTIONS

Typical research questions could include one or more of the following:

- What are the mechanical properties of the upper extremity ligaments?
- Is a pharmaceutical intervention effective at promoting tendon healing in an animal model?
- What is the effect of storage (e.g., freezing, fixation) on mechanical properties of soft tissues?
- Do knee joint ligament mechanical properties infer a stabilizing role in knee joint mechanics?
- Are the mechanical properties of artificial tissues similar to native tissue?
- How does aging or disease affect the mechanical properties of ligaments or tendons?
- etc.

3. METHODOLOGY

3.1 GENERAL STRATEGY

Ligaments and tendons are tested using tensile force on their own, which requires grips to fixate each end, or specimens are prepared with bony attachments. In general, preserving bony attachments is recommended, as this provides an anchor with which to secure the specimen to the testing apparatus. Specimen movement relative to the grips, or specimen slipping, is of particular concern in materials testing. As such, careful attention is made with regards to mounting the specimen into the grips. Once secured, raw force and displacement measurements are recorded throughout the tensile testing protocol. Raw force data are then normalized by the cross-sectional area of the specimen to record stress, and raw displacement data are normalized to the initial length of the specimen to calculate strain. Finally, statistical comparisons are made between the test groups for raw and normalized measurements.

GLOSSARY

✓ **Collagen**. The main structural protein in connective tissue.
✓ **Creep**. Constant applied stress resulting in exponential strain response.
✓ **Ground substance**. Gel-like substance surrounding cells and collagen of connective tissue.
✓ **Hysteresis**. Energy lost with each cycle of applied loading decreases with number of cycles.
✓ **Stress relaxation**. Constant applied strain resulting in exponential stress dissipation.
✓ **Viscoelasticity**. Combined viscous and elastic load response to loading.

SAFETY FIRST

✓ Remember to always wear safety glasses and gloves for protection.
✓ Be careful not to get fingers caught in the test jig during specimen mounting.
✓ Place a plastic protective shield around the soft tissue grip assembly during testing.
✓ Clean the work area and all tools with bleach or disinfectant after testing

3.2 MATERIALS AND TOOLS LIST

- calipers
- custom or supplier soft tissue grips
- dark-colored beads (or lead shot and adhesive)
- digital camera
- hand saw for preparing bone–ligament–bone specimens
- human or animal specimens
- mechanical tester
- scalpel, surgical scissors, cutting board
- tripod for the camera

3.3 SPECIMEN PREPARATION

Step 1. Store the soft tissue specimens. Fresh or fresh–frozen ligament or tendon specimens initially need to be wrapped in plastic bags for proper storage in a freezer prior to tests. The freezer should be held at −20°C or colder.

Step 2. Thaw the specimens. Ligament or tendon specimens should be removed from the freezer, left in their plastic wrappings, and placed on a suitable surface at ambient room temperature or in a warm water bath to fully thaw. After thawing, specimens should be removed from the plastic wrap and sprayed with phosphate-buffered saline solution to prevent dehydration.

Step 3. Cut the specimen. Depending on the research question, the entire soft tissue specimen may be tested, or a portion thereof. If the entire specimen is

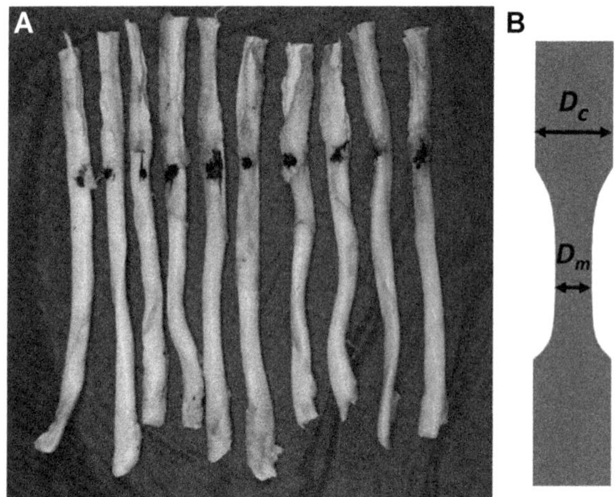

FIGURE 22.2

Soft tissue specimens prepared for testing. (A) Isolated entire tendon specimens cut to length, (B) schematic of a tapered specimen cut to ensure midsubstance failure, where D_c is major diameter of the specimen ends to be placed in the clamp and D_m is major diameter of the specimen at midsubstance, which should be about $D_c/2$. The cross-sectional area perpendicular to the long axis of the specimen is approximately elliptical in shape.

to be tested, then gently use a scalpel to ensure that all extraneous tissue is removed that might confound a cross-sectional area measurement and/or cause the specimen to slip within the grips (Fig. 22.2A). If a portion of the specimen is to be used, isolate a segment of the specimen using a scalpel and a cutting board. In either case, use a scalpel to taper the specimen in the middle, but only in one plane (Fig. 22.2B). This helps obtain a true value of the specimen's mechanical strength by ensuring midsubstance failure instead of in the clamping region. This geometry also allows for measuring cross-sectional area of the specimen at the anticipated true location of failure to obtain an accurate ultimate stress value.

Step 4. Mount the specimen. There are three main gripping mechanisms used for holding ligaments or tendons during tensile testing (Fig. 22.3).

The first type involves the variety of soft tissue grips offered by most mechanical tester manufacturers. These have a serrated or roughened surface that comes into contact with the soft tissue specimen to avoid slippage (Fig. 22.3A). This, however, can create new experimental problems since the roughened surface and force needed to grip the specimen can damage it or create stress concentration factors at the point of contact with the specimen causing premature failure to occur in the specimen at the grip site.

The second type is the preferred method for gripping specimens lacking bony attachments (i.e., tendons) and involves using cryoclamps (Fig. 22.3B). Cryoclamps

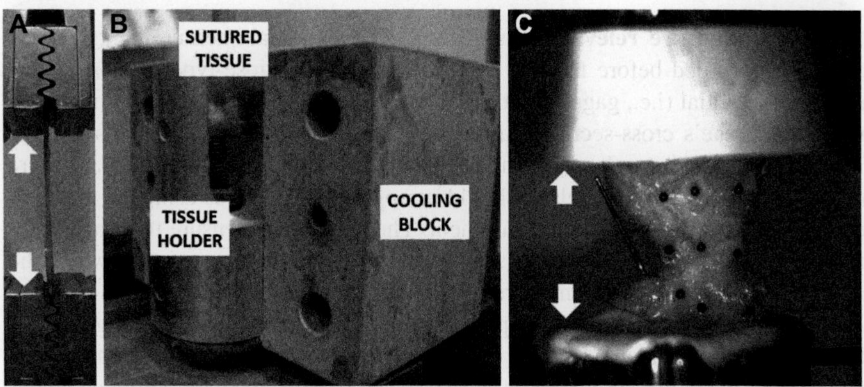

FIGURE 22.3

Ligament and tendon gripping mechanisms. (A) Serrated grips for isolated ligaments or tendons, (B) one-half of a cryoclamp grip for isolated ligaments or tendons, which are sutured around the bar of a tissue holder that is connected to a cooling block that forms a water well for ice block formation, (C) bony end blocks for bone–ligament–bone specimens with small markers (e.g., lead shot) affixed to the tissue surface for optical strain measurement. White arrows indicate the direction of applied tension.

are usually custom-made and involve submersing a low thermally conductive metal (e.g., aluminum) in liquid nitrogen and filling a reservoir around the soft tissue clamp point with water to form an ice block.[1] This methodology is best reserved for failure testing only, since the ice block will have a finite time in which to perform the testing before melting occurs.

The third type involves sophisticated clamping methods used for the bony attachment sites (i.e., in the case of ligaments), which prevent slippage (Fig. 22.3C). For gripping bone, there is little concern of the bone slipping out of the grip; however, there is concern in an uneven strain distribution throughout the soft tissue. Soft tissue insertion sites tend to display larger strains compared to midsection regions located towards the tissue midsection.[2] When gripping is accomplished using bony attachments, it is recommended to also implement a visual noncontact means of calculating the strain in the soft tissue so that the true tissue strain can be measured, which is generally smaller than the strain calculated based on displacement alone.

Step 5. Align the specimen. Another key consideration when mounting the specimen is the specimen alignment. The specimen should be gripped in the upper and lower grips such that it is centrally located in each grip. Failure to accomplish this can lead to uneven loading of collagen fibers that may run the length of the specimen and cause subsequent premature failure. It is recommended to mount the top end of the specimen first, then allow the specimen to hang due to gravity, and then clamp the bottom grip where the specimen would naturally fall. Finally, after mounting it is recommended to ensure the specimen is well-hydrated again using phosphate-buffered saline.

Step 6. Measure relevant dimensions. The essential measurements, which should be recorded before testing, are the specimen's ellipse-type cross-sectional area and the initial (i.e., gage) length. The cross-sectional area should be measured where the tissue's cross-sectional area is smallest for a failure test. This is most frequently done using calipers, although alternative noncontact methods are more accurate. A small 5–10 N preload should be applied to the tissue using the mechanical tester before making this measurement. The preload will ensure that each specimen is under a small consistent tensile force, and therefore, the region in which the cross-sectional area is being measured is also under a small consistent load between specimens. If this is not done, it is possible that the specimen may be "sagging" or deformed slightly, which might introduce inconsistencies into the measurement. Using calipers, make measurements of the minor and major diameter, and then calculate the ellipse-type cross-sectional area. The initial "gage" length of the specimen at the small applied preload should also be recorded since it will be used to calculate the strain value if optical methods are not used. Gage length can be obtained from the recorded displacement value, which will be in the recorded data from the mechanical tester. However, it is always best to record this value in a log book.

Step 7. Prepare specimen for optical strain measurement. If recording strain using optical methods (which is highly recommended), then optical markers need to be affixed to the specimen. Apply a strong adhesive to a number of small, dark-colored beads for contrast, and fix them to the soft tissue surface in a regular pattern (Fig. 22.3C). Mount a digital camera on a tripod, and set it up to record images at a desired frequency based on the duration of the testing protocol. (Note: Optical methods of measuring strain offer a number of distinct advantages. First, visual inspection will help determine exactly where the failure occurred, and therefore, ultimate strain will be more accurately calculated. Second, strain distribution can be quantified throughout the tissue, whereas using the displacement output from the mechanical tester assumes an even strain distribution.)

TIPS AND TRICKS

- ✓ Take all cross-sectional area measurements multiple times and average the results.
- ✓ One researcher performing repeated cross-sectional area measurements improves reliability.
- ✓ Two researchers performing cross-sectional area measurements improves reproducibility.

THE "GOLD STANDARD"

There is no international standard for biomechanical testing of native ligaments and tendons. Thus, researchers should consult peer-reviewed journal articles that have studied this orthopaedic biomechanics application. However, the American Society for Testing and Materials (ASTM) provides potentially useful guidelines in several documents, namely, ASTM D5035 (Standard test method for breaking and elongation of textile fabrics), ASTM F2258 (Standard test method for strength properties of tissue adhesive in tension), and ASTM D638 (Standard test method for tensile properties of plastics).

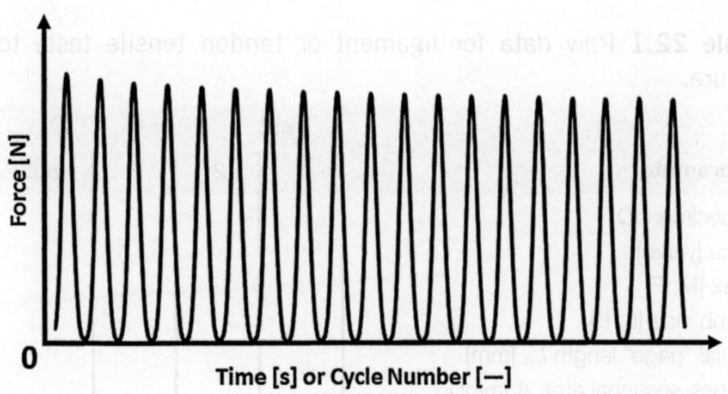

FIGURE 22.4

Typical preconditioning regime for ligaments and tendons. Note how the peak force plateaus and remains constant after approximately five to eight cycles, indicating that the specimen has undergone appropriate preconditioning.

3.4 SPECIMEN TESTING

Step 1. Choose the testing parameters. Most tests to be conducted will define a simple uniaxial tensile failure test. Tests can also be conducted to quantify viscoelastic response, including a stress relaxation or creep test. Parameters that need to be chosen prior to testing include the strain rate (e.g., ~10 mm/min), the maximum strain required for any nondestructive testing (e.g., ~3–5%), and the number of preconditioning cycles required to reach a stable or repeatable response (e.g., ~10 cycles).

Step 2. Perform preconditioning. Preconditioning is done in order to generate a stable viscoelastic response within the tissue during testing (Fig. 22.4). Typically, this involves applying a small subfailure tensile load (~5% tissue strain) over several sinusoidal cycles at an extension rate of approximately 10 mm/min.[3,4]

Step 3. Conduct the test. Place a plastic protective shield around the test assembly. Execute the manual or computer-controlled test sequence to initiate tensioning of the ligament or tendon at the desired speed and to the desired max strain or until specimen failure. If applying optical methods, leave a slight delay to allow time to begin image acquisition with the camera prior to the start of the test.

Step 4. Inspect the specimen. Stop the test sequence after failure occurs or the required duration of loading is completed. Remove the specimen from the test jig carefully, visually inspect the specimen, and take a sufficient number of photos for later inspection. Ensure that the specimen is free of macroscopic tears.

3.5 RAW DATA COLLECTION

Step 1. Record specimen characteristics. Enter human or animal demographic information (i.e., age, sex, and left or right limb) and any other relevant clinical information (Table 22.1).

Table 22.1 Raw data for ligament or tendon tensile tests to failure.

Parameter	Test 1	2	3	etc.
Specimen ID				
Age [years]				
Sex [M, F]				
Limb side [L, R]				
Initial "gage" length L_o [mm]				
Cross-sectional area A [mm^2]				
Failure displacement or length L_F [mm]				
Failure force F_F [N]				
Failure mode				
Other remarks				

Step 2. Record specimen dimensions. Enter the initial "gage" length and the cross-sectional area of the ligament or tendon (Table 22.1).

Step 3. Record failure data. The ultimate or failure force and failure displacement should be noted during the tests and be recorded (Table 22.1).

Step 4. Record failure modes. Some final remarks may also be entered about the nature of failure (Table 22.1). This may include answers to several questions: Did the failure occur in the tissue midsubstance? Is there any evidence of the failure originating at the grips? Is it a clean break or is there visible fiber tearing? etc. Another advantage of incorporating optical strain measurement is the capture of the failure on camera, which will allow for proper review.

3.6 RAW DATA ANALYSIS

Step 1. Create appropriate graphs. Graphs should be created of force vs. displacement, as well as stress vs. strain. From these graphs, mechanical properties can be determined.

Step 2. Compute stress and strain. Compute average stress σ by dividing raw force F by the average cross-sectional area A of the specimen obtained from the initial measurements made prior to testing such that $\sigma = F/A$. Compute strain ε using the initial length or gage length L_o of the specimen such that $\varepsilon = (L - L_o)/L_o$. This should be computed in a spreadsheet with column headings of force, stress, displacement, and strain. However, if determining strain using optical methods, this can be achieved using different methods. In its simplest application, strain is calculated by tracking two markers aligned with the direction of applied force and applying the equation above to determine strain. More complex methods involve tracking a grid of markers and determining a tissue's strain profile in two dimensions. These more complex methods involve tensor algebra and typically require either custom or purchased software to implement.

Step 3. Calculate Young's modulus. Young's modulus E is defined as the slope of the linear elastic region of the stress vs. strain curve. However, it may be difficult to objectively determine which data points should be included when calculating the slope. Best practice dictates that the more data points included in the slope calculation, the more accurate the result will be. In order to maximize the number of points, a linear regression of the stress vs. strain data should be performed. It should be performed iteratively, changing the window on the start and finishing points of the linear region while calculating an R^2 value or measure of best fit of the data. The R^2 value should be maximized while still including the largest number of data points in the slope calculation.

Step 4. Perform statistical analysis. The criterion for statistical difference first needs to be chosen (e.g., $P < 0.01$ or < 0.05). There are software programs for comparing two test groups (e.g., unpaired t-test) or two or more test groups influenced by multiple factors (e.g., analysis of variance, ANOVA). Then, if appropriate, use a statistical test (e.g., Bonferroni, Tamhane, Tukey, etc.) to compare different test subgroups (e.g., treatment vs. control, men $\leq$60 years old vs. men $>$60 years old, etc.).

Step 5. Compute statistical power. Power analysis can be done after the study to ensure there were enough specimens per test group to detect all statistical differences that were actually present (i.e., was type II statistical error avoided?). Statistical power $>$80% is usually considered to indicate there were enough specimens per test group. Note that if good predictions of averages and standard deviations are available from prior studies, then the number of specimens and/or tests can be chosen before the study begins to ensure a power $>$80%.

ENGINEER'S TOOLBOX

An elliptical shape can be used to approximate the typical cross-section of a ligament or tendon specimen. Two measurements, namely, the major axis diameter D_1 and the minor axis diameter D_2 of the specimen, should be made using calipers. The cross-sectional area can then be calculated using the formula for the area of an ellipse using $A = \pi(D_1D_2)/4$.

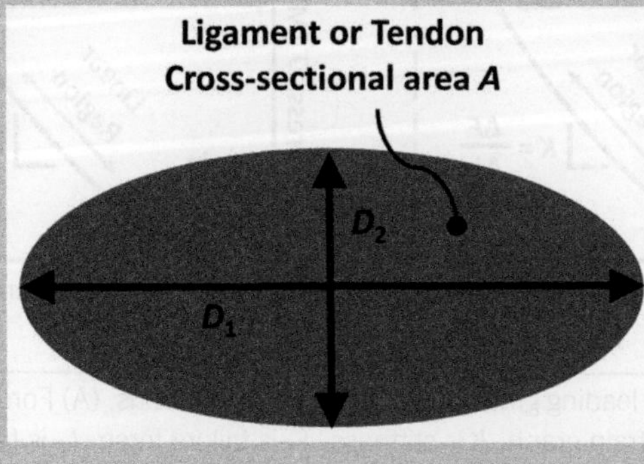

4. RESULTS

Once all ligament and tendon biomechanical tensile testing, data collection, and data analysis have been performed, it is then important to communicate and present the primary results in an understandable and concise manner to the reader of a journal article, conference paper, technical report, or book chapter.

Step 1. Show raw data curves. The force vs. displacement and stress vs. strain curves of most ligaments and tendons have a characteristic shape (Fig. 22.5). The initial region immediately following tensile load application has a curved concave-up shape, as well as a lower slope than the subsequent linear region of the curve. The rationale for this "toe-in" region has been accepted to be due to an "uncrimping" of the zigzag-like orientation of collagen fibers that is characteristic to ligaments and tendons. The crimping is due to hydrogen bonds that form between the fibers. After the collagen fibers have all been uncrimped, the curve enters a linear region when all collagen fibers are being pulled evenly and load distribution is even. A second convex nonlinear region is reached when individual fibers, which are the smallest in cross-sectional area, begin to break. The resultant decrease in force results from fibers each breaking at unique load levels. The stress vs. strain curve generally mirrors the force vs. displacement curve. Finally, the length of the toe-in region may also be of significance. Due to the difficulty in defining the "zero strain" position mentioned above, it is important to maintain consistency of the starting conditions between specimens if a comparison is to be made. In particular, the ultimate strain value may be influenced if standardization of the zero strain position is not achieved.

Step 2. Show data for numerical comparison. Generally, a comparison of some sort is made between test groups (Table 22.2). The key items to be tabulated are the

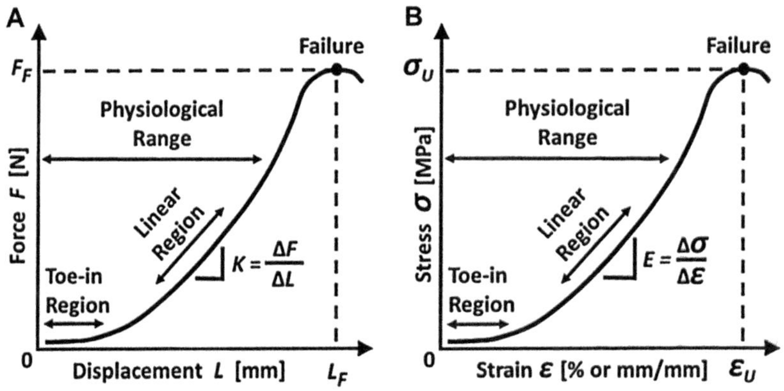

FIGURE 22.5

Representative tensile loading graphs for ligaments and tendons. (A) Force vs. displacement graph, (B) stress vs. strain graph. K is stiffness, F_F is failure force, L_F is failure displacement or length, E is Young's modulus, σ_U is ultimate stress, and ε_U is ultimate strain.

Table 22.2 Ligament and/or tendon test results. Each result should be an average ± 1 standard deviation based on multiple specimens. *P* values indicating statistical differences (e.g., $P<0.01$ or <0.05) between test groups should be calculated and indicated by symbols like asterisks (*), etc.

Test Group	Cross-Sectional Area *A* [mm^2]	Stiffness *K* [N/mm]	Ultimate Stress σ_U [MPa]	Ultimate Strain ε_U [% or mm/mm]	Young's Modulus *E* [MPa]
1					
2					
3					
etc.					

ultimate failure force, Young's modulus, ultimate stress, and ultimate strain, which are easily determined quantities from the raw graphs (Fig. 22.5).

Step 3. Show failure modes. Photos of typical failures should be included to ensure that the failure of the specimen did occur in the mid-region of the tissue, away from the clamps or any other confounding measures (Fig. 22.6).

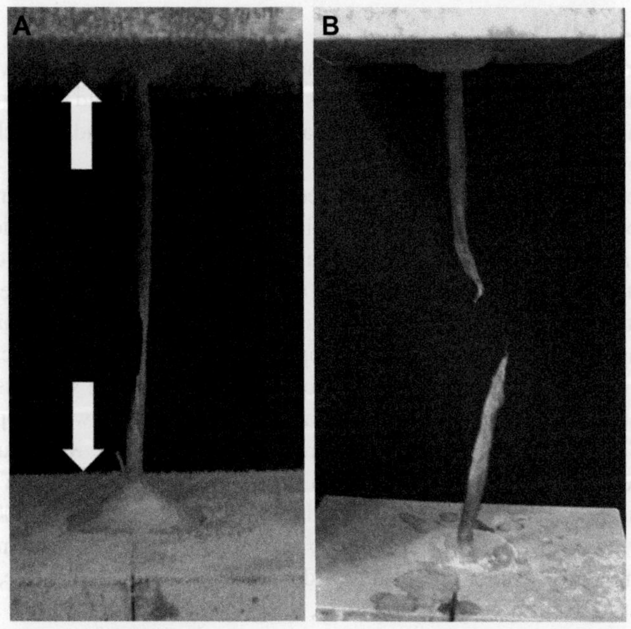

FIGURE 22.6

Tendon specimen tensile testing. (A) During failure, (B) after failure. Failure occurs away from the cryoclamps and is not in frozen tissue regions where brittle failure might be a confounding factor that would influence the ultimate stress value. White arrows indicate the direction of applied tension.

ALTERNATIVES AND ADAPTATIONS

- ✓ **Cyclic tensile tests**. Dynamic viscoelastic properties can be examined with these tests. Specimens are dynamically loaded between two strain or force (stress) values (e.g., 0 and 20 N force) at a set frequency (e.g., 1 Hz). Loss and storage moduli may be determined by examining the hysteresis and comparing the loading and unloading curves.
- ✓ **Stress relaxation tests**. To investigate viscoelastic parameters of soft tissue, these tests can be used. Specimens are quickly loaded to a set strain (e.g., 5%) and remain strained for either a set amount of time (e.g., 10 min) or until there is negligible change in load. Curve fitting to particular viscoelastic models can then be achieved to determine parameters of interest.
- ✓ **Torsional tests**. Ligaments like the scapholunate ligament of the wrist are particularly prone to failure during falls onto the wrist/hand, in which a torsional failure mechanism may be at work. Thus, experimental test regimes for these ligaments should include nondestructive and destructive torsional loading.
- ✓ **Biaxial tests**. Specimens that are small (~5–7 mm square) and thin (<2 mm) are gripped on four edges and load is applied simultaneously in perpendicular directions by two actuators. This loading protocol more closely simulates physiologic conditions for some tissues, such as some planar ligaments or joint capsule tissue. Mechanical properties can be determined in different directions for a more representative model.

5. DISCUSSION

After completing all ligament and tendon tensile tests, data collection, data analysis, and data presentation, then final results can be considered and interpreted in the broader context of some important clinical, biomechanical, and/or technological considerations, as follows.

Characterizing the mechanical properties of ligaments and tendons can be useful for many applications, including, but not limited to, developing synthetic or tissue engineered alternates (e.g., anterior cruciate ligament replacements),[5] comparing clinical treatment or disease/injury groups (e.g., autograft options for anterior cruciate ligament reconstruction, carpal tunnel syndrome of the upper extremity, rotator cuff injury and repair), and determining properties for use in biomechanical models.

The key parameters that are regularly quantified include the stiffness, Young's modulus, ultimate stress, ultimate strain, and failure force, as well as potentially the work of failure (i.e., area under the stress vs. strain curve). Most human ligaments and tendons have ultimate strains in the range of 8–40% with an accompanying ultimate stress of 1–120 MPa and Young's modulus of 5–700 MPa.[6–9] Widely varying results may be attributed to many factors, including, but not limited to, storage methods, gender and health of donor specimens, loading rate, and variations in testing conditions (e.g., humidity, temperature, etc.).

Uniaxial testing of ligaments and tendons remains the simplest method of obtaining mechanical properties. However, it is recognized that there are a number of limitations associated with uniaxial testing. Ligaments and tendons experience multiaxial directional loading under regular physiological use, and therefore, uniaxial biomechanical testing can provide only a narrow look at the function of these tissues. Instead, biaxial biomechanical testing can been conducted for more planar ligaments and tendons, or ligaments may be tested *in situ* using robotic testing systems to determine the tissue's functional contribution to joint biomechanics.

6. SUMMARY

- Ligament and tendon testing has unique experimental parameters related to gripping, viscoelasticity, and quantification of mechanical properties.
- Cryoclamps are suggested for consistent results in specimens without bony attachments.
- Viscoelastic properties of ligament and tendons require a preconditioning protocol.
- Young's modulus, ultimate strain, and ultimate stress can be measured with a tensile test.
- The force vs. displacement and stress vs. strain curves are nonlinear.

7. QUIZ QUESTIONS

1. What is the function of ligaments and tendons in the musculoskeletal system?
2. What are the advantages and disadvantages for including bony attachments in tensile testing?
3. Why should preconditioning protocols remain within physiological stress and strain ranges?
4. How can the measurement of the cross-sectional area influence the calculated results?
5. A set of five repeated measurements obtained using calipers on a ligament showed thicknesses and widths, respectively, of 2.44 and 2.76 mm, 2.50 and 2.69 mm, 2.43 and 2.78 mm, 2.45 and 2.75 mm, and 2.44 and 2.77 mm. The cross-sectional shape of the ligament is elliptical. If the peak force obtained during tensile testing of this specimen was 445 N, what are the cross-sectional area and the ultimate stress? (answer: 5.30 mm^2, 84 MPa).

REFERENCES

1. Rincon L, Schatzmann L, Brunner P, Staubli HU, Ferguson SJ, Oxland TR, et al. Design and evaluation of a cryogenic soft tissue fixation device – tolerances and thermal aspects. *Journal of Biomechanics* 2001;**34**(3):393–7.
2. Woo SL, Gomez MA, Seguchi Y, Endo CM, Akeson WH. Measurement of mechanical properties of ligament substance from a bone-ligament-bone preparation. *Journal of Orthopaedic Research* 1983;**1**(1):22–9.
3. Miller KS, Edelstein L, Connizzo BK, Soslowsky LJ. Effect of preconditioning and stress relaxation on local collagen fiber re-alignment: inhomogeneous properties of rat supraspinatus tendon. *Journal of Biomechanical Engineering* 2012;**134**(3):031007.

4. Woo SL, Orlando CA, GOmez MA, Frank CB, Akeson WH. Tensile properties of the medial collateral ligament as a function of age. *Journal of Orthopaedic Research* 1986; **4**(2):133–41.
5. Leong NL, Petrigliano FA, McAllister DR. Current tissue engineering strategies in anterior cruciate ligament reconstruction. *Journal of Biomedical Materials Research (Part A)* 2014;**102**(5):1614–24.
6. Johnson GA, Tramaglini DM, Levine RE, Kazunori O, Choi NY, Woo SLY. Tensile and viscoelastic properties of human patellar tendon. *Journal of Orthopaedic Research* 1993;**12**(6):796–803.
7. Weber J, Augur AM, Fattah AY, Gordon KD, Oliver ML. Tensile properties of human forearm tendons. *Journal of Hand Surgery European* 2015;**40**(7):711–9.
8. Shani RH, Umpierez E, Nasert M, Hiza EA, Xerogeanes J. Biomechanical comparison of quadriceps and patellar tendon grafts in anterior cruciate ligament reconstruction. *Arthroscopy* 2016;**32**(1):71–5.
9. Jung HJ, Fisher MB, Woo SLY. Role of biomechanics in the understanding of normal, injured, and healing ligaments and tendons. *Sports Medicine, Arthroscopy, Rehabilitation, Therapy & Technology* 2009;**1**:1–17. article ID# 9.

CHAPTER

Measuring the Contraction Force, Velocity, and Length of Skeletal Muscle

23

Radovan Zdero[1], Matthew M. Borkowski[2], Catherine Coirault[3]

Western University, London, ON, Canada[1]; Aurora Scientific, Aurora, ON, Canada[2]; INSERM, Paris, France[3]

1. BACKGROUND

Skeletal muscles are actuators that generate forces to move limbs, transfer load across joints, and stabilize body position or posture (Fig. 23.1).[1,2] Muscle is organized into basic units of ascending size, namely, sarcomeres (i.e., main contractile units), which comprise the myofibrils (i.e., delicate strand-like filaments), which

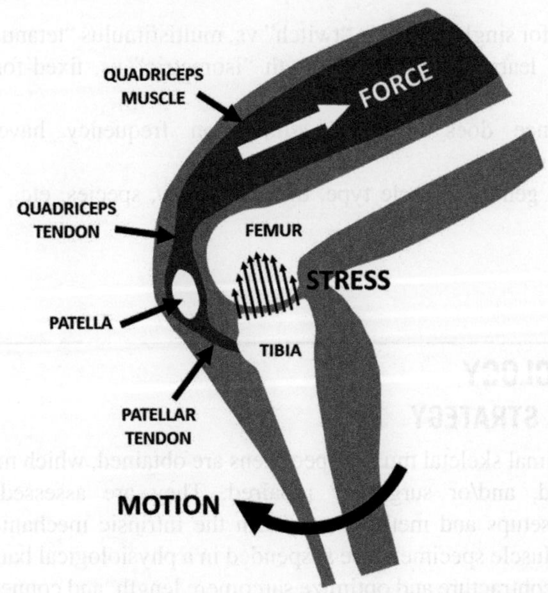

FIGURE 23.1

Knee joint anatomy showing how quadriceps muscle force causes motion of the tibia and stress transfer across the joint.

Experimental Methods in Orthopaedic Biomechanics.

comprise the fibers (i.e., long cylindrical multinucleated cells), which comprise the main muscle belly. Muscle's passive tension arises from elastic spring-like elements stretched beyond their resting length, while active tension is generated by processes within the sarcomere. Active tension is produced in different ways: concentric (i.e., muscle length is shortened), eccentric (i.e., muscle length is extended), isometric (i.e., muscle length is fixed), isotonic (i.e., muscle tension is fixed), isokinetic (i.e., muscle shortening or lengthening velocity is fixed), and isoinertial (i.e., muscle contraction encounters a fixed external resistance). To simplify experiments and computer models in orthopaedic biomechanics research, muscle action is often treated like a system of force vectors acting on limbs or joints. Yet, the magnitude and direction of these force vectors depend on muscle's intrinsic mechanical properties during contraction, especially the interrelationship between force, velocity, and length.[3–16] Therefore, this chapter explains how to experimentally perform *in vitro* measurements of contraction force, velocity, and length of skeletal muscle, as well as how to analyze, present, and interpret results.

2. RESEARCH QUESTIONS

Typical research questions might include one or more of the following:

- What is the interrelationship of contraction force, velocity, and length for muscle?
- Do data vary for single-stimulus "twitch" vs. multistimulus "tetanus" conditions?
- What can be learned from fixed-length "isometric" vs. fixed-force "isotonic" tests?
- What influence does electrical stimulation frequency have on muscle characteristics?
- How do age, gender, muscle type, disease, injury, species, etc., affect muscle properties?
- etc.

3. METHODOLOGY

3.1 GENERAL STRATEGY

Human or animal skeletal muscle specimens are obtained, which may be normal, diseased, injured, and/or surgically repaired. They are assessed using well-established test setups and methods to obtain the intrinsic mechanical properties of muscle.[3–20] Muscle specimens are suspended in a physiological bath, held in pretension to avoid contracture and optimize sarcomere length, and connected to a lever apparatus. Muscle specimens are electrically stimulated using electrodes to induce contraction with single-stimulus "twitch" and multistimulus "tetanus" excitation while under fixed-length "isometric" and fixed-force "isotonic" constraints. Muscle contraction force F, velocity V, length L, and other parameters are measured. Statistical comparisons are then made of F, V, L, and other parameters between test groups to determine the influence of experimental conditions (e.g., twitch, tetanus, isometric, isotonic, stimulation frequency, temperature, etc.) and specimen characteristics (e.g., age, gender, muscle type, disease, injury, species, etc.).

GLOSSARY

✓ **Contraction.** Muscle tightening due to interaction of sliding filaments in the sarcomere.
✓ **Isometric.** Muscle contraction tests done when muscle is kept at a fixed length.
✓ **Isotonic.** Muscle contraction tests done when muscle is kept at a fixed tension.
✓ **Tetanus.** Muscle is stimulated using multiple overlapping electrical excitations.
✓ **Twitch.** Muscle is stimulated using a single brief electrical excitation.

SAFETY FIRST

✓ Remember to always wear a lab coat, safety goggles, and gloves for protection.
✓ Ensure that cutting muscle with the scalpel is always done away from oneself.
✓ Turn off the electrical stimulator to avoid electrical shock during muscle mounting.
✓ Clean the work area and all tools with bleach or disinfectant after testing is done.

3.2 MATERIALS AND TOOLS LIST

- computer and software
- electrical stimulator and electrodes
- forceps (straight or bent) and/or tweezers
- lever arm controller-transducer
- Krebs–Ringer bath
- muscles or muscle strips
- scalpel
- steel wire
- sutures and/or spring clips
- Vernier calipers

3.3 SPECIMEN PREPARATION

Step 1. Recruit study subjects. Obtain institutional ethics approval for the study protocol before recruiting human or animal subjects. Human subjects may be either living or recently deceased within the past 1 h. Animal subjects may be euthanized mechanically by cervical dislocation or chemically using a suitable substance (e.g., sodium pentobarbital), but should be tested within 1 h.

Step 2. Perform muscle biopsies. For human subjects and larger animals (e.g., bovine, ovine, porcine), remove the fascia and then dissect out a strip from the muscle of interest (e.g., quadriceps). Ensure that scalpel cuts are parallel to the direction of the fibers in order to minimize fiber damage. Muscle strips to be tested should be 1–2 mm in thickness, 3–10 mm in width, and 10–20 mm in length. For smaller animals (e.g., frog, mouse, rat), the entire muscle from tendon-to-tendon may be removed in order to maximize the working length of these relatively small muscles, as long as they have a maximum thickness of 1–2 mm. Use Vernier calipers to confirm these dimensions.

Step 3. Secure the muscle ends. Use forceps and/or tweezers for the following procedures when possible. Tie off both ends of the muscle with a 4-0 to 6-0 silk suture to preserve blood content, making sure to leave a little extra suture length for later mounting into the test setup. If beneficial, attach rigid spring clips to the muscle ends with or without a small drop of cyanoacrylate adhesive since spring clips may be easier to handle than sutures. Then, to avoid muscle contraction before the specimen is placed in the main test setup, hang a small weight to the muscle's bottom suture or spring clip to temporarily apply a preload of 1–10 mN or, alternatively, place the muscle on a glass dish with a silgar bottom, and pin both ends to its pre-biopsy length.

TIPS AND TRICKS

- ✓ Make scalpel cuts parallel to the muscle fibers to prevent any fiber damage.
- ✓ Muscles should not be directly touched, excessively pulled, or permitted to dry out.
- ✓ Spring clips are easier to handle than sutures for mounting muscle in the test setup.
- ✓ Periodically do a visual check to ensure the muscle remains properly mounted.
- ✓ The same researcher(s) should perform all tests for consistency in results.

THE "GOLD STANDARD"

No internationally recognized standards exist for *in vitro* testing of skeletal muscle contraction force, velocity, and length. However, the Treat-NMD Neuromuscular Network has produced the documents DMD_M.1.2.002 (Measuring isometric force of isolated mouse muscles *in vitro*) and SMA.M_1.2.002 (Muscle function evaluation through isometric force measurement in mouse models of spinal muscular atrophy) that describe standard operating procedures in their organization for testing methodology and data analysis. In addition, peer-reviewed scientific journal articles and equipment manufacturer instructions should be followed closely.

3.4 SPECIMEN TESTING

Step 1. Mount the muscle in the test setup. Immediately postbiopsy, place the muscle specimen into the test setup, which consists of chemical, mechanical, and electrical components (Fig. 23.2). Components can be individually fabricated, purchased, and assembled in-house;[3–8,10–12] however, the entire setup can also be purchased from a scientific equipment supplier, resulting in a more efficient, standardized, and optimized testing system.[17,18]

The *chemical* components consist of a Krebs–Ringer bath of NaCl (118–140 mM), $NaHCO_3$ (0–25 mM), KCl (4.7–5.9 mM), $CaCl_2$ (1.5–2.5 mM), KH_2PO_4 (1–1.2 mM), $MgSO_4$ (0.5–1.2 mM), $MgCl_2$ (0–1.2 mM), glucose (0–11 mM), pyruvate (0–1 mM), and D-tubocurarine chloride (0–0.3 mM) contained in a glass chamber rigidly fixed to prevent movement. The bath is kept at a fixed temperature typically in the range of 20–37°C (depending on the research question) by using warm water circulating through the outer jacket of the chamber. The bath is bubbled continuously with a mixture of 95% O_2 and 5% CO_2 (or 100% O_2), maintained at a pH of 7.3–7.6, and replaced with fresh solution for each new muscle specimen tested. This mimics an *in vivo* physiological environment and prevents the damaging effects of oxygen deprivation (i.e., hypoxia) on muscle contractile properties. (Note: If it becomes difficult to bubble directly into the small chamber, then use an external reservoir of bubbled solution and pour or pump the media into the chamber.)

The *mechanical* components consist of a horizontal lever arm connected using a double-hooked steel wire to the silk suture or rigid spring clip of the muscle's top end, while the muscle's bottom end is held by its suture or spring clip in an adjustable stationary clamp. The lever arm holds the muscle at its resting and/or optimal length using a low preload of 1–10 mN prior to testing. The mechanical stiffness of the lever apparatus (~5–15 N/mm) should be about two orders of magnitude larger than the stiffness of the elastic component of the muscle (~0.08 N/mm) to avoid force and displacement measurement artifacts during muscle contraction tests.

The *electrical* components consist of a personal computer with suitable software that controls the electrical stimulator and the lever arm apparatus equipped with a force–displacement controller–transducer. The electrical stimulator activates a pair of platinum electrodes (e.g., 2 mm thick × 5 mm wide × 50 mm long) that are placed in the bath at 1–5 mm away from, but parallel to, the muscle to induce muscle contraction. The force–displacement controller–transducer (range, 0–500 mN and 0–5 mm) can control muscle forces and displacements via the lever arm, as well as measure force and displacement during muscle contraction.

Step 2. Equilibrate the specimen. After applying a small preload of 1–10 mN to keep it taught, let the muscle equilibrate in the bath for 10–15 min. The micropositioner may need to be used to periodically stretch the muscle since it may keep relaxing until it equilibrates.

Step 3. Measure the resting length. Once the muscle is equilibrated, lower the bath chamber to expose the muscle, and use a pair of Vernier calipers to measure its resting length L_{rest}. This provides a reference length for future length change measurements. Do not accidentally jab the muscle with the sharp caliper tips.

Step 4. Perform isometric "twitch" tests. These tests determine the muscle's contraction properties at a fixed length (i.e., isometric) during a single excitation

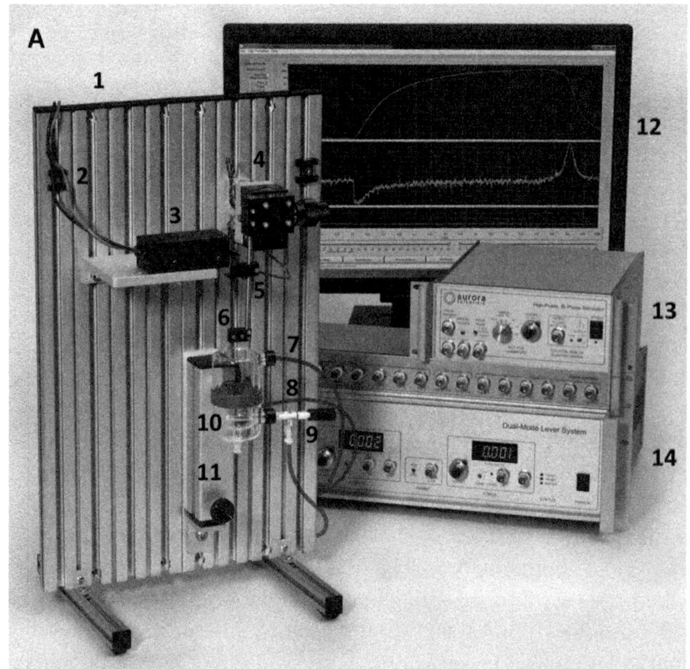

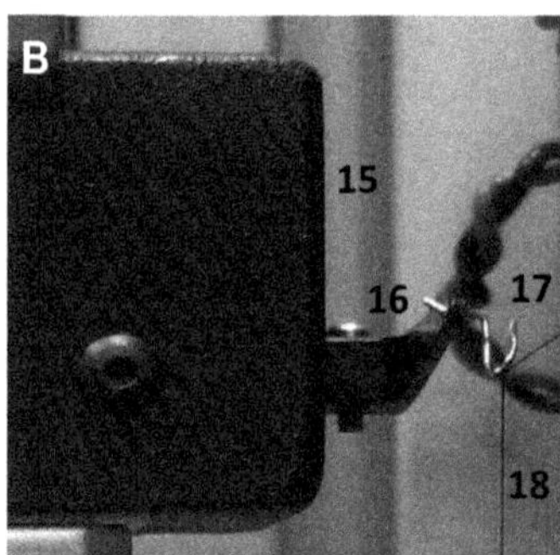

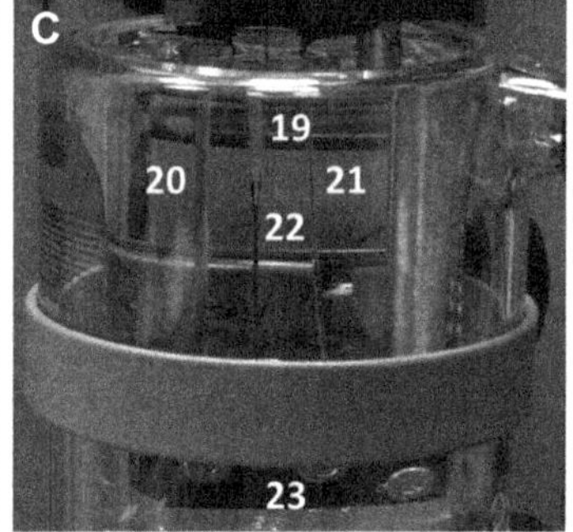

FIGURE 23.2

Muscle contraction test setup. (A) Overall test setup showing (1) metal support platform, (2) force–displacement transducer wiring, (3) force–displacement controller–transducer sensor with lever arm, (4) vertical micropositioner for adjusting muscle length, (5) electrode holder, (6) bottom muscle clamp, (7, 8) inlet and outlet tubes that allow warm water to circulate through the chamber's outer jacket, (9) inlet tube and bubbler for O_2/CO_2, (10) bath chamber, (11) vertical translation slide for bath chamber, (12) computer with software, (13) electrical stimulator, (14) electronic controller for the lever apparatus. (B) Lever apparatus showing (15) controller–transducer sensor, (16) lever arm, (17) stainless steel hook, (18) silk suture connected to the muscle's top end. (C) Bath chamber showing (19) muscle specimen, (20, 21) electrodes, (22) silk suture connecting the muscle's bottom end to the bottom clamp, (23) bottom muscle clamp.

Photos courtesy of www.AuroraScientific.com.

pulse (Fig. 23.3A). To do so, activate the electrical stimulator to apply the first single brief excitation (e.g., rectangular wave, 40 V amplitude, and 5 ms total duration). Next, allow a rest period of 5–10 s, increase muscle length by a small increment (e.g., 0.1 mm), repeat another twitch test, etc., until the muscle's force response curve reaches a maximum amplitude. This muscle length is the optimal length L_o, which allows the muscle to generate the maximum possible force F_{max} during contraction. Then, double-check that this is the true value of L_o by increasing the length using several more increments (e.g., 0.1–0.3 mm), which should cause the muscle force curve to slightly drop. Finally, use L_o to repeat the twitch test 3–5 times to obtain an average value for the various parameters measured from the

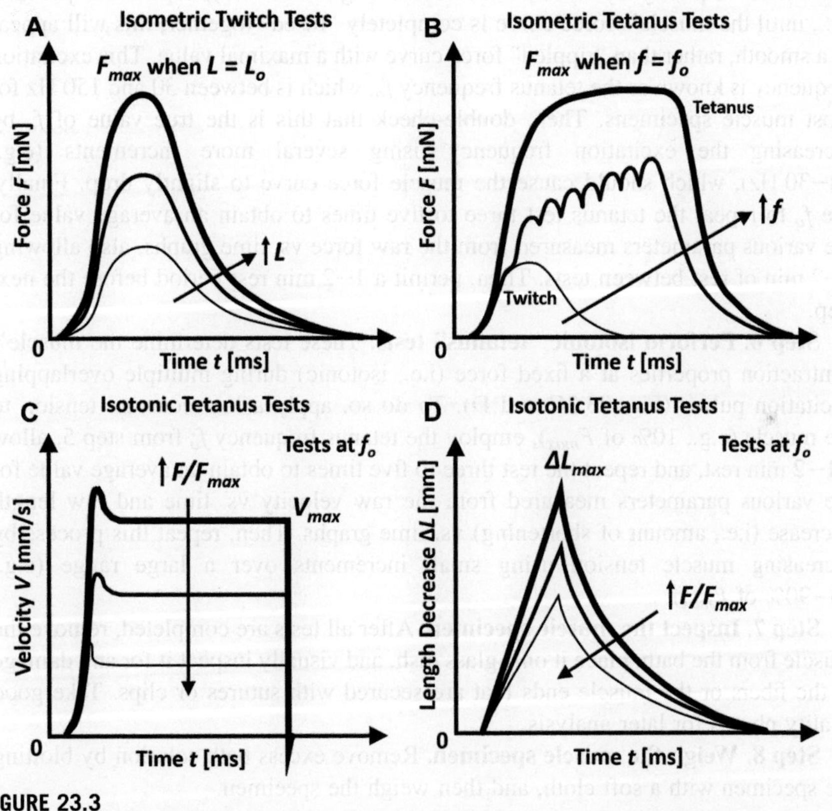

FIGURE 23.3

Raw graphs for muscle contraction tests. (A) isometric twitch tests, (B) isometric tetanus tests, (C) isotonic tetanus tests (velocity), (D) isotonic tetanus tests (length decrease, i.e., shortening). The velocity graph may show a small jump at the start of the isotonic test caused by the finite response time of the lever arm apparatus, while a small subzero drop at the end of the isotonic test indicates the sudden onset of muscle relaxation. f is excitation frequency, f_o is tetanus frequency.

raw force vs. time graphs, also allowing 5–10 s of rest between tests. Then, permit a 30 s rest period before the next step. (Note: L_o is often about 5–10% longer than L_{rest}; thus, they are not to be confused with each other. In addition, prior to any tests, the electrical stimulator may need to be calibrated according to the manufacturer's instructions to find the maximum electrical voltage that will stimulate all the muscle fibers, but without damaging them.)

Step 5. Perform isometric "tetanus" tests. These tests determine the muscle's contraction properties at a fixed length (i.e., isometric) during multiple overlapping excitation pulses (Fig. 23.3B). To do so, activate the electrical stimulator to apply the first series of multiple excitations (e.g., rectangular wave, 40 V amplitude, 10 Hz, and 250–1000 ms total duration). Next, allow a rest period of 1–2 min, increase excitation frequency by a small increment (e.g., 5–10 Hz), repeat the process, etc., until the muscle's force curve is completely "fused" together; this will appear as a smooth, rather than "rippled" force curve with a maximal value. This excitation frequency is known as the tetanus frequency f_o, which is between 30 and 150 Hz for most muscle specimens. Then, double-check that this is the true value of f_o by increasing the excitation frequency using several more increments (e.g., 10–30 Hz), which should cause the muscle force curve to slightly drop. Finally, use f_o to repeat the tetanus test three to five times to obtain an average value for the various parameters measured from the raw force vs. time graphs, also allowing 1–2 min of rest between tests. Then, permit a 1–2 min rest period before the next step.

Step 6. Perform isotonic "tetanus" tests. These tests determine the muscle's contraction properties at a fixed force (i.e., isotonic) during multiple overlapping excitation pulses (Fig. 23.3C and D). To do so, apply a low constant tension to the muscle (e.g., 10% of F_{max}), employ the tetanus frequency f_o from step 5, allow a 1–2 min rest, and repeat the test three to five times to obtain an average value for the various parameters measured from the raw velocity vs. time and raw length decrease (i.e., amount of shortening) vs. time graphs. Then, repeat this process by increasing muscle tension using small increments over a large range (e.g., 10–90% of F_{max}).

Step 7. Inspect the muscle specimen. After all tests are completed, remove the muscle from the bath, place it on a glass dish, and visually inspect it for any damage to the fibers or the muscle ends that are secured with sutures or clips. Take good quality photos for later analysis.

Step 8. Weigh the muscle specimen. Remove excess bath solution by blotting the specimen with a soft cloth, and then weigh the specimen.

3.5 RAW DATA COLLECTION

Step 1. Record qualitative characteristics. Before muscle contraction testing, note the muscle's characteristics, such as species, anatomical location, status, sex, and age (Table 23.1).

Table 23.1 Raw data collected from isometric and isotonic muscle tests.

	Muscle 1	Muscle 2	etc.	Avg	SD
Characteristics					
Species				–	–
Anatomical location				–	–
Status [N, I, D, R]				–	–
Sex [male, female]				–	–
Age [years]					
Mass m [g]					
Volume Q [mm^3]					
Resting length L_{rest} [mm]					
Cross-sectional area A [mm^2]					
Isometric Twitch Tests					
Max force F_{max} [mN]					
Optimal length L_o [mm]					
Rise time t_{max} [ms]					
Half-relaxation time $t_{1/2}$ [ms]					
Isometric Tetanus Tests					
Max force F_{max} [mN]					
Tetanus frequency f_o [Hz]					
Rise time t_{max} [ms]					
Half-relaxation time $t_{1/2}$ [ms]					
Isotonic Tetanus Tests					
Max velocity V_{max} [mm/s]					
Max length decrease ΔL_{max} [mm]					
Half-relaxation time $t_{1/2}$ for length decrease [ms]					
Posttest Visual Inspection					
Damage to specimen middle?				–	–
Damage to specimen ends?				–	–

Avg, *average;* SD, *standard deviation;* N, *normal;* I, *injured;* D, *diseased;* R, *repaired surgically.*

Step 2. Record quantitative characteristics. If possible, also measure other aspects of the muscle specimen, like mass, volume, resting length, and cross-sectional area (Table 23.1).

Step 3. Record muscle contraction properties. Isometric twitch and tetanus tests produce raw data graphs of force vs. time collected by the computer software from which ultimate values for each individual muscle and/or test group can be obtained, like maximum contraction force, optimal length, tetanus frequency, rise time (i.e., the time to reach maximum force), and half-relaxation time (i.e., the time when maximum force drops by 50%) (Fig. 23.3A and B) (Table 23.1). Similarly, isotonic tests generate raw data graphs of velocity vs. time and length decrease vs. time graphs from which ultimate values for each individual muscle and/or test group can be obtained, like maximum contraction velocity, maximum length decrease, and half-relaxation time (i.e., the time when maximum length decrease drops by 50%) (Fig. 23.3C and D) (Table 23.1).

Step 4. Record damage to specimens. Enter some final remarks about any damage that may have occurred to the fibers in the middle of the specimen, or at the ends or tendons used to secure specimens to the test setup (Table 23.1).

3.6 RAW DATA ANALYSIS

Step 1. Perform statistical comparisons. Choose the criterion for statistical difference (e.g., $P < 0.01$ or < 0.05). Then, compare muscle properties (e.g., force, velocity, length, etc.) across different test conditions (e.g., twitch, tetanus, isometric, isotonic, stimulation frequency, temperature, etc.) or specimen characteristics (e.g., age, sex, muscle type, disease, injury, species, etc.). Use statistical software programs to compare two test groups (e.g., paired *t*-test) or two or more test groups influenced by multiple factors (e.g., analysis of variance, ANOVA).

Step 2. Compute statistical power. Perform power analysis after the study to ensure there were enough muscle specimens per test group to detect all statistical differences that were actually present (i.e., was type II statistical error avoided?). Statistical power >80% is usually considered to indicate there were enough specimens per test group. Note that if good predictions of averages and standard deviations are available from prior studies, then the number of muscle specimens and/or tests can be chosen before the study begins to ensure a power >80%.

ENGINEER'S TOOLBOX

Experimental measurements of the maximum mechanical contraction stress generated by individual fibers or fiber bundles (i.e., biopsied muscle strips) can be used to estimate the maximum muscle contraction force produced by a whole muscle. Assume the following: (1) muscle fibers are long, uniform cylinders, tightly packed to form whole muscle; (2) muscle fibers may be oriented at a "pennation" angle with respect to the tendon's proximal origin and distal insertion points in bone; (3) whole muscle force is the sum of the forces generated by all individual fibers; and (4) average contraction stress by definition is the same for a single fiber, fiber bundle, and whole muscle, since stress is a mathematically normalized value computed as force divided by cross-sectional area. Consequently, whole muscle contraction force can be estimated using different variables as,

$$F_w = \sigma_f A_w \cos\theta = \sigma_f \left(\frac{Q_w}{L_w}\right) \cos\theta = \sigma_f \left(\frac{m_w/\rho_w}{L_w}\right) \cos\theta$$

where w is whole, f is fiber, F is force [N], σ is stress [MPa or N/mm^2], A is cross-sectional area perpendicular to the muscle's long axis [mm^2], θ is pennation angle [°], Q is volume [mm^3], L is optimum length (~resting length) for maximum contraction force [mm], m is mass [g], and ρ is density [g/mm^3].

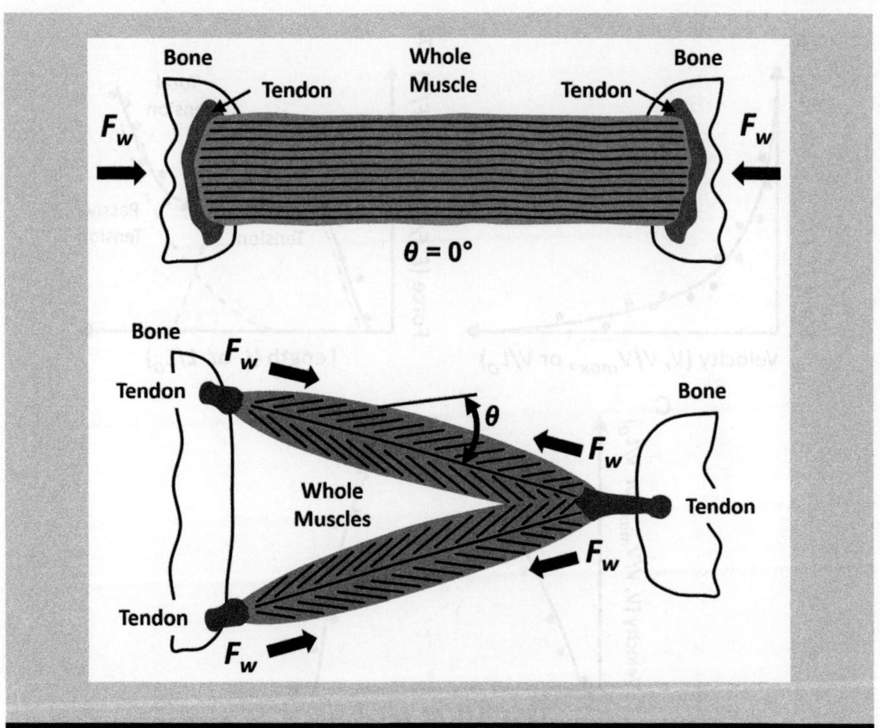

4. RESULTS

Once all muscle contraction data collection and analysis have been performed, it is then important to communicate and present the primary results in an understandable and concise manner to the reader of a journal article, conference paper, technical report, or book chapter.

Step 1. Show *F*–*V*–*L* relationships. Raw data graphs from isometric and isotonic tests provide absolute quantities, which are highly dependent on the muscle specimen's individual characteristics and intrinsic mechanical properties. To allow for proper interspecimen and interstudy comparison, it is sometimes useful to normalize raw data with baseline values. Consequently, the complex interrelationships between force F, velocity V, and length L can be presented graphically using lines of best fit for either the absolute or normalized experimental data points for an individual muscle specimen or muscle tests group (Fig. 23.4).

Several findings are of particular interest. First, the F–V graph (Fig. 23.4A) is commonly called Hill's curve. It takes the general form $(aV + FV) = b(F_{max} - F)$, where a and b are constants and F_{max} is the maximum contraction force.[11,13,14] This curve shows that during muscle shortening, aV is the rate of

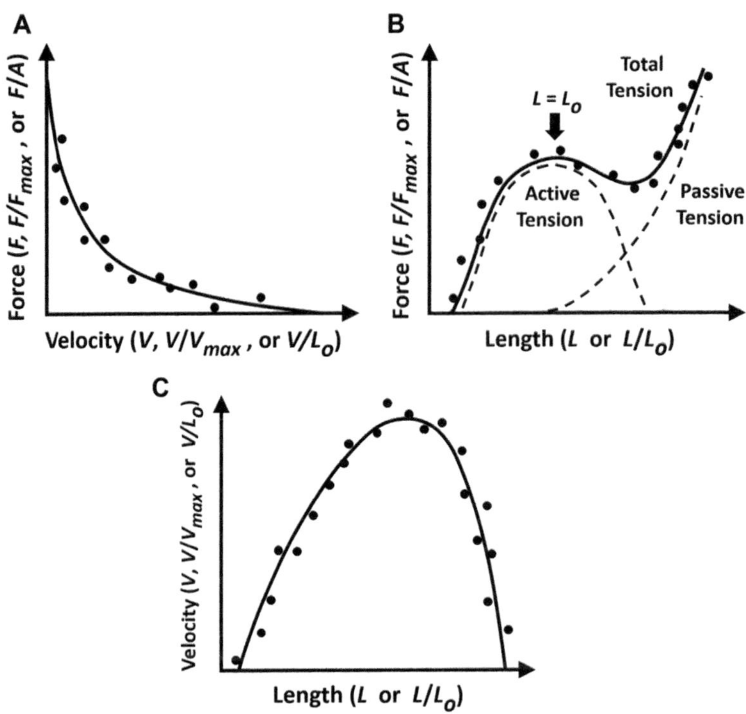

FIGURE 23.4

Isometric or isotonic test results for an individual muscle specimen or test group undergoing twitch or tetanus. (A) *F–V* graph, also known as Hill's curve, (B) *F–L* graph, also known as the Blix curve, shows the contribution of active contractile and passive elastic elements, (C) *V-L* graph. Black dots are experimental data. *F* is force, *V* is velocity, *L* is length, *A* is cross-sectional area. Standard deviation error bars are not present if data points are for an individual muscle, but are present if data points are average values for a test group.

heat production and *FV* is the rate of mechanical work performed (i.e., power). Second, the *F–L* graph (Fig. 23.4B) is often called the Blix curve. It is similar to a force vs. displacement or stress vs. strain curve for a passive material under external tension whose slope at a given point, respectively, is mechanical stiffness or Young's elastic modulus. However, muscle has both active contractile elements and passive elastic elements that resist external tension; thus, the slope of the curve at any point is a kind of resistance stiffness or modulus, but is not to be equated with mechanical stiffness or Young's elastic modulus. Third, the *V–L* graph (Fig. 23.4C) shows that each muscle specimen has an ideal length at which contraction velocity is maximized.

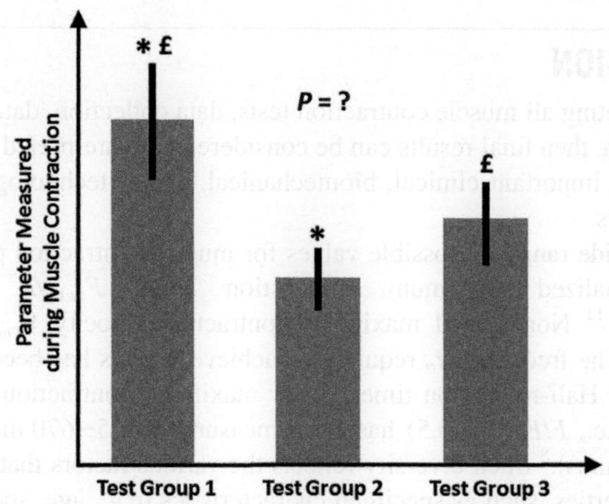

FIGURE 23.5

Representative bar graph for statistical comparisons of muscle contraction data. Results show the average ± 1 standard deviation. *P* values are the statistical differences (e.g. $P<0.05$) between test groups indicated by symbols like asterisks (*), pounds (£), etc.

Step 2. Show statistical comparisons. Present bar graphs of statistical comparisons of any parameter(s) of interest that were measured (e.g., F_{max}, V_{max}, L_o, f_o, t_{max}, $t_{1/2}$, etc.) between test groups to determine the influence of experimental conditions (e.g., twitch, tetanus, isometric, isotonic, excitation frequency, temperature, etc.) and/or specimen characteristics (e.g., age, sex, species, muscle type, disease, injury, etc.) (Fig. 23.5).

ALTERNATIVES AND ADAPTATIONS

- ✓ ***In situ* muscle testing.** Live animals are anesthetized during tests and then sacrificed after data collection. Muscles are surgically exposed (but not biopsied) and then tested using an apparatus (with no Krebs–Ringer bath) similar to *in vitro* testing. Although *in situ* tests are technically more challenging, benefits include preserving the nerve and blood supply to allow physiologically relevant measurements of the absolute magnitude of contractile properties, direct stimulation of the nerve, and repeated muscle stimulation to investigate fatigue.
- ✓ ***In vivo* muscle testing.** Live animals are anesthetized during tests, but then revived and repeatedly tested over an extended period of time. Live human subjects are not anesthetized, but may receive local freezing in the skin. Muscles do not need to be surgically exposed for electrical stimulation since surface electrodes are mounted on the skin or transcutaneous electrodes are inserted through the skin to induce muscle contraction. However, there is no guarantee that individual muscles can be isolated, or that all muscle motor units can be electrically stimulated in order to produce accurate measurements.

5. DISCUSSION

After completing all muscle contraction tests, data collection, data analysis, and data presentation, then final results can be considered and interpreted in the broader context of some important clinical, biomechanical, and/or technological considerations, as follows.

There is a wide range of possible values for muscle contraction properties. For instance, normalized maximum contraction force F_{max}/A is typically 28–300 kPa.[1,8–11] Normalized maximum contraction velocity V_{max}/L_o is about 2–8 s^{-1}.[1,10,11] The frequency f_o required to achieve tetanus has been measured as 30–100 Hz.[10,11] Half-relaxation time $t_{1/2}$ for maximum contraction force F_{max} to drop by 50% (i.e., $F/F_{max} = 0.5$) has been measured at 55–670 ms (twitch) and 90–104 ms (tetanus).[8] Such diversity reflects the various factors that affect muscle contractile properties, such as specimen characteristics (e.g., age, species, slow vs. fast twitch muscles, etc.) and experimental conditions (e.g., isometric, isotonic, twitch, tetanus, excitation frequency, temperature, etc.).

Experimental measurements of maximum mechanical contraction stresses generated by individual fibers or fiber bundles (i.e., muscle strips) can be used to predict maximum muscle contraction forces produced by whole muscles.[1,9] Several factors must be considered, such as the long uniform cylindrical fiber geometry, the "pennation" angle between fibers and the tendon's origin and insertion points, and the equivalence of fiber contraction stress and whole muscle contraction stress since both are computed as force divided by perpendicular cross-sectional area. Once calculated, whole muscle contraction forces can then be incorporated as a system of force vectors acting on limbs and joints for test setups and computer models in orthopaedic biomechanics research. The formula for estimating whole muscle contraction force is $F_w = \sigma_f \times A_w \times \cos\theta$, where F_w is whole muscle contraction force, σ_f is single fiber or fiber bundle contraction stress, A_w is whole muscle perpendicular cross-sectional area, and θ is fiber pennation angle.

There are advantages to *in vitro* contraction testing of isolated skeletal muscle.[8,11,19,20] An important benefit is that experimental conditions and measurements can be strictly defined and controlled to answer a variety of research questions, which may be more difficult to achieve with *in vivo* tests on live humans or animals using electromyography. In addition, confounding factors like nerve or blood supply are eliminated, allowing for more precise characterization of intrinsic mechanical properties of muscle. Moreover, the general shape of the data curves from *in vitro* tests are similar to *in vivo* results. However, some concerns exist about *in vitro* testing.[8,12,19,20] Muscle damage can occur during surgical excision and handling, which affects contractile performance. If muscle fibers are not oriented inline with the sensing direction of the force–displacement transducer, then contraction force, velocity, and length can be underestimated. If muscle specimens are too thick, tissue can be damaged due to a lack of oxygen (i.e., hypoxia) since the Krebs–Ringer bath may not always adequately perfuse the tissue. Inconsistent levels of hypoxia can also occur if muscle specimens are not all immersed into the bath within the same timeframe after biopsy. Finally, although general guidelines exist for *in vitro* muscle testing in the scientific literature[5–8,10–12] and some standardized in-house protocols have been developed by organizations,[19,20] there is an inconsistent application of these procedures by the wider scientific community.

6. SUMMARY

- Muscle activity along limbs or around joints is often modeled or treated as a vector.
- Muscle mechanical properties are characterized by contraction force, velocity, and length.
- Muscle mechanical properties can be experimentally determined using *in vitro* tests.
- Muscle can be tested under isometric, isotonic, twitch, and tetanus conditions.
- Muscle properties are influenced by age, sex, species, disease, injury, etc.

7. QUIZ QUESTIONS

1. What is the difference between the terms "twitch" and "tetanus"?
2. What is the difference between the terms "isometric" and "isotonic"?
3. What is the relationship between muscle contraction force, velocity, and length?
4. What are some specimen characteristics that influence muscle mechanical properties?
5. Estimate the maximum contraction force and the mass of a human whole muscle in a young healthy adult. Assume the muscle has an average diameter of 3 cm, a resting length of 20 cm, a fiber pennation of 20°, a maximum fiber contraction stress of 750 kPa, and a density of 1000 kg/m^3 (answer: 498 N, 141 g).

REFERENCES

1. Mow VC, Huiskes R, editors. *Basic orthopaedic biomechanics and mechano-biology.* 3rd ed. Philadelphia (PA, USA): Lippincott, Williams, and Wilkins; 2005. p. 29–121.
2. Nordin M, Frankel VH, editors. *Basic biomechanics of the musculoskeletal system.* 3rd ed. Philadelphia (PA, USA): Lippincott, Williams, and Wilkins; 2001. 148–171.
3. Wilkie DR. The mechanical properties of muscle. *British Medical Bulletin* 1956;**12**(3): 177–82.
4. Wilkie DR. Measurement of the series elastic component at various times during a single muscle twitch. *Journal of Physiology* 1956;**134**:527–30.
5. Brust M, Cosla HW. The contractility of isolated human skeletal muscle. *Archives of Physical Medicine and Rehabilitation* 1967;**48**:543–55.

6. Eberstein A, Goodgold J. Slow and fast twitch fibers in human skeletal muscle. *American Journal of Physiology* 1968;**215**:535–41.
7. Stevens JC, Jones NB, Fay DF. Apparatus of uniaxial testing of soft human tissues in vitro. *Medical and Biological Engineering and Computing* 1978;**16**:226–8.
8. Stevens JC, Dickinson V, Jones NB. Mechanical properties of human skeletal muscle from in vitro studies of biopsies. *Medical and Biological Engineering and Computing* 1980;**18**:1–9.
9. Winters JM, Stark L. Estimated mechanical properties of synergistic muscles involved in movements of a variety of human joints. *Journal of Biomechanics* 1988;**21**(12): 1027–41.
10. Coirault C, Chemla D, Pery N, Suard I, Lecarpentier Y. Mechanical determinants of isotonic relaxation in isolated diaphragm muscle. *Journal of Applied Physiology* 1993; **75**(5):2265–72.
11. Coirault C, Riou B, Pery-Man N, Suard I, Lecarpentier Y. Mechanics of human quadriceps muscle. *Journal of Applied Physiology* 1994;**77**(4):1769–75.
12. Croes SA, von Bartheld CS. Measurement of contractile force of skeletal and extraocular muscles: effects of blood supply, muscle size and in situ or in vitro preparation. *Journal of Neuroscience Methods* 2007;**166**(1):53–65.
13. Graber TG, Kim J-H, Grange RW, McLoon LK, Thompson LV. C57BL/6 life span study: age-related declines in muscle power production and contractile velocity. *Age* 2015;**37**: 1–16, #36.
14. Hill AV. *First and last experiments in muscle mechanics*. Cambridge, UK: Cambridge University Press; 1970.
15. Herzog W, editor. *Skeletal muscle mechanics*. New York (NY, USA): John Wiley & Sons, Ltd; 2000.
16. Lieber RL. *Skeletal muscle structure, function, and plasticity*. Philadelphia (PA, USA): Lippincott, Williams, & Wilkins; 2009.
17. Instruction Manual: In vitro Test Apparatus for 300C and 305C Muscle Lever Systems. Aurora, Canada: Aurora Scientific Inc., www.AuroraScientific.com.
18. Instruction Manual: Dual-Mode Lever Arm Systems. Aurora, Canada: Aurora Scientific Inc.; www.AuroraScientific.com.
19. Barton ER, Lynch G, Khurana TS. Measuring isometric force of isolated mouse muscles in vitro. July 31, 2008. Treat-NMD Neuromuscular Network, Report No. DMD_M.1.2.002.
20. Ko C-P, Lin M-Y. Muscle function evaluation through isometric force measurement in mouse models of spinal muscular atrophy. July 16, 2010. Treat-NMD Neuromuscular Network, Report No. SMA.M_1.2.002.

About the Editor and Authors

ABOUT THE EDITOR

Radovan Zdero, *Ph.D.*, *C.Eng.*, *M.I.Mech.E.*, obtained a doctorate in mechanical engineering from Queen's University (Kingston, Canada) with a focus on biomechanics and biomaterials. He is the Research Director of the Orthopaedic Biomechanics Lab at Victoria Hospital (London, Canada). He is also a Professor in both the Department of Surgery and the Department of Mechanical Engineering at Western University (London, Canada). His research involves measuring mechanical properties of biological tissue, stress testing of orthopaedic implants, and developing new biomaterials.

ABOUT THE AUTHORS

Ameet Aiyangar, *Ph.D.*, obtained a doctorate in mechanical engineering from the University of Wisconsin–Madison (Madison, WI, USA) with a focus in biomechanics. He is a scientist (SNF, Ambizione) in the Mechanical Systems Engineering Department (Structural and Biomedical Engineering Group) at EMPA, Swiss Federal Laboratories for Materials Science and Technology (Duebendorf, Switzerland). His research interests lie in orthopaedic biomechanics, specifically in the biomechanics of healthy, disease-altered, and surgery-altered human lumbar spines.

Tarik Attia, *Dipl.-Ing.*, obtained a master's degree in mechanical engineering with a focus on biomechanics and biomaterials from Leibniz University (Hanover, Germany). He is pursuing a doctorate in biomedical engineering at the University of Toronto and is associated with the Lunenfeld Tanenbaum Research Institute at Mount Sinai Hospital (Toronto, Canada). His research objective is to develop a technology to improve the mechanical properties of irradiated sterilized human bone allografts for orthopaedic reconstruction.

Mina S.R. Aziz, *M.D.*, *M.B.Ch.B.*, *M.Sc.*, obtained medical certifications from the Faculty of Medicine at Fayoum University (Fayoum, Egypt) and the Medical Council of Canada (Ottawa, Canada). He followed that with a master's degree in orthopaedic biomechanics from the Institute of Medical Science and completion of the Collaborative Program in Musculoskeletal Sciences at the University of Toronto (Toronto, Canada). He is a resident-in-training in the Internal Medicine program at the University of British Columbia (Vancouver, Canada).

Zahra S. Bagheri, *Ph.D.*, *E.I.T.*, obtained a doctorate in mechanical engineering from Ryerson University (Toronto, Canada) with a focus on composite biomaterials. She did a postdoctoral fellowship in the Department of Mechanical Engineering at McGill University (Montreal, Canada) with a research interest in orthopaedic biomechanics. She is a Research Fellow at the Toronto Rehabilitation Institute (Toronto, Canada). She has also been an associate editor and reviewer for several biomedical engineering and medical journals.

Matthew M. Borkowski, *B.A.Sc.*, obtained a bachelor's degree in engineering science from the University of Toronto (Toronto, Canada). He is the Sales and Support Manager, as well as a design engineer, for Aurora Scientific Inc. (Aurora, Canada), which serves researchers in nearly 40 countries.

Habiba Bougherara, *Ph.D.*, *P.Eng.*, obtained a doctorate in mechanical engineering from Ecole Polytechnique (Montreal, Canada). She held a postdoctoral fellowship in biomedical engineering at the National Research Council (Ottawa, Canada). She is Associate Professor in the Department of Mechanical and Industrial Engineering at Ryerson University (Toronto, Canada). Her research focuses on the development of new orthopaedic implants and biomaterials.

Allan Brett, *M.A.Sc.*, *B.Eng.*, obtained a master's degree in biological engineering from the University of Guelph (Guelph, Canada), where his research focused on tissue testing of the transverse carpal ligament, a ligament known to play a key role in aetiology of carpal tunnel syndrome. He is a member of the Running Injury Clinic research group (Calgary, Canada), which applies big data analysis techniques to the study of musculoskeletal injuries.

Martin Browne, *B.Sc.*, *Ph.D.*, *C.Eng.*, *M.I.Mech.E.*, obtained a bachelor's degree in materials science from the University of Manchester (Manchester, UK) and spent time in industry researching carbon fiber composite materials. He obtained a doctorate from the University of Southampton (Southampton, UK) for research into improving the corrosion resistance of titanium medical devices. He has been a senior research fellow, lecturer, and reader. He has a personal chair as Professor of Applied Biomaterials and is head of the University of Southampton's Bioengineering Science research group.

Tony Chen, *Ph.D.*, obtained a doctoral degree in biomedical engineering from the University of Rochester (Rochester, NY, USA). He performed his postdoctoral training at the Hospital for Special Surgery in the Department of Biomechanics (New York, NY, USA). His research interests are in understanding changes in joint contact mechanics after injuries. He is also interested in cellular and tissue responses to changes in joint contact mechanics.

Catherine Coirault, *M.D.*, *Ph.D.*, obtained a doctorate in medicine and a doctorate in science from the University of Paris 11 (Paris, France). She is a research scientist at INSERM (Paris, France), where she leads the team investigating the pathophysiology of contractile dysfunction of striated muscle caused by various muscular disorders.

Meghan C. Crookshank, *Ph.D.*, *M.D.*, obtained a doctorate in biomedical engineering from the Institute of Biomaterials and Biomedical Engineering, received a doctorate in medicine from the Faculty of Medicine, and is a resident-in-training in the orthopaedic surgery program, all at the University of Toronto (Toronto, Canada). She has also done research at the Martin Orthopaedic Biomechanics Lab at St. Michael's Hospital (Toronto, Canada) with a focus on orthopaedic biomechanics and biomaterials.

Kathleen Denis, *Ph.D.*, *M.Sc.Eng.*, obtained a doctorate in mechanical engineering from KU Leuven (Leuven, Belgium) with a focus on robot assistance in knee surgery. She is head of the Mechanical Engineering Technology Cluster at Campus Group T (Leuven, Belgium) and Assistant Professor in the Faculty of Engineering Technology at KU Leuven. Her research involves the development and accuracy analysis of smart instrumentation.

Alex Dickinson, *M.Eng.*, *Ph.D.*, *C.Eng.*, *M.I.Mech.E.*, obtained master's and doctoral degrees in mechanical engineering from the University of Southampton (Southampton, UK). He has spent a semester studying computer-aided engineering at ESTACA (Paris, France) and has worked in the medical device industry. He lectures on Engineering Design and Computing, holds a Royal Academy of Engineering (R.A.Eng.) Research Fellowship in biomechanics of lower limb prosthetics, and is a Fellow of the Higher Education Academy.

Alejandro A. Espinoza Orías, *Ph.D.*, obtained a licentiate degree in mechanical engineering from the Universidad Mayor de San Andrés (La Paz, Bolivia) in addition to master's and doctoral degrees in aerospace and mechanical engineering at the University of Notre Dame (Notre Dame, IN, USA). He later completed a postdoctoral fellowship in spine biomechanics at the McKay Orthopedic Research Laboratory at the University of Pennsylvania (Philadelphia, PA, USA). He is an Assistant Professor in the Department of Orthopedic Surgery at Rush University Medical Center (Chicago, IL, USA). His research interests include joint biomechanics, the study of structure–function relationships in orthopaedic materials/tissues, and applications of 3-D printing in orthopaedics.

Trevor C. Gascoyne, *M.Sc.*, *P.Eng.*, obtained a master's degree in biomedical engineering from the University of Manitoba (Winnipeg, Canada). He is the Manager of clinical and biomedical research at the Orthopaedic Innovation Centre (Winnipeg, Canada). His research experience spans patient-centered clinical trials, failure analysis of retrieved orthopaedic implants, implant testing and evaluation, as well as development of new products and materials for the orthopaedic industry.

Karen Gordon, *Ph.D.*, *P.Eng.*, obtained a doctorate in biomedical engineering from Western University (London, Canada) with a focus on upper extremity orthopaedic biomechanics. She is an Associate Professor in the School of Engineering at the University of Guelph (Guelph, Canada) and has conducted extensive soft tissue testing research related to upper extremity tendons, cartilage, and meniscal tissue.

Peter Goshulak, *M.H.Sc.*, obtained a master's degree of health science in clinical engineering with a focus on orthopaedic biomechanics from the Institute of Biomaterials and Biomedical Engineering at the University of Toronto (Toronto, Canada).

Leah E. Guenther, *M.Sc.*, *E.I.T.*, obtained a master's degree in mechanical engineering from the University of Manitoba (Winnipeg, Canada), specializing in the wear testing and failure analysis of orthopaedic bearing materials. She is the Manager of biomedical engineering and services at the Orthopaedic Innovation Centre (Winnipeg, Canada), where her responsibilities include pin-on-disk wear testing, joint simulator testing, and helping develop orthopaedic products for industry.

Jade He, *B.Sc.*, obtained an undergraduate degree in bioengineering (premedical) from the University of California, San Diego (La Jolla, CA, USA) and has also completed two internships in the Department of Biomedical Engineering at Doshisha University (Kyoto, Japan). She is a graduate student in biomechanics at Rush University (Chicago, IL, USA). Her research interests include joint biomechanics, human motion, and exercise and sport science.

Bernardo Innocenti, *Ph.D.*, obtained a doctorate in mechanical engineering from the University of Florence (Florence, Italy). He is Professor of Biomechanics at BEAMS Department (Bio-Electro and Mechanical Systems) at Université Libre de Bruxelles (Brussels, Belgium). His main research field is orthopaedic biomechanics, in particular, the analysis and investigation of the human knee joint in healthy conditions and with a prosthesis, in terms of musculoskeletal loading, bone stress distribution, and the effect of different implant designs.

Farrokh Janabi-Sharifi, *Ph.D.*, *P.Eng.*, obtained a doctorate in electrical and computer engineering from the University of Waterloo (Waterloo, Canada). He is a Professor of Mechanical and Industrial Engineering and the Director of the Robotics, Mechatronics, and Manufacturing Automation Laboratory at Ryerson University (Toronto, Canada). His major research focus is optomechatronic systems in biomedical engineering applications, such as teleoperation, image-guided surgery and intervention, human operator modeling and workspace optimization, and haptic interfaces.

James A. Johnson, *Ph.D.*, *P.Eng.*, obtained a doctorate in biomedical engineering from the Technical University of Nova Scotia (Halifax, Canada). He is the Co-Director of the Bioengineering Laboratory at the Hand and Upper Limb Clinic at St. Joseph's Hospital (London, Canada). His research includes the biomechanics of the shoulder and elbow, as well as the contact mechanics of hemiarthroplasty and total joint arthroplasty techniques of the upper limb.

G. Daniel G. Langohr, *Ph.D.*, obtained a doctorate in biomedical engineering from Western University (London, Canada) with a focus on upper limb biomechanics, wear, and biomaterials. He is a researcher at the Hand and Upper Limb Clinic at St. Joseph's Hospital (London, Canada). His research includes the biomechanics of the shoulder and elbow, as well as the contact mechanics of hemiarthroplasty and total joint arthroplasty techniques of the upper limb.

Steven Leuridan, *M.Sc.Mech.Eng.*, obtained a master's degree in mechanical engineering from KU Leuven (Leuven, Belgium). He is a researcher at the Biomechanics Section of the Department of Mechanical Engineering at KU Leuven. His research focuses on the development of smart instruments for use in orthopaedic surgery. He has broad experience in experimental and computational orthopaedic biomechanics.

Yu-Heng (Vivian) Ma, *B.A.*, obtained a bachelor's degree in molecular and cellular biology with a specialization in immunology from the University of California, Berkeley (Berkeley, USA). She is working on a doctorate in the Cellular Biomechanics Lab at the University of Toronto (Toronto, Canada), where her research focuses on the interaction between osteocyte mechanosensitivity and bone metastasis.

Troy MacAvelia, *B.Eng. (Hons.), M.A.Sc.*, obtained several degrees in mechanical engineering from Ryerson University (Toronto, Canada). His master's work focused on computational and experimental work on surgical bone drilling. He has also worked as a mechanical engineer in industry contexts involving mechanical design, lean manufacturing, and process planning and design.

Ziauddin Mahboob, *M.A.Sc.*, obtained a master's degree in aerospace engineering from Ryerson University (Toronto, Canada) with a focus on the biomechanics of total knee replacements. He is pursuing a doctorate in mechanical engineering at the same institution with an interest in developing polymer-based composite materials for industrial and medical applications.

Kevin A.J. Middleton, *B.Sc.E.*, obtained a bachelor's degree in mechanical engineering with a material's science specialization from Queen's University (Kingston, Canada). He is working on a doctorate in the Cellular Biomechanics Lab at the University of Toronto (Toronto, Canada), where his research focuses on developing microfluidic systems to study bone cell interactions during fluid flow stimulation.

Bruce Nicayenzi, *B.Eng.*, obtained a bachelor's degree in aerospace engineering from Ryerson University (Toronto, Canada). He has also done research at the Martin Orthopaedic Biomechanics Lab at St. Michael's Hospital (Toronto, Canada) with a focus on orthopaedic biomechanics and biomaterials.

Leonard C. Pastrav, *Ph.D.*, *M.Sc.Eng.*, obtained a doctorate in mechanical engineering from KU Leuven (Leuven, Belgium) with a focus on vibration analysis to assess the fixation of orthopaedic implants. He obtained a master's degree in mechanical engineering from Technical University "Gheorghe Asachi" (Iasi, Romania). He is Senior Researcher in the Smart Instrumentation research group and teaches in the Faculty of Engineering Technology at KU Leuven.

Cheryl E. Quenneville, *Ph.D.*, *P.Eng.*, obtained a doctorate in mechanical engineering from Western University (London, Canada) with a focus on injury biomechanics. She is Director of the McMaster Injury Biomechanics Lab and an Assistant Professor in the Department of Mechanical Engineering and the School of Biomedical Engineering at McMaster University (Hamilton, Canada). Her research focuses on developing injury limits and evaluating safety devices using a combined experimental and numerical approach.

Kathryn Rankin, *M.Eng.*, *Ph.D.*, *A.M.I.Mech.E.*, obtained master's and doctoral degrees in mechanical engineering from the University of Southampton (Southampton, UK). Her doctoral work investigated the cemented fixation of advanced polymeric orthopaedic implants by employing experimental testing at coupon and full construct levels, as well as a variety of noninvasive surface and volume strain measurement techniques. She is also a STEM (Science, Technology, Engineering, and Maths) Ambassador.

Jacob Reeves, *M.E.Sc.*, obtained a master's degree in mechanical engineering from Western University (London, Canada) with a focus on wrist fractures. He is a doctoral candidate at the Hand and Upper Limb Clinic at St. Joseph's Hospital (London, Canada). His research focuses on design optimization and the use of in vitro and in silico models for the quantification of bone response to shoulder reconstruction.

Emil H. Schemitsch, *M.D.*, *F.R.C.S.C.*, obtained medical training from the University of Toronto (Toronto, Canada) and a national certification from the Royal College of Surgeons of Canada (Ottawa, Canada). He is an award winning and internationally recognized orthopaedic surgeon interested in developing new devices and techniques for joint replacement and fracture fixation. He is past President of the Canadian Orthopaedic Association and is Chief of Surgery at the London Health Sciences Centre (London, Canada).

Suraj Shah, *M.A.Sc.*, *M.A.*, obtained a master's degree in mechanical engineering with a focus on biomedical imaging techniques and biomechanics, as well as a graduate degree in patent law, both from Ryerson University (Toronto, Canada). He is pursuing a doctorate in orthopaedic biomechanics from the Institute of Medical Science at the University of Toronto (Toronto, Canada).

Joshua Slane, *Ph.D.*, obtained a doctorate in materials science from the University of Wisconsin–Madison (Madison, WI, USA) with an emphasis in orthopaedic polymeric biomaterials. He is a postdoctoral researcher at the Institute for Orthopaedics Research and Training at KU Leuven and University Hospitals Leuven (Pellenberg, Belgium). His research interests lie in the fields of orthopaedic biomaterials, total joint replacement, and ligament reconstruction techniques.

Matthew R.S. Tsuji, *B.Sc.H.*, *M.Sc.*, *M.D.*, *F.R.C.S.C.*, obtained an orthopaedic surgery degree from the University of Toronto and specializes in arthroplasty, trauma, and upper extremity reconstruction. He has also done research at the Martin Orthopaedic Biomechanics Lab at St. Michael's Hospital (Toronto, Canada) with a focus on orthopaedic implant testing and the incorporation of new biomaterials and technologies into orthopaedic surgery.

Juan F. Vivanco, *Ph.D.*, obtained a doctorate in materials science (biomechanics emphasis) from the University of Wisconsin–Madison (Madison, WI, USA) with a focus in biomaterials and biomechanics. He is an Assistant Professor in Bioengineering and Civil Engineering at the University Adolfo Ibáñez (Viña del Mar, Chile). He also has experience in experimental biomechanics and biomaterials to evaluate artificial bioscaffold structures for bone regeneration at the macro-, micro-, and nanolevel.

Stephen D. Waldman, *Ph.D.*, *P.Eng.*, is a Professor of Chemical Engineering at Ryerson University (Toronto, Canada) and an Affiliated Scientist at the Li Ka Shing Knowledge Institute of St. Michael's Hospital (Toronto, Canada). His laboratory focuses on cartilage tissue engineering and the surgical repair of cartilage for both orthopaedic and otolaryngological (ear, nose, and throat) applications.

Hongsheng Wang, *Ph.D.*, obtained a doctoral degree in mechanical engineering from the University of North Carolina at Charlotte (Charlotte, NC, USA). He received his postdoctoral training at the Hospital for Special Surgery in the Department of Biomechanics (New York, NY, USA). His research interests are development and evaluation of methodologies and technologies for prevention and intervention in musculoskeletal injuries and disorders.

Mei Wang, *Ph.D.*, obtained master's and doctoral degrees in mechanical engineering from Queen's University (Kingston, Canada). She is an Associate Professor and Director of Biomechanics Research in the Department of Orthopaedic Surgery at the Medical College of Wisconsin (Milwaukee, WI, USA). She is also an Adjunct Professor in Biomedical Engineering and Orthopaedic and Rehabilitation Engineering Center at Marquette University (Milwaukee, WI, USA). Her research interests include spine biomechanics, biomechanics of orthopaedic trauma, and sports injuries.

Joanna F. Weber, *M.A.Sc.*, obtained bachelor's and master's degrees in biological engineering from the University of Guelph (Guelph, Canada). She is a doctoral student in the Department of Mechanical and Materials Engineering at Queen's University and in the Collaborative Biomedical Engineering Program (Kingston, Canada).

Christian Wight, *B.Eng.*, obtained a mechanical engineering degree from Dalhousie University (Halifax, Canada). He is pursuing a doctorate in biomedical engineering with clinical concentration at the University of Toronto (Toronto, Canada), studying fretting corrosion in modular hip replacements. He has industrial experience in developing various hip, knee, spine, trauma, and soft tissue reconstruction systems for international markets.

Thomas L. Willett, *Ph.D.*, received a doctorate in biomedical engineering from Dalhousie University (Halifax, Canada) with a focus on the mechanics of collagenous biomaterials and tissues. He is a skeletal tissue mechanist and biomedical engineering professor at the University of Waterloo (Waterloo, Canada). His research interests include understanding the role of collagen and the effects of its modifications (due to disease or other causes such as irradiation sterilization) on the mechanical performance of cortical bone, as well as engineering new materials for skeletal reconstruction.

Mitchell Woodside, *B.A.Sc.*, *M.A.Sc.*, obtained a master's degree in mechanical engineering from the University of Toronto (Toronto, Canada). His research focused on the role of collagen structure and cross-linking on the fracture toughness of cortical bone.

Lidan You, *Ph.D.*, *P.Eng.*, *F.C.S.M.E.*, is an Associate Professor in the Department of Mechanical and Industrial Engineering and the Institute of Biomaterials and Biomedical Engineering at the University of Toronto (Toronto, Canada). She is the director of the Cellular Biomechanics Laboratory, where her research focuses on solving biomechanical questions about the musculoskeletal system at the cellular level, such as antiresorptive effects of mechanical loading on bone tissue, pressure effects on bone cell mechanotransduction, mechanical loading effects on bone metastasis, osteocyte mechanosensitivity for diabetic conditions, and the microfluidic system for bone cell mechanotransduction.

Radovan Zdero, *Ph.D.*, *C.Eng.*, *M.I.Mech.E.*, obtained a doctorate in mechanical engineering from Queen's University (Kingston, Canada) with a focus on biomechanics and biomaterials. He is the Research Director of the Orthopaedic Biomechanics Lab at Victoria Hospital (London, Canada). He is also a Professor in both the Department of Surgery and the Department of Mechanical Engineering at Western University (London, Canada). His research involves measuring mechanical properties of biological tissue, stress testing of orthopaedic implants, and developing new biomaterials.

Index

U

V

W

X

Y